The Complete Handbook of Electrical & House Wiring

No. 913
$10.95

The Complete Handbook of Electrical & House Wiring

By S. Blackwell Duncan

TAB BOOKS
Blue Ridge Summit, Pa. 17214

FIRST EDITION

FIRST PRINTING— MARCH 1977

Copyright © 1977 by TAB BOOKS

Printed in the United States
of America

Library of Congress Cataloging in Publication Data

Duncan, S   Blackwell.
  The complete handbook of electrical & house wiring.

  Includes index.
  1. Electric wiring, Interior—Amateur's manuals.
2. Electric wiring—Amateur's manuals. I. Title.
TK9901.D84   621.319'24   77-1770
ISBN 0-8306-7913-8
ISBN 0-8306-6913-2 pbk.

Contents

Introduction

Residential electrical wiring includes both new and old homes. It ranges from the design and installation of a complete new system in a new house, to the addition or updating of existing wiring, and to the complete rewiring of an older home. Unfortunately, though, residential wiring is an area often shied away from—or shunned altogether—by the homeowner and do-it-yourselfer. There are a good many excuses offered, but in the final analysis few make much sense, so you are left with only two valid reasons for not doing it yourself:

(1) You are just plain scared to death of electricity. You have no confidence at all in your abilities as a do-it-yourselfer to overcome that fear and develop your knowledge of the electrical field. Perhaps you are convinced that electricity is some kind of a malevolent or occult force known only to a select few tradesmen who always seem to speak in a mysteriously unidentifiable jargon. If this is the case—forget it! Call an electrical contractor and unload all your problems on him. I have known several folks who were reduced to a quivering jelly at the mere thought of grafting a new plug on the end of an extension cord; the malady seems to be an irreconcilable phobia.

(2) You can't fight City Hall. This is an even more impossible reason to cope with, not because the desire to be a do-it-yourselfer isn't there, but because in a good many places, particularly municipalities, the homeowner is strictly enjoined from doing any electrical work of any sort in his own

home. All wiring must be done under permit by licensed (and more often than not, union) professionals under the direction of a master electrician and the scrutiny of a team of official inspectors. In many respects this is an unfortunate situation, but one that the private citizen can do absolutely nothing about, except perhaps to bemoan yet another example of bureaucratic balderdash and further misguided and misbegotten efforts of a striving government to protect us all from ourselves, at tremendous cost and with questionable effect.

Now then, if you are not afflicted with either of these two conditions, there is really nothing to keep you from doing your own residential electrical work if you want to—even if you have no prior experience with electricity. The work is not all that hard, nor is the time consumed more than for any other major project. The costs are reasonable enough and the savings over having the same work done professionally are considerable. Even the small effort you must expend in learning how to do the work probably won't be too terribly unbearable. On top of that, the results are likely to be better than you might otherwise expect to get.

The level of skill needed to wire up a complete house is so rudimentary that even a would-be do-it-yourselfer who has never hefted a toolbox can readily rise to the occasion. If you can learn to turn a screw with a screwdriver, drive a nail (even a bit crooked) with a hammer, bore a hole with a drill, and cut a piece of wire with a pair of diagonals, you've about got it made. And strange though it may sound, residential wiring rarely gets more complicated than that. Oh, it can, of course, and rewiring an old house sometimes calls for the use of numerous common shop tools and their attendant skills. But there is nothing there that you won't be able to handle, rest assured.

Wiring is not especially taxing from a physical standpoint, either, especially in new construction. There's usually not much in the way of heavy work, and all of the material is quite easy to handle. Exceptions to this might arise in the erection of a complete entrance raceway or mast, or in boosting heavy appliances into place, but you can always yell for help on such infrequent occasions.

The essential factor in electrical wiring is *knowledge*. Difficulties arise mainly from a lack of knowledge. You should have a reasonable understanding of what electricity is and how it works. You should know what material and equipment is available to you for installation in your own system, so that you can intelligently choose your components and tailor them to your own special needs—always with an eye toward costs, ease of installation, and effective and efficient usage. You will need to know how to go about laying out and calculating your own system. And above all, you must have a working knowledge of the rules and regulations that govern electrical installations, specifically those parts of the National Electrical Code and any local codes in your area that apply to residential wiring.

And that is what this book is all about. The information presented here is sufficient to enable you to conceive, lay out, calculate, install, and test a complete wiring system, or any part thereof, in your own home. Unfortunately, there is no way to give you a single standard blueprint that will fit *every* electrical installation; there are literally thousands of combinations and possibilities. Each job is specific in itself. And this means, in turn, that you will have to adjust and reshape some of the more general information to suit your own needs and local conditions.

You may also find that this book is not complete. It may not answer *all* of your questions, or treat *all* of your problems, or explain *every* aspect of electrical wiring that faces you in your own installation. But the residential electrical wiring field is so immense that I doubt whether any single volume could possibly cover every point. The basic stepping stones, however, are here, and hopefully broad enough that you won't stumble along the way. The path should be clear enough for you to fill in any specific details applicable to your own work by doing only a little bit of research.

In using this book, quickly read through the entire volume first in order to get the overall picture. Then return to those sections that apply specifically to your own projected electrical system, and mull them over thoroughly. In order to avoid redundancy, and a lot of unneccesary pages, very little information has been repeated. A point discussed under *Lighting*, for instance, might be equally important to *Modernization and Updating*, but may be mentioned only briefly there. This will entail a certain amount of skipping back and forth on your part in order to assemble all of the details for a given installation.

Then too, the electrical wiring field is to some extent blessed with a dubious distinction of requiring what might be called "circular knowledge" for full understanding of the subject. This makes matters difficult in trying to explain everything in neat little blocks of material that progress through the subject in straightforward and orderly fashion. For instance, a certain amount of knowledge about making circuit calculations is necessary for a complete understanding of how to make up an electrical layout. But on the other hand, a certain amount of design knowledge is necessary to understand how to go about making the calculations. So, round and round we go, like a dog chasing its tail. And this too will necessitate some jumping about from place to place as you read.

But once you dive in and become involved, all of this will become clear. When you have digested the contents of these pages, correlate them with the latest National Electrical Code, and your own local code if there is one. If necessary, add in specialized detail information from other sources and proceed with your project. You will soon find that electrical wiring tasks are not tasks at all; they are like any other good do-it-yourself projects. Challenging, satisfying, enjoyable, and well worth the doing.

Good luck, and may your circuits never be short.

S. Blackwell Duncan

Part 1:
Fundamentals

Chapter 1
Basics of Electricity

Fortunately, you don't have to be an electrical engineer to design and install a home wiring system. In fact, if you had a complete step-by-step set of instructions listing every move, you probably could install a system with no knowledge whatsoever of electricity. But in reality, all installations differ in specific details, so each must be tailored to fit existing needs and conditions. Then they must be checked and tested, and sometimes analyzed for troubles. This means that you must be equipped with some fundamental facts that can be applied as necessary. So a relatively painless dose of basic electricity (as it is used in the general wiring field) should be sufficient for you to see the job through.

It is not likely that you will need to know—or even want to know—all the material presented in this chapter. But every effort has been made to provide, in simplified form, all the rudiments needed to design and install a complete electrical system by yourself. It supplies a reasonable foundation upon which you can build your knowledge of electricity as need, interest, or curiosity dictates. You should therefore read this chapter through to acquaint yourself with the basics of electricity, even if you do not intend to study it in depth at this time. A little knowledge is better than none, at least when it comes time to do your own electrical wiring.

Of course it is only possible to barely scratch the surface here, and the material must of necessity be much simplified and geared only to the general requirements of residential wiring. If you need to expand your knowledge, or simply become interested in learning more, your library will have a whole shelf of books to which you can refer.

CURRENTS AND VOLTAGES

To begin with, electricity is a force. As you know, everything around us is made up of atoms. Atoms are in turn composed of tiny elementary particles called *protons, electrons*, and *neutrons*. The electrons are always on the move in an orbit around the nucleus of the atom, and under the right conditions can be kicked out of orbit to move from one atom to the next. Electrons have a negative charge while protons are positive. Neutrons are neutral, having no charge.

A continual transference of electrons from one atom to the next is called *electron flow*, and constitutes an *electric current*. The relative ease with which the transference can be made to take place is called *conductivity*. Copper, for instance, because of its makeup conducts an electric current well, while wood does not and hence can be used as an insulator.

Electric current also has *polarity*—referred to as plus and minus, or positive and negative. The electron flow is from the negative pole through the atoms of the conductor to the positive pole.

In order to create an electric current, a pressure called *electromotive force* (emf) has to be applied. This force can be created by chemical reaction (as with a battery) or mechanically (as with a generator) and is called *voltage*. As electric current passes through the conductor under the pressure of the emf, a weak magnetic field proportional to the current is set up that surrounds the wire. As we will see later, this magnetic field can be amplified and used in a number of ways to good advantage. There are times, too, when the presence of a magnetic field is a disadvantage.

Electricity comes in two varieties—*alternating current* (AC) and *direct current* (DC). For the time being we will consider direct current since it is the easiest to fathom. We will tangle with AC later. DC travels at any given level of emf and strength of current. The flow is constant and always in the same direction through the conductor from minus to plus. If you had to represent it graphically, you would draw a straight line as in Fig. 1-1.

OHM'S LAW

This brings us to the most important and widely used fundamental of basic electricity, one that you will need time and again, which is expressed as *Ohm's law*. There are three primary aspects of electricity. One is emf, measured by the *volt* and expressed as *voltage*.

To find the current, I = E/R, so:

    amps = volts/ohms
    I = 100/10 = 10 amps

To find the power, P = EI, so:

    watts = volts × amps
    P = 100 × 10 = 1000W

Fig. 1-1. Direct current graph. When the circuit is completed, direct current rises to its full strength instantaneously from X, the point at which it is turned on. It then continues at the same level until turned off at Y. The distance, or value, of X to Z is called the amplitude.

The second aspect is the electron flow, or current, measured by the *ampere* and expressed as *amperage*. This does not refer to the current flow itself, but to the *rate* of flow. An ampere is 6.24×10^{18} electrons—they're really pretty tiny—flowing past a given point in one second. This mass of electrons, all of them taken together as a lump (they do have size and weight, though you can't see them), is called a *coulomb*. The coulomb, then, is a unit of measure like a gallon, while the ampere is a measure of flow rate, like gallons per minute through a pipe (Fig. 1-2).

The third aspect is *resistance*. When current flows through a conductor, it meets a certain resistance to its flow—regardless of how good the conductivity of the material is. This resistance is measured in *ohms* and symbolized by the sign Ω.

Fig. 1-2. Current flow. When the mass of electrons comprising one coulomb passes point P in the interval of one second, the resulting current flow is one ampere.

George Simon Ohm, a German physicist, proved years back that for any circuit the electric current is directly proportional to the voltage and inversely proportional to the resistance. This is usually expressed in a mathematical formula of remarkable simplicity: where E stands for emf in volts, I stands for current in amperes, and R stands for resistance in ohms. Using the formula, we can find any one value if the other two are known. Thus, emf is equal to current times resistance, also stated as volts equals amperes (usually abbreviated *amps*) times ohms. (This may be written as $E = I \times R$, or without the times sign as $E = IR$.) If you know that the resistance of a circuit is 20 ohms, and that the current flow is 10 amps, then the voltage, or emf is $20 \times 10 = 200$ volts (Fig. 1-3).

Similarly, current equals emf divided by resistance, or stated another way, amps equals volts divided by ohms. Using the same example as above, if we know that the emf of a

Table. 1-1. Basic Formulas of Ohm's Law

TO FIND	FORMULA	EXPLANATION
emf or voltage	$E = I \times R$	The emf equals current times resistance, or stated another way, the volts equals amps times ohms. If $R = 10\,\Omega$ and $I = 20A$, then $E = 20 \times 10 = 200V$.
current	$I = E/R$	The current equals the emf divided by the resistance, or amps equals volts divided by ohms. If $E = 200V$ and $R = 10\,\Omega$, then $I = 200/10 = 20A$.
resistance	$R = E/I$	The resistance equals the emf divided by the current, or ohms equals volts divided by amps. If $E = 200V$ and $I = 10A$, then $R = 200/10 = 20\,\Omega$.

Fig. 1-3. Formula triangle for Ohm's Law. By blanking out the unknown factor, the proper combinations of known factors appear.

circuit is 200 volts and the resistance is 20 ohms, then the current is 200 divided by 20, or 10 amps.

The third combination of the formula states that resistance is equal to emf divided by current, or ohms equals volts divided by amps. Again using the above example, if we know that the voltage is 200 and the current is 20 amps, then the resistance equals 200 divided by 10, or 20 ohms.

If you find memorizing mathematical formulas difficult, memorize instead the Ohm's law triangle shown in Figure 1-3. If you can envision the triangle well enough to draw it out whenever you need the formulas, it will serve you well. Just put the tip of your finger over the factor symbol that you need to find, read the rest of the formula from the triangle, and make the necessary figure substitutions and calculations.

CIRCUITS

Electricity must flow in a circuit, out from the negative pole and back to the positive pole. This is called a *closed* circuit, or *complete* circuit since the conductor is a continuous loop. If there is a break anywhere along the line, it becomes an *open* circuit—no current flows; it has nowhere to go and the circuit is "dead."

A closed circuit is composed of three elements. The first is the emf source, the second is the conductor through which the current flows. The third is called the load, or load resistance. An electrical current cannot simply loaf along through a circuit without doing anything; it must have a load which converts its energy into heat, light, or some other form of power.

If the load is absent and the circuit is closed, or if the load malfunctions in such a way as to provide for current path

Fig. 1-4. A simple electrical circuit, where E_S is the source of electromotive force.

which is electrically little different from the conductor itself, the result is a *short* circuit—and usually creates a certain amount of fireworks.

Figure 1-4 shows the simplest possible closed circuit, while Fig. 1-5 shows a more advanced circuit that can be opened and closed, or turned off and on.

Fig. 1-5. Two simple electrical circuits. In A, the switch is open (OFF position), so the circuit is open and the load is inoperative. In B, the switch is closed (ON position), so the circuit is closed and the load is operating.

To find the current, I = E/R, so:

amps = volts/ohms

I = 100/5 = 20 amps

Fig. 1-6. Determing current flow in an electrical circuit.

To give you a practical idea of how you may have use for Ohm's law, let's consider Fig. 1-6. Assume that the emf (E_s) is 100 volts and the load is a lamp with a resistance of 5 ohms. You have to provide a conductor heavy enough to handle the current safely, but what is the current? Simple. Current equals voltage divided by resistance, remember? The load current is 20 amps.

POWER

There is another factor that we need to consider too, and that is power. Power equations tie in closely with the Ohm's Law equations, and you will need them frequently. There are several different forms of power, such as mechanical, heat, and light, which we will look into later. Our concern for the moment is electrical power, which is expressed in *watts*.

One watt is equal to the power in a circuit where one ampere of current flows at one volt of emf. The formula is $P = EI$; power equals volts times current. Correspondingly, $P = E^2/R$, which is the voltage squared, (multipled by itself), divided by the resistance. Also, $P = I^2 R$, which is the current squared, multiplied by the resistance.

Referring to Fig. 1-7, you can see that with the stated values and using the same process as in Fig. 1-6, a current of 10 amperes flows in the circuit. Now, to find the power of the circuit, simply multiply volts times current, or 100 times 10 equals 1000 watts.

By using one or another of the twelve equations shown in Table 1-2, you will be able to solve the bulk of the ordinary electrical problems that will confront you in laying out, installing, testing, and troubleshooting your own electrical

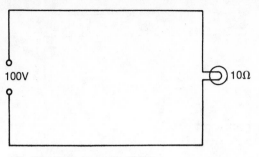

To find the current, I = E/R, so:

amps = volts/ohms
I = 100/10 = 10 amps

To find the power, P = EI, so:

watts = volts × amps
P = 100 × 10 = 1000W

Fig. 1-7. Determining power consumption of an electrical circuit.

installation. As long as you can discover two of the factors of each equation, whether from a specification sheet or by actually taking measurements, you can then find the remaining unknown factor.

CIRCUIT MEASUREMENTS

Once you get past the stage of flashlights and simple doorbells, electrical circuits become more complex than those we have looked at so far. There are three kinds of circuits with their attendant formulas and characteristics that we must consider.

Table. 1-2. The Twelve Basic Formulas Used in Electrical Work

TO FIND	FORMULAS		
amperes (I)	$I = E/R$	$I = P/E$	$I = \sqrt{P/R}$
volts (E)	$E = IR$	$E = P/I$	$E = \sqrt{PR}$
ohms (R)	$R = E/I$	$R = P/I^2$	$R = E^2/P$
watts (P)	$P = EI$	$P = I^2R$	$P = E^2/R$

Fig. 1-8. Simple series circuit showing current flow and polarity.

Series Circuit

The *schematic* (circuit diagram) in Fig. 1-8 is of a series circuit. Notice that the current flows from the negative pole of the voltage source (E) through each load and back to the positive pole, and that the same current flows everywhere in the circuit. The conductor between the negative pole of the source and the first terminal of the first load is negative. So is the first terminal of the load with respect to its second terminal, making the second terminal positive in polarity. The next load's terminals are negative, then positive again, and so on.

Knowing the polarity is often important. When you apply the leads of a DC voltmeter across the terminals of a load to measure the voltage drop, for example, you must know the polarity of each terminal. Voltmeter actions are unidirectional (one direction). If you reverse the leads, the indicating needle will try to move in the wrong direction, down past zero instead of upward along the scale.

Each of the loads in a circuit has a certain resistance. In a series circuit, the total resistance of the circuit is equal to the sum of the individual load resistances. In the circuit in Fig. 1-9, there are two loads, making the total resistance in the circuit 10 ohms. Using the same method as in Fig. 1-6 again, the current flowing in the circuit proves to be 10 amps. It is also apparent that the circuit consists of two parts, load 1 and load 2. There is a 10-amp current flowing through the circuit, and the emf across the two source terminals is 100 volts. Let's calculate the voltage across the first section of the circuit, across the terminals of load 1. The formula is $E = I \times R$, or $10 \times 5 = 50$ volts. What happened to the remaining 50 volts? That is across load 2, and you can calculate this in the same way.

The preceding example points up two more elementary principles of electricity. The first is that Ohm's law is just as applicable to *parts* of a circuit as it is to an entire circuit. The second is known as Kirchoff's law, which states in part that the sum of the voltages across the loads in a closed loop is equal to the electromotive force in the loop. The voltage we found across load 1 is called a *voltage drop*, or *IR drop* (because it is found by multiplying current *I* times resistance *R*). In other words, the sum of the voltage drops around a circuit is equal to the voltage at the source.

To bring this series circuit a little closer to home, let's assume that load 1 and load 2 are incandescent light bulbs screwed into sockets and attached to the conductor. The current flows through load 1 and lights it, then through load 2 and does the same. Now, a light bulb is designed to operate at a specific voltage for maximum efficiency. If the emf of the circuit is 100 volts and you put in two 100-volt bulbs, what happens? Each bulb glows dimly because they are splitting the voltage between them and operating at only 50 volts each. If you series more bulbs in the circuit, they will all glow more and more feebly. But if you put in a pair of bulbs rated at 50 volts each, they would glow at full strength because each is receiving its rated voltage.

To find the voltage drop across each load,
the formula is $E = I \times R$.

$$E_1 = 10 \times R_1 = 10 \times 5 = 50V$$
$$E_2 = 10 \times R_2 = 10 \times 5 = 50V$$

Fig. 1-9. Determining the voltage drop in a series circuit.

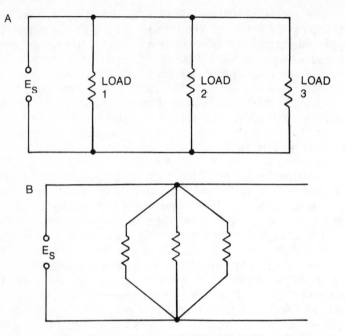

Fig. 1-10. (A) Simple parallel circuit. (B) Electrically identical circuit.

If you wanted to use five bulbs in series, each would use one-fifth of the source voltage, so that each would have to be rated at 20 volts for full output. Some strings of Christmas tree lights operate on this principle. If you unscrew any one of the bulbs, all the rest will go out. You have opened the circuit so that no current flows. Note too that if you were to put in two bulbs rated at only 25 volts each, they would burn out rapidly because they would be operating at 50 volts each—well over their rated capacity.

In a series circuit all the loads are hooked up end to end, so to speak. They all have a proportionate voltage drop, the sum of which equals the emf source voltage. Either all loads function at once, or none do since the same current flows through the whole circuit. Load resistances are added to find total resistance of the circuit; load wattages are added to find the total wattage.

Parallel Circuit

The parallel circuit presents a somewhat different situation. The circuit shown in Figure 1-10 is a parallel type with three separate loads. Note that, unlike the loads in the series circuit, each load is connected directly to the source voltage, and each receives that voltage directly. If one load is removed, the others continue to function normally, through the total current flow and resistance of the circuit change.

Each load in a parallel circuit is actually a *branch* circuit coming from the source. For purposes of illustration, though, we will call them *legs* of the circuit, since branch circuits have a different meaning in the general wiring field (though the term is electrically correct). This will avoid some confusion later.

There are several methods of figuring the resistance of a parallel circuit. Unlike resistance in a series circuit, which *increases* as loads are added, the total resistance of a parallel circuit *decreases* as more loads are added. Referring to Fig. 1-11, let's assume that load 1 has 5 ohms of resistance, load 2 has the same, and load 3 has 10 ohms. In a parallel circuit we can find the total resistance by using reciprocals. To start with, we know that $I = E/R$. In a parallel circuit, as in a series circuit, the total current equals the voltage divided by the total resistance. Put another way, this would be IE/R_T, where R_T is the total resistance. In this example,

$$E/R_T = E/R_1 + E/R_2 + E/R_3$$

The E's cancel out, leaving

$$1/R_T = 1/R_1 + 1/R_2 + 1/R_3$$

Using the values given,

$$1/R_T = 1/5 + 1/5 + 1/10$$

$$\frac{E}{R_T} = \frac{E}{R_1} + \frac{E}{R_2} + \frac{E}{R_3} \qquad \frac{1}{R_T} = \frac{1}{5} + \frac{1}{5} + \frac{1}{10} = \frac{1}{2}$$

$$\frac{E}{R_T} = \frac{E}{5} + \frac{E}{5} + \frac{E}{10} \qquad R_T = 2\Omega$$

Fig. 1-11. Determining total resistance of parallel loads.

$$R_T = \frac{R_1 \times R_2}{R_1 + R_2}$$

$$R_T = \frac{5 \times 10}{5 + 10} = \frac{50}{15} = 3\ 1/3\ \Omega$$

Fig. 1-12. Determining total resistance of two unlike parallel loads.

a series of simple fractions. The common denominator is 10, so

$$2/10 + 2/10 + 1/10 = 1/2$$

So, $1/R_T = 1/2$, so $R_T = 2$ ohms.

Another method of finding resistance applies in the case of two loads that are unequal. This is a further offshoot of the Ohm's Law formulas, and is written

$$R_T = \frac{R_1 \times R_2}{R_1 + R_2} .$$

In other words, in Fig. 1-12 you would simply multiply $5 \times 10 = 50$, then add $5 + 10 = 15$, and divide 50 by 15 to get the answer of 3 1/3 ohms.

A quick way to find the total load resistance of any number of equal loads connected in parallel (like a line of light bulbs) is to divide the size of any one load by the total number of loads in the circuit. Thus, from Fig. 1-13, 20 ohms divided by 4 loads equals 5 ohms for a total circuit resistance.

When you determine the total resistance of a parallel circuit (Fig. 1-14A), you can then draw another circuit which, though it may not be physically the same, is an *electrical equivalent* (Fig. 1-14B). Most circuits, especially in the field of general electrical wiring, can be reduced in this way despite their apparent complexity. Using this equivalent circuit, it is a

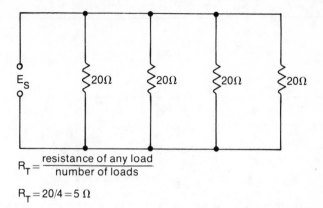

$$R_T = \frac{\text{resistance of any load}}{\text{number of loads}}$$

$$R_T = 20/4 = 5 \ \Omega$$

Fig. 1-13. Determining total resistance of any number of like parallel loads.

simple matter to find the current in the circuit when the voltage is known. If the voltage is 100, as shown in Fig. 1-14B, then the current will be E/R, or $100/6.66$, or 15.8 amps.

Note, that while the total current (I_T flows through the entire simplified circuit, the same is not true of the original parallel circuit. Figure 1-15 is electrically the same as the circuit in Fig. 1-11, but is drawn a different way, though the

$$R_T = \frac{R_1 R_2}{R_1 + R_2} = \frac{10 \times 20}{10 + 20} = \frac{200}{30} = 6.66 \ \Omega$$

To find the total current of the equivalent circuit, $I = E/R$.

$I = 100/6.66 = 15.8A$ approximately

Fig. 1-14. (A) Reducing a parallel circuit to an equivalent total resistance (B) Determining current flow in the equivalent circuit.

27

Fig. 1-15. Breakdown of current flow in a parallel circuit.

same load resistance values are used in each diagram. But since each leg of this parallel circuit is actually a little circuit in itself, the current flowing in each leg is entirely dependent upon the load in that leg ($I = E/R$). So,

$$I_1 = E/R_1 = 100/5 = 20 \text{ amps}$$

Load 2 is the same. For load 3,

$$I_3 = E/R_3 = 100/10 = 10 \text{ amps}$$

Adding the three individual leg currents gives us the total of 50 amps, just as reducing to an equivalent simple series circuit and then figuring the current would result in a flow of 50 amps.

To turn things around, suppose you know that in that same circuit (Fig. 1-15) the current flowing in leg 1 is 20 amps, leg 2 is 20 amps, and leg 3 is 10 amps. You could then easily find the resistance value of each individual load by applying the formula $R = E/I$, or find the wattage of each with $P = E \times I$, or solve for any of the various combinations shown in the formula chart of Table 1-2.

Series-Parallel Circuit

The last category of circuits that we will discuss here is called the series-parallel. As the term implies, these circuits are made up of various combinations of series and parallel arrangements. They are quite common in electrical work of all sorts. Figure 1-16 shows a series-parallel circuit that is simply a combination of Figs. 1-9 and 1-10. The first leg of the circuit, with loads A and B, is a series circuit. This series circuit is then in parallel with loads C, D, and E, which in turn are in parallel with one another.

The various electrical values of the circuit can be found as a whole, or in individual legs, or in various combinations. Simply apply Ohm's law and the power equations as in

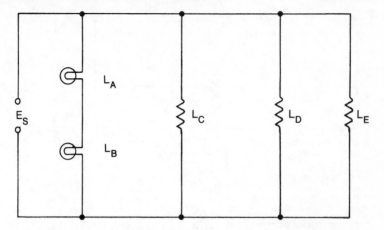

Fig. 1-16. Simple series-parallel circuit. Load A and load B are in series with one another. Loads C, D, and E are in parallel with one another and with load AB.

previous examples. To reduce the circuit to its simplest form, add the two series loads to one equivalent load, with the result shown in Fig. 1-17A. This four-leg parallel circuit can then be reduced by reciprocals to the one equivalent load and series circuit shown in Fig. 1-17B.

A more complex series-parallel circuit is shown in Fig. 1-18. Though it may look impossible, this circuit still can be broken down in the same way. Loads E and F are obviously in series with one another and so can be added to form EF. EF is then in parallel with D, and both are in parallel with B. Using reciprocals, an equivalent value for EFDB is found. This then puts EFDB in series with C, and the two are added together to become EFDBC. EFDBC is then in parallel with A, and their reciprocals are added in computing a total load resistance that can be used to diagram a simple one-load circuit.

Fig. 1-17. Reducing series-parallel circuit in Figure 1-18 to equivalent circuits.

E and F in series.
EF in parallel with D.
EFD in parallel with B.
EFDB in series with C.
EFDBC in parallel with A.
EFDBCA = total resistance of circuit.

Fig. 1-18. Reducing a complex series-parallel circuit.

Figure 1-19 shows two examples of how the same circuit can be analyzed when some of the factors are known. It is important to realize that such complex circuits first have to be *completely* reduced and then worked backward leg by leg toward E_S in order to arrive at the correct values. At times this can be a confusing jigsaw puzzle, and it is usually worthwhile to check your results. For instance, in Fig. 1-19A the total current is approximately 35.7 amps. In Fig. 1-19B the current through R_{2T} and R_3 is 21.3 amps. The current through R_1 is 14.3 amps. The two currents together equal 35.6 amps, or about the same as the total in Fig. 1-19A. Other values can be calculated as well, throughout the circuit.

Circuit With Conductor Resistance

There is another type of series-parallel circuit that is not nearly so obvious as the preceding, but is nonetheless quite common. You'll recall that all electrical conductors present some resistance to current flow; you will never encounter any such condition as zero R. The amount of resistance in a conductor, insofar as we are concerned here, is primarily a function of the cross-sectional area and the length of the conductor. For the most part, in short runs of conductors of the types used for general wiring, this conductor resistance is negligible. But if the conductors are quite long, the effect is that of an unseen load connected in series with the circuit.

In Fig. 1-20A we have three loads connected in parallel at the end of a very long run. The load shown in series, load A, represents the total resistance of the conductor. This resistance, by the way, can be computed for any size and shape of any material, but in practice in the field the needed figures are generally taken from tables already compiled for the purpose.

This circuit would represent an operational problem for the electrician, as you will see when we finish dissecting it. The first step is to reduce parallel-connected loads R_B, R_C, and R_D to R_T, as shown. R_T is then added to R_A, since they are in series (Fig. 1-20B). Using the formula $I = E/R_T$ you can see that the current flowing in the equivalent, or total, circuit is 25 amps. But since there is a series load involved (the R of the conductor), the same voltage does not appear in all parts of the circuit. Instead, there is a voltage drop across each load. Since $E = IR$, the drop for R_A is 25 × 2, or 50 volts. The voltage drop for R_T is the same. What this means, practically, is that the three loads (B, C, and D) will operate at a line voltage of only 50 volts instead of the source voltage of 100, while the conductor itself "uses up" the other 50 volts by conversion to heat through its own resistance. If the three loads happened to be standard 120-volt light bulbs, for instance, they would barely glow.

A. To find the total current:

R_5 and R_6 are in series, so $R_5 + R_6 = 10\,\Omega$.

10Ω is in parallel with R_4, so $10/2 = 5\,\Omega$.

5Ω is in parallel with R_2, so $\dfrac{5 \times 6}{5 + 6} = \dfrac{30}{11} = 2.7\,\Omega$

$2.7\,\Omega$ is in series with R_3, so $2.7 + 2 = 4.7\,\Omega$.

$4.7\,\Omega$ is in parallel with R_1, so $\dfrac{4.7 \times 7}{4.7 + 7} = \dfrac{32.9}{11.7} = 2.8\,\Omega$.

Then $I = E/R$, so $I = 100/2.8 = 35.7A$.

B. To find other individual values, for instance:

$P = E^2/R$, to find the wattage of R_3. $E =$ voltage drop across R_3; R_3 is known. From above, R_{EQ} of R_2, R_4, R_5, and R_6 is 2.7Ω. R_{2T} is in series with R_3, with a voltage of 100. So,

$I = E/R = 100/4.7 = 21.3A$ flowing in that leg. $E = IR$, so for R_3,

$E = 21.3 \times 2 = 42.6V$. P of R_3 is $(42.6 \times 42.6)/2 = 907.38W$.

Fig. 1-19. Determining total current draw of a complex series-parallel circuit is shown in (A). Determining other values is shown in (B).

$$\frac{1}{R_{BCD}} = \frac{1}{5} + \frac{1}{5} + \frac{1}{10} = \frac{2}{10} + \frac{2}{10} + \frac{1}{10} = \frac{5}{10} = \frac{1}{2}$$

$R_{BCD} = 2$

$R_T = R_{BCD} + R_A = 2 + 2 = 4$

$I_T = E/R_T = 100/4 = 25A$

Voltage drop across R_A is $IR_A = 25 \times 2 = 50V$

Voltage drop across R_{BCD} is $IR_{BCE} = 25 \times 2 = 50V$.

Fig. 1-20. (A) A type of series-parallel circuit that might be found in electrical wiring. (B) Determining the voltage drop in the circuit. R_A represents the resistance of a long conductor.

Now consider Figure 1-21. Though this may appear to be the same as Fig. 1-20, it is not. In this case the series resistance is between loads B and C instead of at the head of the circuit, and the resulting figures are quite different. This could represent a circuit where load B is connected across the circuit close to the voltage source, while C and D are connected a long distance away, so that the bulk of the resistance in the line is between load B and loads C and D. The first step is to reduce the circuit to the equivalent shown and find the value of R_{CD} (Fig. 1-21B). Then, since R_{CD} is now in series with R_A, we add the two, for a total of 5.33 ohms. This total, as shown in Fig. 1-21C, is now in parallel with R_B.

The next step is to find what current flows in the circuit leg containing R_{ACD}. With this ascertained, we refer back to Fig. 1-21B, where R_A and R_{CD} are separated, to find the voltage drop across each. This is again a series leg, and we know that the same current flows in all parts of the leg. Thus, IR_A equals 37.52 volts. Similarly, the voltage drop across R_{CD} equals IR_{CD} or 62.47 volts. The two voltage drops together equal the total voltage entering the leg of the circuit, or nearly so. Now, going back to Fig. 1-21A, you can see that while load B remains unaffected, loads C and D will operate at only 62.47 volts.

$$\frac{1}{R_{CD}} = 1/5 + 1/10$$
$$= 2/10 + 1/10$$
$$= 3/10$$

So $R_{CD} = 10/3 = 3.33\ \Omega$

$R_{ACD} = R_A + R_{CD} = 2 + 3.33 = 5.33\Omega$
I through R_{ACD} is $E/R_{ACE} = 100/5.33 = 18.76A$
Voltage drop across R_A is $IR_A = 18.76 \times 2 = 37.52V$
Voltage drop across R_{CD} is $IR_{CD} = 18.76 \times 3.33 = 62.47V$

Fig. 1-21. Breaking down a series-parallel circuit where R_A represents a long conductor between load B and loads C and D, and determining the voltage drop.

33

If you have decided by now that series-parallel circuits can be tricky, obscure, and even downright ornery, you are by no means alone in your opinion. Most electricians, when confronted by some strange occurance in a circuit, like a dim bulb or a groaning motor, check to see if something has gotten "seriesed" in what should have been a parallel circuit, or test for excessive voltage or *IR* drop. So difficult or not, a working knowledge of such circuitry is useful and sometimes necessary.

If you are having some problems understanding series-parallel circuits, or any of the material that we have discussed so far, this would be a good time to stop and do a bit of experimenting. One of the best ways to learn is to dig out a pad of paper and a pencil and, perhaps using some of the example schematics on the previous pages, start making up circuits of your own. Draw the same circuits in different ways, assign varying values, work through whole series of equations, cross-checking all the way. To help even more, you can devise some simple circuits with a lantern battery, a few lengths of bell wire, and some bulbs and sockets. Draw the circuit first, assemble it, work out the calculations, and then measure the results in the actual circuit with a meter to see if they compare.

ALTERNATING CURRENT

The next step is to investigate AC, or alternating current. Though what you have learned about DC will hold true, there are some important differences, additions, and intricacies that we will have to look at.

First, alternating current continuously reverses itself —anywhere from one or two, to millions of times per second. This is graphically represented in Figure 1-22. When the circuit is closed, the voltage and the current gradually rise, headed toward plus, until both reach a peak and then start to diminish. They pass the zero line, and at this point reverse and start toward minus, reach a peak equal in minus value to the plus peak, and start back up again only to do the same thing over and over again until the circuit is opened. Each pair of reversals is called a cycle, and alternating current is said to function at so many cycles per second—thus the abbreviation *cps*, which you have probably noticed many times in the term *60 cps*. Most of our electrical service in this country operates at 60 cps. The expressions *60 cy.* or *60 c.* mean the same thing. Now we have a new designation as well—*hertz*, abbreviated *Hz*, so you will now see the expression *60Hz*.

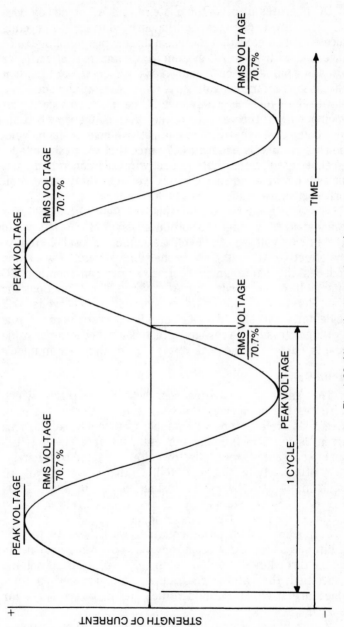

Fig. 1-22. Alternating current graph.

35

AC Voltages

DC operates at a continuous emf, but AC obviously does not. The level changes continuously during the cycling process. This means that there are three types of voltage to be concerned with in an AC system—*instantaneous, effective* or *rms*, and what is variously known as *peak, crest,* or *maximum* voltage. Instantaneous voltage is the voltage at any one given point in the cycle, or at any given time. Peak voltage is that which occurs at the very top of the cycle, or the very bottom, there being one negative and one positive peak in each cycle. The rms voltage is equal to 70.7 percent of the peak voltage, and that is the value that we are primarily concerned with. Unless otherwise indicated, all AC voltages that you read of, work with, or measure are rms voltages.

There is good reason for this odd figure of 70.7%. For instance, an AC voltage that produces peaks of 156 volts would have an rms voltage of 110 volts, a familiar household figure. This effective AC voltage does the same amount of work and produces the same amount of energy as an identical SC voltage. In other words, this AC voltage of 110 functions the same, as far as power is concerned, as 110 volts DC. Incidentally, this should not be considered as an *average* value of voltage; it is not. Average voltage is another specific value equal to 0.636 times the peak voltage, or maximum amplitude.

Circuit Values

The same three categories hold for current, which alternates the same way the voltage does. There is a peak current as well as an instantaneous current, and the rms current is 70.7% of the peak current. And since power (P) is equal to current (I) multiplied by voltage (E) and both I and E are continuously variable, it follows that power in an AC circuit may also be peak, instantaneous, or rms. Note, however, that resistance (R) is not cyclical, since it is a function of the properties of the conductor or load.

Ohm's law and the power equations work the same way for AC circuits as for DC, except that you must remember not to mix the three kinds of values in any given problem. If nothing is specified, the values are assumed to be rms. When other values are used, all are identified, as E_{RMS} or I_{PEAK}, for instance.

You may see effective or rms voltage, current, or power referred to as being *one half* of the peak voltage, current, or power. This might seem ridiculous at first glance, but it is true nonetheless. Using the same figures as above, if a peak voltage is 156 volts, then the rms voltage must be 110 volts. How can 110 be half of 156? In terms of numbers, it cannot. Instead, this is the expression of a mathematical relationship. A *root mean square*—from whence the abbreviation *rms*—is the square root of the arithmetic mean of the squares of the numbers in a given set of numbers, which is quite a mouthful. When applied in this case, the formula is

$$E_{\mathrm{RMS}} = E_{\mathrm{PEAK}} / \sqrt{2} = 0.7071 E_{\mathrm{PEAK}}$$

The square root of 2 is 1.4142, and half of this is 0.7071, which is rounded to 0.707 and expressed as a percentage, 70.7%. Rounded off, $1.41E_{\mathrm{RMS}}$ equals E_{PEAK}, and $0.707E_{\mathrm{PEAK}}$ equals E_{RMS}. You can see the two-to-one relationship.

Phase

The next characteristic of AC to consider is phase. As you can see from Fig. 1-23, each complete cycle of AC, in effect, is a full circle, and can be marked off into degrees. The positive peak occurs at 90°, the negative peak at 270°, and reversals occur at 180° and 360° (or 0°). The degrees themselves are called *phase* degrees, and each degree represents 1/360 cycle. Now, if you apply an AC voltage to a load of theoretically pure resistance, the current cycle will exactly match the voltage cycle, so that on graphing them the lines cannot be told apart. Current and voltage are then in step with one another, or *in phase*.

In Fig. 1-24 another voltage, B, has been applied, beginning 1/8 cycle later than voltage A. In this case, voltage B is said to lead voltage A by 45°. If, as is usually the case in AC circuits, the resistance to the emf consists of pure resistance plus other influencing factors, then A might represent the alternating voltage and B the alternating current, leading by 45°. In practice though, the B peaks representing current flow would likely be lower than shown. In fact, sine-wave representations such as these can be worked out for an infinite number of circuits with an infinite number of conditions imposed upon them, and many of the results present a rather odd and often complex appearance.

You will often hear the term *phase* used in a somewhat different context in practical electrical work. This is in reference to a single-phase service, or a three-phase service, or a two-phase four-wire system, or some such. This has to do with the manner in which a particular commercial source of

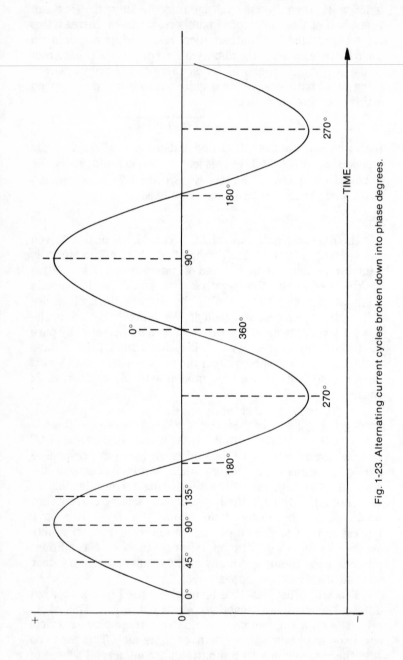

Fig. 1-23. Alternating current cycles broken down into phase degrees.

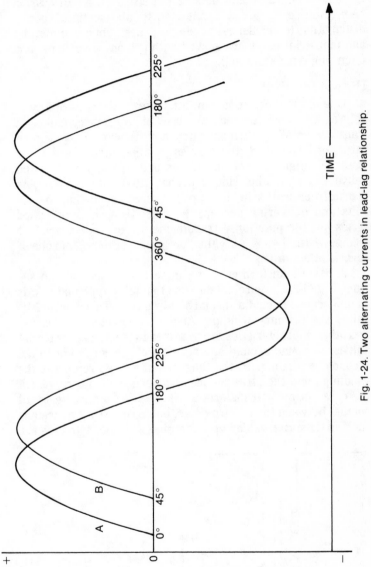

Fig. 1-24. Two alternating currents in lead-lag relationship.

39

power is presented to the consumer by way of the powerhouse, transmission network, and assorted transformers. The details of how these systems are put together and operate is extremely complex and outside the scope of our discussion here. Suffice it to say that nearly all electrical power distribution to residences is single phase, while commercial and industrial users are served with polyphase systems, and often single-phase as well.

TRANSFORMERS

One of the most important attributes of AC is its ability to be readily transformed by use of a device called a transformer. With a transformer, a particular voltage can be raised or lowered to other values. A *step-up* transformer, for instance, might accept an emf on one side of 100 volts and deliver it on the other side as perhaps 1000 volts. A *step-down* transformer works in the opposite direction, taking the 100 volts and delivering perhaps 10 volts. Using a complicated series of computations, transformers can be designed to change virtually any AC value, or several of them, into one or several other values.

A basic transformer is a simple device, as you can see from Fig. 1-25. A wire conductor of suitable type and size is wound around part of a special soft iron core. This input side is called the primary winding. Another conductor, whose size and other characteristics are matched to the necessary output values, is wound around another part of the core to form the secondary winding. In the diagram, the loops represent the windings, and the straight lines between the loops represent the core. In most transformers there is no direct electrical contact between the two windings, but in some special designs such as the *autotransformer*, common windings with taps are

100V

1000V

Fig. 1-25. Simple step-up transformer.

PRIMARY   SECONDARY

Fig. 1-26. Primary and secondary relationships of a simple transformer.

used. The general operating principles and the results obtained, however, are much the same.

When an emf is applied to the primary leads, or taps, magnetic lines of force are immediately set up and form a magnetic field around the primary winding. The lines of force cut across the secondary winding and induce another separate voltage there at the same instant. A half cycle later the applied voltage reverses its direction, and the magnetic field also reverses. When it does, another surge of voltage having opposite polarity is induced in the secondary winding.

The *turns ratio* is based on number of times the conductors are coiled around the core. The ratio of primary to secondary windings determines the ratio of primary to secondary voltage. The frequency of the source voltage has no effect upon the voltage, current, or power relationships. In the interests of efficiency, though, the voltage will affect the mechanical design of the transformer.

In theory, the consumed power at the primary side of a transformer is equal to the power consumed at the secondary side. Suppose, for instance, that you have a transformer such as the one shown in Fig. 1-26. The primary side is designed for an emf of 100 volts and a maximum current of 10 amps. The windings are in a one-to-five step-up ratio, and the secondary delivers 500 volts. The secondary current, however, is not stepped up by five times to 50 amps, but rather is divided by five, to 2 amps. The primary power equals IE, or $100 \times 10 = 1000$ watts. The secondary power is theoretically the same, so the secondary current equals P/E, or

1000/500 = 2 amps. But this is an ideal situation never realized in practice.

Actually, transformers are somewhat less than 100% efficient because of leakage flux, internal resistance, and other factors. Part of the power is lost in the transforming process. A poor transformer might deliver as little as 80% of the possible 1000 watts in the preceding example. All transformers have a certain power rating too, and they should not be operated above or beyond that capacity. The result of this "lost" power in the transformer appears in the form of heat. A little heat means lost efficiency, but a lot of heat means a burn-out.

INDUCTANCE

Resistance in an AC circuit is sometimes "pure," or nearly so, much like the resistance of a DC circuit. But more often than not, there are other factors that go into the makeup of AC resistance. For instance, we noted that in a transformer a magnetic field from one conductor, or winding, cuts across another winding to create a magnetic field. This magnetic process is called *inductance*, and it produces a force that opposes any change in current flow. Inductance is present in many AC circuits, either by design or by accident. Inductance, symbolized by the letter L, is measured by a unit called a *henry*, abbreviated H, which equals a current flow increase at the rate of one ampere per second with one volt applied as emf.

Inductance may be introduced purposely into a circuit, as in certain types of lighting control. It may also be a byproduct of some device like a transformer, motor, or fluorescent light ballast—not necessarily wanted, but present anyway. Or, inductance can result through inadvertant miscalculation in circuit design or installation.

Though inductance is often ignored, if present to any great degree, it can have a notable effect upon the current flow and power capabilities of a circuit. Inductance opposes the flow of alternating current, and this condition is called *reactance*, or more properly, *inductive reactance*. Pure inductive reactance in a circuit causes the current to lag 90° behind the applied voltage. Since this is a form of resistance in the circuit, inductive reactance is measured in ohms and generally is specified as AC ohms, symbolized by X_L. Ohm's law applies, but as you can see in Fig. 1-27, the formula for inductors is a little more complicated than those we have used so far.

You will notice that the inductive reactance of the circuit in Fig. 1-27 is responsible for a current of 0.58 amps. You might

110V 60Hz — 0.5H

$X_L = 2\pi fL$

$I = \dfrac{E}{2\pi fL}$ OR $I = \dfrac{E}{X_L}$

Fig. 1-27. Determining the reactance of a simple inductive circuit.

I = AC amperes
E = AC volts
$\pi = 3.1416$
f = frequency in hertz
L = henrys of inductance

$I = \dfrac{110}{2 \times 3.1416 \times 60 \times 0.5} = \dfrac{110}{188.5} = 0.58$ amps

Reactive power $E_{RMS}I_{RMS} = 110 \times 0.58 = 63.8$ vars

assume that the circuit therefore consumes a proportionate amount of power. But it does not. In fact, it consumes no power at all, despite the current draw, so the source does no work. Current flow lags the voltage by 90°, meaning that the average power component of the inductance is half negative and half positive, so they cancel each other. The power actually flows back and forth between the conductor and the magnetic field in time with the cyclical alternations of the emf, and the effective result is that each alternation cancels the other.

This peculiarity of reactance gives rise to new terms—*reactive power* and *true power*. True power is that power which actually does work, and in this example that value is zero. Reactive power doesn no work and is measured in volt-amps reactive, abbreviated *vars*, also written as reactive volt-amperes. In a purely reactive circuit as in Fig. 1-27 the number of vars is, by Ohms's law, equal to the product of the volts and amps, or *EI*, but this is not true for circuits that also contain resistance because the phase of the voltage and current is no longer 90° apart. We will continue our discussion of calculating power a little later.

When the inductance in henrys in any particular circuit is known, it is easy to compute the inductive reactance and then plug the answer into equations for total circuit values. But inductance can be an elusive rascal, especially in those frequent cases where specific inductances are not deliberately designed into a circuit. There are a number of ways to determine inductive reactance, mostly difficult, but there is one simple backdoor approach. As you know, you can use a *wattmeter* to register real power in a circuit; the power company reads one at your house every month. There is a similar instrument, called a *varmeter*, that will measure directly the vars of reactive power in a circuit or system. The reading can then be used to figure the current flow attributable to reactive power. You can also use an AC ammeter to assist in the proper sizing of conductors and overcurrent protection.

CAPACITANCE

Capacitance is another factor that is fairly common to AC circuits. By definition, this is the property of a system of conductors and *dielectrics* (nonconductors) that permits the storage of electricity when a potential (voltage) difference exists between conductors. Capacitance is measured in *farads* (F) or, since the farad is such a large unit, more often in

$$X_C = \frac{1}{2\pi f C}$$

$$I = E/X_C \quad \text{or} \quad I = 2\pi f C E$$

I = AC amperes
E = AC volts
π = 3.1416
f = frequency in hertz
C = farads of capacitance

$$I = 2 \times 3.1416 \times 60 \times .05 \times 110 = 2073 \text{ amps}$$

Reactive power = $E_{RMS} I_{RMS}$ = 110 × 2073 = 228,080 vars

110V 60Hz   0.05F

Fig. 1-28. Determining the reactance of a simple capacitive circuit.

microfarads. One microfarad equals one-millionth of a farad, and is written mfd or more commonly μF, where the Greek letter μ (mu) is a standard prefix meaning one-millionth. In practice, capacitance is usually introduced into a circuit by means of a specially made device called a *capacitor* or, to use the old term, a *condenser*. Capacitors are widely used in commercial and industrial work, but not often in residential wiring. Some motors, though, are equipped with them.

Capacitance in an AC circuit will also cause reactance, but in this case it is called *capacitive* reactance and symbolized as X_C. Its presence causes current flow in a circuit to *lead* the voltage by 90°. But like an inductor, a capacitor does not consume power. The reactive power is measured in vars and the effective resistance is measured in ohms, just as with inductance. Figure 1-28 shows a purely capacitive circuit, and as you can see, the method of finding capacitive reactance values is similar to that for inductive reactance. In addition, capacitive reactance also can be directly measured with a varmeter.

COMBINING REACTANCES

Some circuits may contain only reactance, either of a capacitive or an inductive nature. Combined values are symbolized as X, which may be converted either to X_C for capacitive reactance or X_L for inductive reactance when differentiation is necessary.

There may be several reactances in a circuit, and arriving at the total values for each type is handled in much the same way as for resistance. Series inductances are added:

$$X_T = X_{L1} + X_{L2} + X_{L3}$$

Parallel inductances are combined by reciprocals:

$$1/X_T = 1/X_{L1} + 1/X_{L2} + 1/X_{L3}$$

Note that with inductive components one special rule must be observed: they must be physically far enough apart so that their respective magnetic fields do not overlap and cause mutual inductance or coupling.

Series capacitances are added:

$$X_T = X_{C1} + X_{C2} + X_{C3}$$

while parallel capacitances are also combined by reciprocals:

$$1/X_T = 1/X_{C1} + 1/X_{C2} + 1/X_{C3}$$

A

Total resistance = impedance = Z.
$$Z = \sqrt{R^2 + X^2}$$
$$= \sqrt{2^2 + 4^2}$$
$$= \sqrt{4 + 16}$$
$$= \sqrt{20}$$
$$= 4.472 \ \Omega$$

R 2 Ω

100V

X   4 Ω

B

Impedance Z = 4.472 Ω.
To find current: I = 110/4.472 = 24.6A
To find true power: P = I²R = 24.6² × 2 = 1210W
To find apparent power: U = EI = 110 × 24.6 = 2706 volt-amperes
To find reactive power: Q = I²X = 24.6² × 4 = 2420.6 vars

Fig. 1-29. Determining the impedance of a circuit containing resistance, capacitance, and inductance.

IMPEDANCE

A common combination in the electrical wiring field is reactance and resistance together in the same circuit. This is called *impedance*; it is defined as the square root of the sum of the squares of the resistance and the reactance of a series AC circuit. More simply, it is the total opposition to AC by an electric circuit, expressed in terms of AC ohms and represented by the symbols Z.

Series Circuit

If a circuit contains a capacitive or inductive reactance in series with a resistance, use the formula

$$Z = \sqrt{R^2 + X^2}$$

to find the value for AC ohms in the circuit. In this formula, R is the resistance in ohms and X is the reactance in ohms. So, if R in a particular circuit is known to be 2 ohms and X is 4 ohms, then impedance Z will equal the square root of $4 + 16$, which is the square root of 20, equal to 4.472 ohms. Figure 1-29A gives an example of how this is done.

Once Z is determined, then the Ohm's law formulas can be used in Fig. 1-29B for the circuit values by simply substituting

Z for R; they are effectively the same. If the circuit contains a capacitive reactance in series with a resistance, the formula and procedure is just the same, except that the symbol X_C is substituted for X_L.

Parallel Circuit

The possibilities are equally good for finding an inductive reactance in parallel with a resistance. In this case, the formula to apply is

$$Z = \frac{RX_L}{R^2 + X_L}$$

Again, the formula and procedure for finding Z when a capacitive reactance is in parallel with a resistance is the same except for the substitution of X_C for X_L.

Complex Circuits

Whenever inductive reactance *or* capacitive reactance occurs in a circuit along with resistance, follow the preceding rules to find impedance Z. However, if *both* inductive reactance and capacitive reactance are present in the same circuit, with or without resistance, the situation changes. You will recall that the current lags the voltage by 90° in an inductive circuit, while the current leads the voltage by 90° in a capacitive circuit. If X_L should happen to be equal to X_C, this means that in a series circuit they will cancel each other, and in a parallel circuit the resulting reactance can be infinitely large.

After the total values for X_C and X_L are found individually, then two new formulas come into use. To determine the total reactance in a series circuit, subtract X_C from X_L by the formula

$$X = X_L - X_C$$

In the parallel case, the formula is

$$X = \frac{-X_L X_C}{X_L - X_C}$$

Note that a negative answer is possible, so in practical electricity, values of inductive and capacitive reactance can be purposely balanced so that they cancel out, leaving a given circuit free from their effects.

In many circuits, all three components—inductive reactance, capacitive reactance, and resistance—may occur

$E_S = 110$ volts
$R = 4 \, \Omega$
$X = 16 \, \Omega$
$I = 6.67$ amps
$Z = 16.49 \, \Omega$

Apparent power: $U = EI = 6.67 \times 110 = 733.7$ VA
True power: $P = I^2R = 6.67 \times 6.67 \times 4 = 177.96$ W
Power factor is the ratio of apparent power to true power:
$P_F = U/P = 178/734 = 0.2425$, or 24%.

B

$E_R = IR = 4 \times 6.67 = 26.68$ V. Thus: $P = E^2R = \dfrac{26.68 \times 26.68}{4} = 177.95$ W

$E_X = IR = 16 \times 6.67 = 106.72$ V.

Note that $E_R + E_X = 133.4$ V, which is 23.4V greater than E_S.

However, the proper way to combine the two voltages is this way:

$E_T = \sqrt{E_R^2 + E_X^2} = \sqrt{26.68^2 + 106.72^2} = \sqrt{12101} = 110$ V

Fig. 1-30. Determining the power values of a circuit containing resistance, capacitance, and inductance.

at the same time. All that is necessary then is to solve for X and then use that value in the formula

$$Z = \sqrt{R^2 + X^2}$$

in series circuits, and

$$Z = \frac{RX}{R^2 + X^2}$$

for parallel circuits. Z equals the total impedance in AC ohms of the circuit.

POWER IN AC CIRCUITS

The power in an impedance circuit is measured in three ways. Since resistance is involved, there will be a *true power* figure in watts. Reactance will bring about a *reactive power* figure in vars, and there will also be an *apparent power* in volt-amperes or VA, which is the combination of the true and the reactive power. Referring back to Fig. 1-29B, you can see how the values for each are determined. Note that though apparent power has no formal symbol, it is sometimes designated U, which also happens to be the symbol used for radiant energy. The three different kinds of power are a bit confusing, but perhaps the chart in Table 1-3 will help you to sort them out.

Figure 1-30A shows another circuit that contains both resistance and reactance. In this case, all values are known except for the various power ratings, which are then computed. You can see that the apparent power of the circuit is about 734 volt-amperes, and this is the figure that must be used in sizing the conductor and overcurrent protection device capacity. If you mistakenly used the formula $P = IE$, the power consumption would appear to be the same figure of about 734 watts—but these are only "apparent" watts. Part of that power is *wattless power* since it is reactive and actually does no work. Only the resistance R represents *working watts*, work actually accomplished. Since we know the values of both R and I, we can find the true power comsumption of R, which is 177.96 watts.

The relationship between the working watts and the wattless power in a circuit is called the *power factor*, a ratio that is usually expressed as a percentage. By dividing the true power by the apparent power, you can see that the power factor of this circuit is only about 24%, which is not very good. While the power factor of a pure DC circuit is unity, or 100%, because the current and voltage are always directly in step, the same is normally not true in AC circuits. For one reason or another in AC circuits, the current is always out of phase with the voltage at least a little bit. However, AC circuits can be

Table. 1-3. The Three Types of Electrical Power

POWER	SYMBOL	FORMULA	UNIT	UNIT ABBR.
apparent	U	$U = EI$	volt-ampere	VA
true	P	$P = I^2R$	watt	W
reactive	Q	$Q = I^2X$	volt-ampere reactive	var

Fig. 1-31. Complex circuit reduction to equivalent circuits to determine various values.

$X_A = 2237.2\ \Omega$
$R_B = 90\ \Omega$
$X_D = 1077.9\ \Omega$
$Z_F = 378.4\ \Omega$

Resulting equivalent circuit.

L3 = 4H
L4 = 6H
C1 = 75 µF
C2 = 90 µF
C3 = 125 µF
C4 = 110 µF

E_S = 110V 60 Hz
R1 = 160 Ω
R2 = 20 Ω
R3 = 70 Ω
L1 = 1H
L2 = 3H

Leg A: $X_{L4} = 2\pi fL = 2\times3.14\times60\times6 = 2261\ \Omega$

$X_{C4} = \dfrac{1}{2\pi fC} = \dfrac{1}{2\times3.14\times60\times0.00011} = \dfrac{1}{0.042} = 23.8\ \Omega$

$X = X_L - X_C = 2237.2\ \Omega$

Leg B: $R = R2 + R3 = 90\ \Omega$

Leg C: $X_{C2} = \dfrac{1}{2\pi fC} = \dfrac{1}{2\times3.14\times60\times0.000090} = \dfrac{1}{0.034} = 29.4\ \Omega$

$X_{C3} = \dfrac{1}{2\pi fC} = \dfrac{1}{2\times3.14\times60\times0.000125} = \dfrac{1}{0.047} = 21.3\ \Omega$

$X_C = X_{C2} + X_{C3} = 50.7\ \Omega$

Leg CD: $X_{L3} = 2\pi fL = 2\times3.14\times60\times4 = 1507.2\ \Omega$

X_C (from leg C) = 50.7 Ω

$X = \dfrac{-X_L \, X_C}{X_L - X_C} = \dfrac{-1507.2\times50.7}{1507.2 - 50.7} = \dfrac{-76415}{1456.5} = -52.5\ \Omega$

Leg D: $X_{L2} = 2\pi fL = 2\times3.14\times60\times3 = 1130.4\ \Omega$

X (from Leg CD) = −52.5 Ω

$X = X_{L2} + X_{CD} = 1130.4 + (-52.5) = 1077.9\ \Omega$

Leg E: $X_{C1} = \dfrac{1}{2\pi fC} = \dfrac{1}{2\times3.14\times60\times0.000075} = \dfrac{1}{0.028} = 35.7\ \Omega$

Leg EF: $R_1 = 160\ \Omega$

X (from Leg E) = 35.7 Ω

$Z = \dfrac{RX}{\sqrt{R^2 + X^2}} = \dfrac{160\times35.7}{\sqrt{25600 + 1274.5}} = \dfrac{5712}{163.9} = 34.9\ \Omega$

Leg F: $X_{L1} = 2\pi fL = 2\times3.14\times60\times1 = 376.8\ \Omega$

manipulated by balancing reactance and resistance values to approach a condition of unity power factor.

Another way to find the power consumption of R is to determine the voltage drop across R, as in Fig. 1-30B, and then use the formula $P = E^2/R$. But if you then find the voltage drop across X and add the two, you will notice that the sum is greater than the source voltage. How can this be? The reason is that the voltage and the current in the circuit are out of phase. To take this fact into consideration, you must use the square root of the sum of the squares, and the voltages will then equal the supply voltage E_S.

As is the case with complex DC series-parallel circuits, their AC counterparts must be broken down piecemeal, leg by leg, to arrive at a set of values for the complete circuit. Figure 1-31 shows you how this is done. Once the breakdown is complete, other values such as the current draw or power consumption in each leg or in combinations of legs can be determined in the usual manner.

CONCLUSION

The foregoing fundamentals are simplified, in that much more could be added in the way of still more fundamentals, additional refinements, special cases, exceptions to the rules, side effects, and so on. In the electrical field, there are rules and formulas for about everything imaginable, but most of this information would only serve to confuse the issue for the do-it-yourself home electrician. In fact, some of the material discussed here may not be of immediate practical value to you, such as the calculations for series-parallel AC reactive circuits. Nonetheless, by understanding the major elements of how electricity works for you, you will have a better grasp on how to design and install an efficient, functional, and safe system.

Chapter 2
Rules and Regulations

American inventors were quick to grasp the possibilities of harnessing electrical power, American businessmen were quick to see the potential, and the American people were quick to accept the use of electricity. Well before the year 1900, electrical power was in relatively widespread use in all major population areas of the country, providing energy for production work in the factories and mills and mines, and energy for convenience and pleasure in hotels, shops, and homes. And right along with the tremendous assets of electrical power as a servant came the liabilities. For electricity, like fire, is a force to be handled with great care and treated with respect, as many found to their sorrow. Minor mishaps, major disasters, personal injuries and deaths began to occur with alarming regularity.

Many of the problems were recognizable in those early years, many more were not. The state of the art was not well advanced, technology and electrical expertise was in its infancy. There was no central agency, no guiding hand, little widespread exchange of information and data that would fruitfully amalgamate the young electrical industry and the electrical installation trade, and which would offer parallel possibilities of advancement in technology, application, and safety to all concerned. Nor were there any standards worth noting, or protection of any sort either as to quality, effectiveness, or safety for the consuming public.

THE NATIONAL ELECTRICAL CODE

In 1897 a group called the National Conference of Electrical Rules—established somewhat earlier and consisting of interested and concerned men from various segments of electrical and other industries from around the country—sponsored a publication called the National Electrical Code (NEC). This was the first attempt at any sort of codification in the electrical wiring field, and one can assume that the impact was something less than resounding. In 1911 the National Fire Protection Association (NFPA)—started some 15 years before in the interests of studying and improving upon the methods of fire prevention and protection—took over the task of revising, updating, enlarging, and publishing the National Electrical Code, and has continued to do so ever since (Fig. 2-1).

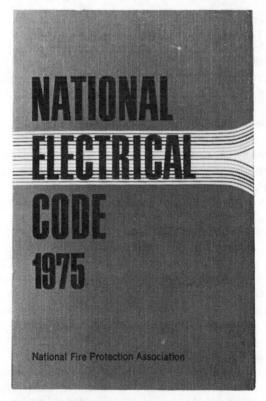

Fig. 2-1. The National Electrical Code, in its latest edition, is an essential reference book for anyone doing residential electrical work of any kind.

The NFPA is not a regulatory or enforcing body, and the National Electrical Code is not a textbook, engineering manual, or installation handbook. But it is a complete and comprehensive set of guidelines for the use and installation of electrical material and equipment, compiled with one particular point in mind—safety. Electricity can be a dangerous force, and the NEC provisions are aimed at minimizing conditions that might lead to fire, explosion, electrical burn and shock, and similar personal hazards, insofar as the present state of the art permits.

The material contained in the NEC is a distillation of field experience, practical application, laboratory research, and exhaustive testing. The process is a continuous one, involving countless thousands of hours of time and hundreds of men, over the past three-quarters of a century. The resulting information should be regarded by anyone working in the electrical wiring field as essential to achieving a minimum safety level.

Nonconformance to this standard does not necessarily mean that any given installation may be hazardous, though that certainly might be the case. If the nonconformance were on the plus side, then the job could well be a bit safer than standard. But if some requirement were disregarded, the result could be the opposite.

For instance, the NEC allows the use of plastic-sheathed, nonmetallic cable in wiring frame houses. This is a widespread practice, and in most areas is considered adequate. You might, however, elect to run all wiring conductors in pipe or tubing. The resulting wiring job would be a safer one, assuming the job was done properly, because of the greatly increased protection afforded the conductors. On the other hand, you might use the same type of cable on a circuit exposed to the weather, or directly buried in the ground. This use, however, is not allowed by the NEC because all information and experience points to the fact that the cable will not stand up under such use for any length of time. Failure and malfunction will occur eventually, a potentially hazardous situation exists, and the wiring is substandard (Fig. 2-2).

The NEC is published in revised editions on a regular basis. The one in use at the time of this writing is the 1975 Edition, with the next one, the 1978 Edition, scheduled to be published in September of 1977. It has also been adopted in its entirety by the American National Standards Institute (ANSI). As far as ANSI and the NFPA are concerned, the provisions of

Fig. 2-2. An illegal usage under the NEC. This outdoor receptacle is not provided with a weatherproof cover, so it is wide open to the ravages of sun, rain, snow, and dust, as well as mud splashing up from the ground.

the Code are informational and advisory. But the NFPA, which holds copyright to the Code, allows any or all of the material to be used by public authorities so that it may be adopted in the form of laws, ordinances, regulations, or orders.

Though the idea was somewhat slow to catch on, today most areas in the country have adopted the NEC as law at state, county, or municipal levels. Despite the fact that in many places enforcement and inspection exist in little more than name alone, the NEC is the recognized authority for electrical wiring standards everywhere. The Codebook has become the "bible," usually sworn by—and at, once in a while—by most electricians, electrical supply dealers, manufacturers, inspecting authorities, insurance companies, and governmental officials at all levels. It only makes sense that you too should familiarize yourself with its provisions, and then follow them.

Your library may have a copy available, or they may be on sale at your building department offices. Failing that, you can order a copy from the National Fire Protection Association, Publications Sales Department, 470 Atlantic Avenue, Boston, Massachussets 02210. The publication number is NFPA #70, but be sure to request the latest edition when you order. Current price is $5.50, but this is subject to change at any time. Another publication, #SPP-6A, the NFPA Handbook of the National Electrical Code, is also available. This book includes the complete text of the NEC and adds explanatory notes, drawings and diagrams, application notes, and interpretations as appropriate. A third book, NFPA #70A, is an extraction from the full NEC, covering only the material applicable to one- and two-family dwellings.

UNDERSTANDING THE CODE

The NEC does indeed cover a great deal of ground, and a certain amount of the material will probably be of no interest to the homeowner who wants to wire his own house. And yet, an overall acquaintance with the whole volume can do no harm, and it may give you a more thorough understanding of that part of the material that directly affects your own installation, by virtue of comparison. The first four chapters, which cover definitions and requirements, methods and materials for general wiring applications, and equipment used in the installation of wiring systems, will be essential. In addition, you might find need for some of the information in following chapters as well, such as for swimming pools or recreational vehicles, or certain hazardous locations.

There have been two principle complaints about the NEC by those who use it. Some say that it is hard to read, and harder to understand. Others say that it is difficult to look anything up and follow through to a solid, well-defined answer. Both complaints are valid in some respects, but on the other hand, the Code was never meant for a training guide, lending itself to easy digestion. Partly of necessity, it is written in a language of its own in order to satisfy the needs of insurance companies, governmental authorities and the legal profession. To most of us, for instance, a *device* is simply a piece of equipment used for some particular purpose, or it might be a scheme of some sort. But in the NEC, a *device* means one thing and one thing only: "A unit of an electrical system which is intended to carry but not utilize electrical energy."

But in other respects, the complaints are not valid. Every field has its language, after all, whether it be auto mechanics, golf, computer programming, or backpacking. And the stranger to any field is likely to be a bit bewildered until he gets himself organized and oriented. The truth is, once you get the hang of it, the NEC is not a difficult book to get through and understand. The two biggest drawbacks probably are the small print and the many pages.

By remembering a few simple points, you will have no trouble in fathoming the NEC. In the first place, much of the material consists of flat statements characterized by the use of the word *shall*. Whenever you see this, the meaning is that this is a *mandatory* requirement—you have no option, you will or will not do thus-and-such, whether you enjoy it or not. The remaining material consists of advisory notes and recommendations, which simply means that while you need not "necessarily" do thus-and-such, you would still be well advised to do so because that is a better or more effective procedure.

Some of the material is judgemental, being left up to the discretion of the inspecting authority, if one exists, or to the user of the Code. For instance, rigid metallic conduit (pipe) is an old standby in general wiring; it can be used under just about any conditions, including direct burial in the earth. But some soils are highly corrosive to various metals, and conduit can't be buried in such locations unless the material is judged suitable for the conditions, or perhaps not at all. So the questions arise: How corrosive is the earth, and to what metals? What type of protective coating is needed? Should another wiring method be used, or perhaps a non-metallic conduit? Can damage result, or malfunction, or deterioration, and if so, over how long a period of time? Questions of judgement, you see, are brought about by local conditions and are capable of being properly answered only by those directly on the scene who have some expertise in the matter.

As far as terminology is concerned, some terms will be strange to you, and ordinary dictionary definitions may or may not be precise. One of the first things to check, however, is Article 100 of the Code, which contains specific definitions of many of the terms used in and, to a degree, peculiar to the Code. This brief section is by no means all-inclusive, but you will find many more definitions, both formal and working explanations, throughout the pages of this text and in the Glossary. If you fail to find a needed definition or explanation either here or in the Code, stop in at your local library and look through their reference material in the electrical field. As a final resort, check with your electrical supplies dealer. Though a few are not very charitable to persons they consider to be "Sunday sparkies" or moonlighters, most are only too happy to help out with any problems you might have, especially if they sense a dollar or two in the offing.

The easiest way to master the NEC is first to look through the table of contents, so that you can see how the book is arranged and what the major information groupings are. Then skip to the back and run through the index, which will give you a more detailed breakdown on just what is covered, and where. Finally, just start at the front and ramble on through the entire book, or at least those parts that apply to your needs, to get the overall picture of what the NEC involves and how it will affect your work.

You will note that everything in the Code is listed in logical groupings. One article covers grounding, another treats with services, a third is on boxes and fittings, and so on—rather than being lined up according to types of installations, or conditions, or classes of service. This means that in order to figure out all of the details for any particular installation, or to find the answer to a particular problem, you will have to jump back and forth in the text a number of times before you gather up all of the bits and pieces.

This is not a difficult process, however, and is mostly a matter of common sense and either knowing or finding out exactly what you are after. Let's suppose that you don't have enough places in your living room to plug things in, so you decide to add an existing circuit. What material in the NEC applies? You look up *plugs* in the index; nothing there. You must have the wrong term. How about *duplex*? Nothing there, either. Well then, *convenience outlet*? Success! And you are referred to *receptacles*. After absorbing all the pertinent information here, you go on to the section on *boxes*, since the receptacle will be mounted in a box. Then you go on to *grounding*, and *conductors*, and *circuit extensions*, following all of the noted referrals to other articles and jumping around among the various elements and procedures required to wire up a convenience outlet.

Sometimes questions or puzzlements arise that have to do not with the requirement itself, but rather with how to accomplish a desired result. For instance, Article 370, which covers boxes and fittings, states in Section 11 that "Plaster surfaces that are broken or incomplete shall be repaired so that there will be no gaps or open spaces at the edge of the box or fitting." This is quite straightforward, but leaves the process up to you. You can apply Spackle or plaster or cement or whatever; the inspector probably doesn't give a hoot how you do it or what you use, within reason, just so long as the gaps and openings are, in his opinion, effectively and permanently sealed.

To take another example, Article 250, Section 32 says that "Metal enclosures for service conductors and equipment shall be grounded." Period. But what exactly is to be grounded? What constitutes an acceptable ground? What hardware or equipment do you use to effect the ground, and how do you go about doing it? Some of that information you can find or deduce by digging around in other parts of the NEC. The rest, however, has to do with accepted trade practice in installation, and acceptable and available material to do the job. Some of these details you can find in this text, others from your supplier. Electrical inspectors are often quite willing to help out with suggestions, but you shouldn't call upon them to engineer your wiring system or teach you the wiring trade. While many would be perfectly capable of doing so, this is not their job and they are generally overworked as it is. Common courtesy dictates that you not pester the inspector unless some unusual problem arises that may include a judgmental decision or a Code interpretation on his part.

Questions sometimes arise over the word *approved*, which frequently appears in the NEC. By definition, this means "acceptable to the authority having jurisdiction," which really doesn't say a whole lot. There are three principle points, though. "Approved" equipment or material is that which has been tested by a private laboratory or testing-and-research firm, and each item usually carries a seal, label, or some notation to the effect that Such-and-Such Testing Labs considers this to be a safe and usable piece of equipment. One

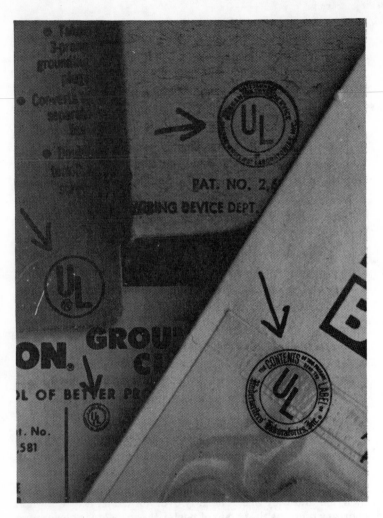

Fig. 2-3. All electrical equipment and material should show the UL or some other recognized seal of approval.

of the most widely known and respected seals is that of the Underwriters Laboratories, the familiar UL listing stamp (Fig. 2-3). This seal affirms that the article—not that particular one, but a representative production run sample—has been tested and holds up to minimum safety and quality standards.

"Approved" methods or manners are simply those methods and procedures that are in common use throughout the general wiring field and have been judged safe, useful, effective, practical, and workmanlike. Such methods may or may not be stated specifically in the NEC, but they are found in study courses for electricians, for instance. Under certain conditions, soldering a joint between two conductors may be an approved method, provided that the solder joint is a good one. But it is up to the solderer to know the difference between a bad solder joint, which would not be an approved method, and a good solder joint, which would constitute an approved method.

An "approved" type of equipment means that some certain design or construction of an item renders it usable under some specific condition. An example of this would be that a lighting fixture to be mounted in a corrosive location must be made and approved for that purpose. Usually such equipment or material is stamped or labeled as being suitable for whatever the special condition might be.

In the final analysis, the ultimate approval rests with the inspecting authority. Just because a method or a piece of gear is approved by somebody, somewhere, does not automatically mean that it will also be approved in every individual case. But by following approved methods and procedures, and using equipment and materials that bear a stamp of approval, the chances are good that you will win any inspector's approval, unless he notices something awry that you do not. And that brings us back again to the judgmental decisions.

LOCAL REGULATIONS

While many areas in the country use and depend upon the National Electrical Code as the sole basis for electrical wiring regulations and inspection, some do not. Instead, they use the NEC as a basic platform and go on from there. In the simplest cases the local jurisdiction may use the entire NEC plus a few additional regulations that they feel will better serve the local conditions. Other locales, especially some of the larger cities, operate under a complete municipal electrical code that refers back to the NEC whenever necessary or advisable (Fig. 2-4). Many things allowable under the NEC may not be under local codes, and there may be many additional requirements as well in areas not touched upon or expanded by the NEC.

The NEC, for instance, says something to the effect that general-use snap switches—the standard type that you have mounted on your walls—can be used with tungsten filament lamp loads if the load doesn't exceed the current rating in amps of the switch at 120 volts. The Los Angeles City

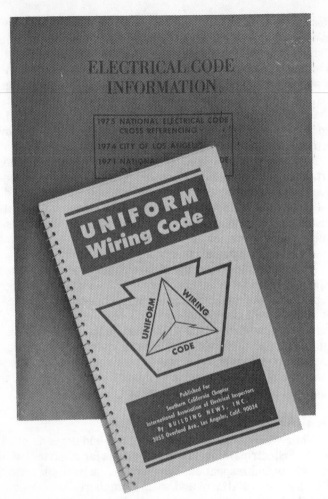

Fig. 2-4. In many areas of the country, local and regional codes are supplementary to the NEC and must be observed.

Electrical Code, however, takes exception to this and goes a step further along the road to safety. They disallow the use of general-use snap switches for tungsten lamp loads, and require the use of "T" rated switches designed to withstand the heavy start-up currents of tungsten lamps.

The NEC says only that temporary wiring services must provide overcurrent protection of appropriate rating. The Los Angeles City Code adds the requirement (undoubtedly based on past experience) that those overcurrent protection devices *shall be* circuit breakers of the tamper-proof type. And though the NEC has much to say about cooking appliances, no minimum circuit capacity is established beyond what is necessary to operate the appliance in question. Los Angeles, though, specifies that the minimum size circuit for an electric range with three or more burners and one or more ovens *shall be* 40 amperes.

Your first task, then, is to familiarize yourself with the National Electrical Code, at least as far as rules and regulations are concerned. This process will be much easier, by the way, if you first become acquainted with electricity and electrical wiring, so that you will have at least some general idea of what you are reading about. Once you have all of this in hand, find out what variations are introduced by local codes by comparing the two. Local codes may be written at city, county, or state level, or all three, but you can get the details from either the city or county building department offices nearest you. If you live in a rural area where there is no formal building department, contact the nearest governmental official of any level and go from there.

Wherever electrical codes are in effect and are enforced, you have three things to do that are solely your responsibility. First you must apply for any necessary permits, which might be simple and all-purpose, covering every phase of building or remodeling, or may be multiple, one for each of the trades including electrical. Second, you must pay any necessary fees. And third, you must arrange with the electrical inspecting authority for inspections. Several may be required, depending upon local codes, the nature of the work, or the desires of the inspector. The usual situation, though, is one inspection when the entire job has been roughed in—no finish coverings have been applied and everything is still out in the open—and a second inspection when the job is completed and ready to function. It is up to the installer to call for inspections, and often the inspector must make his rounds within a certain time period after the call. In many areas, noncompliance with the inspection requirements can result in rather discouraging penalties such as stiff fines, or worse, refusal of an occupancy permit.

SAFETY PRECAUTIONS

The proper and safe use of electricty is governed first by the physical laws of the force itself, the "natural" laws. Electricity acts as it acts, and we must work within those boundaries. Secondly, use is governed—as far as the buyer, user, and installer of electricity and its associated equipment in the residential wiring field is concerned—by the various laws, rules, regulations, and conventional practices that we ourselves have established, such as the National Electrical Code.

There is also a third factor involved that should always be kept firmly in mind, and that is good old horse sense. There is no way to place too much stress upon the fact that electricity is a powerful (and invisible) force that can be downright dangerous to life, limb, and property. And most problems that arise can be traced to ignorance, stupidity, carelessness, corner-cutting, and just plain lack of common sense and disregard of the facts. Certainly, these are not very good reasons for endangering yourself, your loved ones, your friends and guests, and your hard-earned property.

Most people realize that electricity can start a fire. That's the principle behind the operation of the cigarette lighter in the dashboard of your car. Electrical fires can start from a short circuit or from overloading a circuit so that the conductors heat up over a period of time, which can sometimes be quite lengthy. They can start from poorly maintained equipment, overworked appliances, aged and cracked or repeatedly overheated insulation. They can start from high ground impedances, or mice gnawing at the conductor insulation, or a hundred other reasons. The point is, they can start all too easily if both the installer and the user of the electrical system do not make every effort to use their heads and follow the rules.

Most people also realize that unleashed electricity can seriously injure or kill. Few understand how that happens—or how easily it can happen. A good many studies have been made about the tolerance of the human body to electrical force, and the findings are both interesting and revealing. To give you something to compare, note that the usual household electrical supply is 230 or 115 volts, alternating at 60 cycles per second. The current drawn by a standard 100-watt bulb is a bit under 1 ampere, while a toaster might require on the order of 7 amperes. A doorbell operates on about 10 volts.

Tests have shown that some individuals can stand no more than 12 volts AC at around 7 or 8 milliamperes (1 mA equals 1/1000 amp) with good contact before they have to let go of the leads. Others tested could withstand from 20 volts to as much as 40 volts at the same current and frequency, with the leads held in dry hands, but for only a few seconds before they lost voluntary control of their arm muscles. A few with heavily calloused, dry hands could briefly withstand up to 120 volts because of the high initial resistance of the human skin. But only a few seconds of contact was needed to break the skin down and cause blisters at the contact points, greatly increasing conductivity. The sensation of shock occurs at about one milliampere; strong discomfort at about eight milliamperes. Under the right conditions, a jolt of only 30 milliamperes—about 3% of the current draw of a 100-watt bulb—can be fatal.

In the event of electrical shock, either of two things may happen. First, the current may pass through the body in fractions of a second and interfere with the normal function of the nerves at the breathing control center at the back of the neck. The impulses from the nerves are cut off, the breathing muscles no longer respond, so artificial respiration must be started immediately and continued until the nerves recover. They may not. Second, the current may stop the heart entirely, or perhaps cause it to go into fibrillation, which can be just as lethal. Again, artificial respiration is the initial treatment for any serious shock, perhaps followed by more drastic measures.

This, then, is the force that we just take for granted, the servant that is always there at the flip of a switch, and without which our lives would be dull and tedious indeed. With a full knowledge of the harmful potential of electricity as well as the beneficial, you can see that a bit of thought, care, and common sense will result in maximum benefits and safety, and minimize any potential problems. Electricity will never bite back if you don't give it a chance.

Chapter 3

The Residential Wiring System

As a matter of convenience, an electrical wiring system can be broken down into a number of sections, each with its own definition and terminology. In different areas of the country there are sometimes a few variations in terms as well as some colloquial expressions used to define parts of the system, but on the whole you should have little trouble keeping them straight. There is some overlap, too, between the parts of a residential system and the parts of commercial or industrial installations, the latter being much more complex. Everything that follows here, though, will refer particularly to residential systems.

SERVICE

The most obvious part of the wiring system is the *service*. This term includes all of the equipment and conductors necessary to carry the electricity supply, from the nearest source point maintained by your local power company, into the residence itself. If you look out your window you may see power poles with a series of high-voltage lines strung on them. The objects mounted high on the poles—they look like trash cans with knobs—are transformers that supply stepped-down power from the lines, usually in the form of 115 or 230 volts AC at 60 Hz, single-phase, three-wire. The service part of the installation runs from the secondary terminals of this transformer to the distribution equipment in or on the house.

The main wires that go from the transformer to the house are called *service conductors*, or sometimes *supply conductors*. If they droop from the transformer pole to a point of attachment on the house, the span of wire from point to point is called a *service drop* (Fig. 3-1). These conductors may be multiple spans of single wires hanging free in the air. But if the conductors are run together in one protective jacket, this is called a *service cable*. Some types consist of three conductors, two with their own jackets and one bare, twisted together in a continuous spiral; this is sometimes referred to as *triplex*.

Not all service conductors are aerial. In many cities and some subdivisions, service comes from an *underground main* or *transformer vault*. Many individual homeowners prefer to hide the service conductors underground from the distribution pole to the house (Fig. 3-2). In this case, the service conductors are called a *service lateral*, which extends from the point of connection at the transformer to the first junction box inside the house. If there is no junction box, the service lateral ends at the point where it enters the building.

In fact, all service-conductors change, if not actually then at least in name, at some particular point in any system. In the case of the service lateral, the service conductors become *service-entrance conductors* at the first junction box inside the building. That is, from that first junction box to the following service equipment, they become known as service-entrance conductors. If there is no junction box, the name changes to service-entrance conductors at the point where the service enters the building. In the case of an aerial system, the service-entrance conductors are those that are tapped onto the service drop at a point close to the building, running from there into the building and to the service equipment.

The *service equipment* in its strictest meaning is simply a disconnect device in a suitable enclosure, which can be used to disconnect the service from loads inside the building. This can be a pull-out set of fuses, a circuit breaker, or a switch. This is usually tagged as the *main disconnect, main fuse, main breaker, main switch*, or simply "*the main*." Most often it is mounted inside the building, but can also be placed outdoors if in a weatherproof enclosure. Sometimes, as in the case of a long service lateral, an additional circuit breaker is placed near the supply source, but this is not considered a main. The purpose is only to provide overcurrent protection for the service lateral itself or for convenience when the system has to be shut down. Overcurrent protection of the main is also a part

Fig. 3-1. Typical residential service drop.

Fig. 3-2. Typical residential service laterals.

DIST. POLE

8' MIN.
CONDUIT PROTECTION

METER

GRADE LEVEL

DRIVE

WALK

LATERAL—METALLIC CONDUIT—6" MIN.

LATERAL—NONMETALLIC CONDUIT—18" MIN.

LATERAL—DIRECT—BURIED CABLE—24" MIN.

BUSHING

SLEEVE UNDER
CONCRETE AND PAVING

BUSHING

BUSHING

INSIDE
SERVICE
EQUIPMENT

SERVICE-ENTRANCE
CONDUCTORS
FROM POINT OF
ENTRY ON.

of the service equipment. This protection is an integral function of a main breaker or main pull-out fuse set. When a switch is used as a main disconnect, a set of fuses is mounted within the same enclosure, or a close-coupled second enclosure.

In practice, most main disconnects in residential service are located inside the building in a large enclosure that also houses fuses or circuit breakers to handle the distribution of power throughout the house. This is variously called a *main entrance panel, entrance box, main box*, or some similar term.

The service-entrance conductors are sometimes enclosed in a special raceway or in conduit or tubing, and this is called a *service raceway*. A *service mast* would be one example; this consists of a length of pipe connected to the service equipment and rising through the roof of the building, properly anchored and guyed. The service-entrance conductors travel from the service equipment up through the pipe, emerge through a special cap known as a *weatherhead* or a *service-entrance head*. The conductors come out of the weatherhead at a point higher than the end of the service drop, curve downward in an arc, and then curve upward again to join the service drop. This curve of wire is called a *drip loop*, and its purpose is to allow moisture to run to the bottom of the loop and fall away instead of running along the conductors and into the raceway.

Another example of a service raceway occurs when the conduit and weatherhead, often just called an entrance pipe, are mounted on the outside wall of a building. In other cases a direct-burial service lateral enters the building through a conduit that extends to the service equipment. Here, the lateral becomes service-entrance conductors at the point of entry, and they run through a service raceway to the main disconnect enclosure.

Somewhere between the point of supply source and the service equipment, there must be a meter housing, box, or socket, and a meter (Fig. 3-3). These may be located on the nearest power pole, usually at about eye level, with the service drop running down the pole from the transformer to the meter, back up again, and away into the service drop. Or, the service conductors may become a lateral, dropping down from the meter box into the ground and on to the building. Frequently the meter box is located on the service-entrance conductor line at some convenient point, either mounted on the outside wall or, less often, on the inside, next to the service equipment.

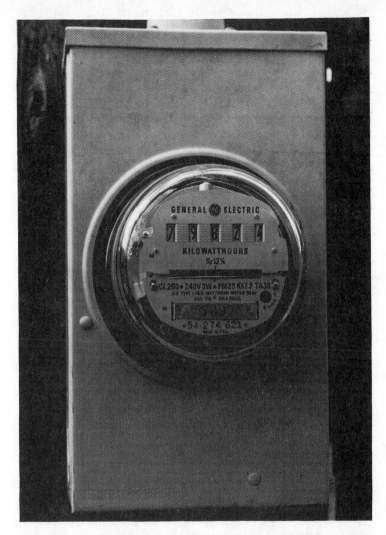

Fig. 3-3. A watt-hour meter, which measures the amount of power consumed, is a part of every electrical system installation.

In nearly all cases the local power company has responsibility for at least a part of the service. In an aerial system, they will supply and install the service drop, complete with all necessary hardware. They will also tap the service-entrance connectors onto the service drop. They will sometimes provide and install the service-entrance conductors down to the top of the meter box, especially when the service is of small capacity and the service cable remains physically unbroken from the pole to the meter box. Individual underground systems are usually the responsibility of the owner of the property. Often the cable may be purchased from the power company, but the owner must install it, from the pole top to the building. In practice, helpful linemen will sometimes offer to secure the service raceway to the power pole when they are making the transformer connections and activating the service. Municipal underground systems are usually the responsibility of the governing authority, though certain installation fees may be charged back to the owner.

There is considerable variation in the matter of meters and boxes. In some areas, the box is supplied to the owner (we'll use the words owner and electrician interchangeably here, in the assumption that you will be both) free of charge and with advice on mounting. The power company then plugs in and seals the meter, again free of charge, when the installation is ready to be "livened up." In other areas, the power company may do the installing of both, either with or without charge. Or, you may be handed the meter box along with a bill, and you might have to buy the meter as well. You will have to contact your serving power company to find out exactly what the situation is. Note, too, that in some instances there may be a need for two meters and their boxes. Both meters may be tapped to one set of service entrance conductors, as for a two-family dwelling or for separately metered water heating. Or, one may be tapped to the service conductors while the other is in an "off-peak" water heating circuit that originates inside the building.

FEEDERS

Once beyond the load-side contacts of the meter, everything in the system is the responsibility of the owner. From the point of the main disconnect, the wiring system spreads out through the building. If the main disconnect is by itself, not part of a main entrance panel, then the conductors to a distribution panel are called *feeders*, or *feeder conductors*. There may be more than one such panel, and all the lines to them are feeders. A line from one distribution panel to another is also called a feeder. The sole purpose of a feeder is to carry "bulk" power from one point to another for further distribution there.

BRANCHES

The circuits that travel out from the main entrance panel or from other distribution panels to the points of use, are called *branch circuits*. By definition, a branch circuit is that portion of a wiring system extending beyond the final overcurrent device protecting that circuit. In other words, the line from one particular overcurrent device, which may be a fuse or a breaker, to its end, and nowhere protected by any other fuse or breaker, is a branch circuit. Through these circuits the power reaches its final points of use.

There are four kinds of branch circuits. First, there is the *general-purpose branch circuit*, which supplies both lighting and small appliances. Second, there is the *appliance branch circuit*, which supplies appliances only, nothing else. Third, there is the *individual branch circuit*, which supplies only one piece of equipment or convenience outlet. And last, there is the *multiwire branch circuit*, which contains two "hot" wires and a neutral. Note that in theory any of these branch circuits, and feeders as well, can be either indoor or outdoor, and either underground or aerial if outdoor, or open or concealed if indoor.

LOADS

Each branch circuit, regardless of its type, either terminates in one outlet or serves a number of outlets. Defined electrically, an *outlet* is any point in the system where power is taken out for use. In other words, these are the spots where the various loads are attached, either by direct-wiring or by attachment plugs. And the loads, which taken together are called the *total connected load* (TCL), are the ultimate parts of the electrical system, the reason for the existence of all that goes before. They are sometimes referred to in specific terms, such as a heating load, lighting load, motor load, air-conditioning load, or in the case of a mixed bag, just a general load. The residential TCL generally consists of three principle loads, plus a lesser fourth category.

Appliances

In this world of conveniences and time-savers, the current draw caused by all the various appliances in a home often makes up the bulk of the TCL. *Appliance* is a general term used for utilization equipment that is usually manufactured in certain shapes, sizes, and types, and installed or connected to perform one or more specific domestic tasks. Familiar examples include the washer, dryer, freezer, and range. All told, there are probably hundreds of items that come under this definition. In wiring systems they are broken down into three groups.

A *stationary appliance* is one that normally stays in one place and is not easily moved around, though it could be if necessary without any disassembly of itself or its surroundings, like a refrigerator. A *fixed appliance*, on the other hand, is fastened into place and cannot be removed without something of a struggle. An undercounter dishwasher would be one example, a countertop food preparation center another. *Portable appliances* are legion—toasters, mixers, vacuum cleaners, blenders, radios, hair dryers, crock pots, and on and on. Most are relatively small and easily storable, and all are designed to be readily moveable from place to place at the user's whim.

Some fixed appliances are wired directly into the electrical system, while others may have plug-and-receptacle connections hidden behind them. Stationary appliances are usually connected by means of an attachment plug, but some can be wired directly. Portable appliances all use cord-and-plug arrangements.

Fixed and stationary appliances must be considered as an integral part of the electrical system; in some cases this is mandatory. Insofar as is possible, current and future needs, both physical and electrical, should be taken into consideration in the planning stages of the residential wiring system. While portable appliances are not an integral part of the system, they are a part of the total load borne by the system and so adequate provision, plus a little extra, should be made for them, too.

Lighting

Lighting accounts for a large part of the TCL, especially in a large residence occupied by a number of people, and where architectural and decorative lighting is an integral part or important design feature of the structure. Chapter 5 deals with the details of lighting in depth.

Heating

Heating constitutes a fair part of the total load in most homes. Of course if the home is electrically heated, that portion of the TCL is substantial, and this is discussed in Chapter 4. But homes heated by other means may also have auxiliary electric heating units in bathrooms, for instance.

In northern climates, heat tapes are commonly used to keep gutters, downspouts, or eaves free from ice build-up. Heat tapes are also used on water pipes, and immersion heaters in stock or poultry watering troughs. Special cables are sometimes buried under driveways and walks or embedded in steps to keep them clear of snow, or used in greenhouses for soil heating purposes. Electric water heating is quite common everywhere.

Other Loads

In some cases part of the TCL may be attributable to various sorts of equipment that do not fall into the above categories. This can occur even in small homes if, for instance, a workshop is powered by the system. Large power tools like planers or heavy-duty table and radial saws use a great deal of power. Built-in air conditioning is also a major consumer, though the smaller units designed to serve one room are usually classed as appliances.

There may also be ventilating fans, oil burner transformers, heat blowers, circulating pumps, well pumps, waste water aerators, or any number of other items, large and small, which together can account for considerable power usage. Farmsteads are almost always equipped with various pieces of machinery, sometimes to the extent that they become a major part of the total load.

And last, there always seem to be a few items that do not fit in anywhere and must be classed as miscellaneous. Examples? Well, a model railroad set-up is one, a home weather installation another. Or perhaps an antenna rotating system, or an intercom system, or electronic gear. There are a lot of possibilities, and miscellaneous or not, they should be taken into account when the electrical system is designed.

MAIN PANEL

The main entrance panel mentioned previously may be the distribution center for power throughout the building (Fig. 3-4). Most main entrance panels consist of a metal enclosure of appropriate size, which contains two heavy copper bars, or *busses*, mounted vertically and insulated from the box. Molded fuse sockets may be permanently attached to these busses, or they may be designed to take removable circuit breakers held in place by a spring-clamp arrangement or screws. In addition, there is another silver-colored *terminal block* mounted by itself either vertically or horizontally. The busses carry the "live" or "hot" lines, while the *neutrals* (which aren't always neutral) or "white wires" are connected to the terminal block.

Fig. 3-4. Typical fused-main, service entrance panels suitable for residential application. Left, a 200-amp panel with 200-amp pull-cover and fused main disconnect (top), and four 60-amp and two 30-amp two-pole pull-out fuse holders (middle), and 24 single-pole plug fuses (bottom). Right, a 100-amp panel with 100-amp main fuse pull-out (top), three 60-amp and one 30-amp two-pole pull-out fuseholders (middle), and 16 plug fuses (bottom). (Courtesy Wadsworth Electric Mfg. Co.)

A *surface-mount* entrance panel has a front cover that is screwed on. A *flush-mount* panel is designed to be set into an opening in the wall and has a cover that overlaps the perimeter of the panel to cover the edges of the hole; it includes a smaller hinged door for access to the overcurrent devices.

Most main entrance panels come with a particular size of main overcurrent protection device already installed at the top of the panel and clearly identified. Thus, a 100-ampere main panel may have a 100-ampere 2-pole breaker in it, or it may have a pull-out fuse holder designed for a pair of 100-ampere fuses (not supplied). Below this main breaker will be a row of spaces for other smaller overcurrent protection devices, from which the power is distributed through the branch circuits.

The spaces are numbered as single circuits, but they can be used in various combinations.

There is another type of main entrance panel sometimes used in residential service, called the *split-bus panel* (Fig. 3-5). Here the two main power-carrying busses are split into sections. The top set of busses holds up to six two-pole circuit breakers. One of these feeds directly to a lower set of busses, which are in turn arranged to hold overcurrent devices for small distribution circuits or branch circuits. The remaining five (or fewer, as needed) breakers at the top, protect feeders that go out to remote distribution panels or load centers. No more than six of these breakers may be used. All of them together act as the main disconnect, even though they are not physically connected together and each must be tripped or disconnected individually. So instead of disconnecting the entire residential load from the supply with one motion, as with a master switch or breaker up to six separate motions are needed. However, this is allowable in residential work, and it

Fig. 3-5. Internal wiring diagrams of two fusible, split-bus, main entrance panels suitable for residential application. Left, with 200-amp main lugs, six two-pole pull-out fuse holders, and 16 single-pole plug fuses. Right, 150-amp main lugs, five two-pole pull-out fuse holders, and 12 single-pole plug fuses. (Courtesy Wadsworth Electric Mfg. Co.)

is usually referred to as the "six movements of the hand" rule. This type of panel is also available for use with fuses instead of circuit breakers.

GROUNDING SYSTEM

The grounding system, along with any necessary bonding between elements of the wiring system, is an essential part of the wiring installation. A *grounding electrode conductor* originates at the service equipment or the meter box, where it is attached to the *grounding electrode* itself.

In many installations this grounding electrode may consist of the interior metallic cold-water piping system, including the street main or well casing. The grounding electrode may also by any substantial underground metallic piping network or other metallic object. One or more ground rods are commonly used, too, especially in rural homes using nonmetallic (plastic) piping in the water supply. Rods made for the purpose are available at electrical supply houses, and 3/4-inch trade size, galvanized pipe with a corrosion-proofed inner surface can also be used. As many rods as necessary are driven into the ground, close to the service equipment location, to provide a low-impedance path to earth for any fault currents that might occur in the wiring system.

OVERCURRENT DEVICES

Overcurrent protection devices do just that; when an overload or short in a circuit or feeder line occurs, they automatically open up and disconnect the circuit or feeder from the supply. The two principal kinds are fuses and circuit breakers.

Fuses

There are two basic types of fuses—*plug* and *cartridge*. Both contain specially designed, soft metal links that will melt away and open the circuit when the current exceeds a certain value. Though fuses "blow" almost instantaneously when a short circuit occurs, they may show a definite time lag before opening up on a circuit overload.

The plug fuse (Fig. 3-6) has a *screw base* (like a light bulb) with the body made of heavy glass or a refractory material, and with a window in the top so that the link is visible. The earlier type is called an *Edison base*, and recently these have been superseded by the noninterchangeable *Type S* fuses.

The most common fuse size are 15, 20, 25, and 30 amperes, though in-between sizes are available for special purposes.

Fig. 3-6. Typical plug fuses. (Courtesy Bussman Manufacturing)

Fig. 3-7. Typical cartridge fuses. (Courtesy Bussman Manufacturing)

Type S fuses are arranged in three groups, 1—15, 16—20, and 21—30 amperes, and the groups are not interchangeable because of different base designs. The principle object is to prevent the dangerous misuse of, say, a 30-amp fuse in a 20-amp circuit—an all too common occurance.

Special *time-lag fuses*, with a longer than normal time delay built into them, are also available. These are often used for protection in motor circuits since they will allow the passage of quick and harmless overloads caused by the starting surge.

All plug fuses are rated for 150 volts or less, and not over 30 amperes of current. They are used singly on single-pole branch circuits.

Cartridge fuses (Fig. 3-7) are tubular with current-carrying end caps. These are designed to be inserted into a double spring-clip holder. They look much like the ones in your automobile except the center sections are not made of glass. There is a wide range of ratings from a small fraction of an amp to several hundred amps.

In residential work, cartridge fuses are used for two-pole circuits, including protection for the service-entrance conductors. There are two styles. One has a built-in metal link, and when this separates, the entire fuse has to be replaced. The other type has removable end caps that secure replaceable links, so that the fuse body can be used again.

Cartridge fuses come in a great range of current capacities and are rated for 600 volts or less. The 0—60 ampere sizes use plain end caps, while those from 61 to 600 amperes are fitted with a short blade on each cap.

Circuit Breakers

A circuit breaker (Fig. 3-8) is far more complex in design but more efficient in operation than fuses. A little black plastic box with a switch handle on the front, it looks a bit like a light switch. Breakers come in a tremendous array of ratings, sizes, and configurations, and in special physical and electrical designs for all manner of purposes. Those generally used in residential work are of three basic types—*single-pole*, *single-pole double* or "piggyback," and *double-pole*.

Single-pole breakers are rated at 20, 25, and 30 amperes for the most part. These are used on branch circuits.

Single-pole doubles are designed to provide single circuit. They fit in the same physical space of one breaker slot (or pole) in the distribution panel. The object is to increase the

Fig. 3-8. Typical circuit breakers. (A) 15-amp single-pole. (B) 50-amp two-pole. (C) 20-amp single-pole piggyback allowing two branch circuits in one circuit slot. (D) two-pole double piggyback, allowing four single-pole circuits in two single circuit slots. (Courtesy Wadsworth Electric Mfg. Co.)

flexibility of general-purpose branch-circuit connection options where loads are light and diverse. These are available only in 15- and 20-ampere sizes.

Double-pole breakers are designed to handle 240 volts, so they can be used for two-wire branch circuits or feeders, or for main disconnects. The 20-amp rating is the smallest, then 30, 40, 70, 100, 150 and 200 amperes, with a few less standard sizes in between. Nothing larger is likely to be found in a residence, though standard ratings continue up through 6000 amps.

Unlike fuses, which are done for after an overload, breakers normally need not be replaced. A burned-out breaker, though perfectly possible, would be an uncommon circumstance.

When a short circuit occurs, the internal mechanisms of the breaker open the circuit almost instantaneously, and the handle flips down to the *off* position. As long as the fault remains, the handle cannot be returned to the on position—it simply won't go. Once the line is cleared the handle can be reset, the internal machinery returns to normal position, and the circuit is again functional. In the event of an overload there is a certain time lag before the mechanism trips out, so harmless overloads or surges of short duration are "overlooked."

Breakers also have the obvious advantage of acting as a simple switch that you can use to turn a circuit off and on at will. Though initially more expensive than fuses, they offer greater protection and convenience. They seldom if ever have to be replaced.

Ground Fault Interrupters

There is a new type of protective device now in use called a ground fault circuit interrupter—usually known as a GFCI or just GFI (Fig. 3-9). These look and react somewhat like a circuit breaker, but afford additional protection. Instead of only tripping on a short or overload, they also detect a leakage of current from a "hot" line to a ground, not the neutral line but some other ground such as the metal frame of an appliance. Their purpose is to guard against shock hazard by immediately detecting a current *difference* between the input and output lines, which normally are equal. They react within tiny fractions of a second to open the circuit before any harm is done. They are quite sensitive and are able to respond to leakage currents on the order of 0.006 amps.

Several types of GFIs are now being manufactured. Some are designed to exactly replace ordinary circuit breakers in a breaker panel, thereby protecting an entire circuit. They are rated at 120 volts for single-pole circuits and come in standard sizes of 15, 20, 25, and 30 amps. Others are joined to receptacles and sized to fit in standard wall boxes, with the GFI protecting only that particular receptacle. Some are similar in nature but installed in weatherproof housings for outdoor use.

DISTRIBUTION PANELS

Often a main entrance panel has an insufficient number of circuit possibilities for a particular installation, or perhaps the

DO NOT USE WHERE EXPOSED TO RAIN
TEST BEFORE A PLUG IS INSERTED

Fig. 3-9. One type of GFI protective device. (Courtesy Harvey Hubbell Incorporated)

service entrance equipment may consist only of a main disconnect and associated main overcurrent protection device. In such cases, distribution panels are necessary.

Large distribution panels are usually called *load centers*. They may provide anywhere from 8 to 16 more single-pole circuit spaces. Smaller panels with a lesser number of circuits are usually referred to as *branch circuit panels* or *subpanels*, but their purpose is the same—distribution.

Distribution panels are available in a wide number of capacities for either fuses or circuit breakers. They are made along the lines of a main entrance panel, with busses for the "hot" legs of the feeder circuit and a neutral bus terminal block, but they have no provision for a main disconnect. As with a main entrance panel, the gear is all mounted in a suitable metal enclosure with removable cover, and the panel can be either surface or flush mounted.

BOXES

In terms of sheer numbers, boxes of various sorts are used more in an electrical installation than any other item.

Fig. 3-10. A few of the many types of standard wall or device boxes available. (Courtesy Appleton Electric Company)

Wherever there is a termination of the conductors in a circuit—whether for tap connections, splices, or for devices—there must be a box. In some instances, boxes are used even where there are no breaks in the wiring, simply for convenience in pulling wires through a raceway.

Wall Boxes

The most common box in residential work is called a *switch box, wall box,* or *outlet box*; they are all for the same purpose. There are many different kinds as shown in Fig. 3-10. They come in several depths and configurations and with minor design differences, but they all measure three inches high by two inches wide on the face and are made for flush or concealed mounting. These are the boxes used to house wall switches, convenience outlets, dimmers, and the like. But they are also often used for mounting fixtures, thermostats, low-voltage switches, and TV-FM jacks.

Some boxes are equipped with mounting brackets or flanges for securing them to studs or joists. Others have a good number of holes in them for mounting directly with nails or screws. All have *mounting ears* on the face of the box, which are usually adjustable fore and aft so that the face of the box

Fig. 3-11. Some types of standard wall or device boxes can be partly disassembled and then joined or "ganged" to provide space for two or more standard devices, one per box section.

can be brought out flush with the finished surface when mounted on a recessed support.

The depth of the boxes varies from 2 to 3 1/2 inches. Some are perfectly rectangular while others have beveled top and bottom back corners for easier installation and wire routing. Slotted knockouts are provided; by inserting a screwdriver in the slot and prying back, you can remove a small chunk of the box to create a hole through which the cables will run. Most boxes, except for the deep ones, have only four knockouts, and the rule is one cable per hole. Many boxes are also fitted with clamps, which are tightened down with a screw to hold the cables in place.

The sizes most commonly used are probably the 2 1/4-inch and 2 1/2-inch depth variety—they are generally referred to as *standard wall boxes*. Those deeper than 2 1/2 inches are called *deep boxes*. Those shallower than 1 1/4 inches are called *shallow boxes*, but these are used only when there is no other way out of a special problem, such as in a furred-out wall over concrete block with only a small space between the block and the finished surface.

In addition, while some of these boxes are of solid construction, many kinds are "gangable." That is, by loosening a screw you can remove a side plate, then join another box (also minus a side plate) to the first as in Fig. 3-11. Tighten the screw again and you have a *double width* or *two-gang* wall box. In theory you can gang as many boxes as you like, but triples or an occasional four are about the practical maximum. Their use is for mounting a pair of duplex convenience outlets and a switch, or three outlets, or whatever combination of devices is necessary. Larger supply houses also stock solid-construction wall boxes manufactured in two, three, and more gang sizes.

Most wall boxes are made of steel, but there are some types of plastic boxes. These are nonconductive and inexpensive but less utilitarian and rugged than steel, so they see rather limited use. Most electricians settle upon the particular brand and type of wall box that they like best, then use that for everything possible; they buy something different only out of necessity.

Junction Boxes

The *junction box* (Fig. 3-12) is also widely used, mainly for two purposes—as a connection point for two or more pairs of conductors, and for mounting and connecting lighting fixtures. Some types are designed for surface mounting, while others have mounting brackets which, when properly positioned, allow them to be flush mounted. When used solely for connections, they are fitted with a cover secured by screws. When used to hold most types of lighting fixtures, a special *fixture mounting bar* is secured across the box opening with the cover screws, and the fixture is then attached to the bar. There is also a type of fixture that can be mounted directly onto certain junction boxes.

Like wall boxes, junction boxes come in a variety of sizes and shapes. The most popular seems to be the 4-inch *octagonal* (eight-sided) type, which is made in depths of 1 1/2 and 2 1/8 inches. Next in line is the 3 1/4-inch octagonal, then the 3 1/2-inch octagonal, both of which are 1 1/2 inches deep. These are the sizes generally used for fixture mounting, and for which special lampholders called *porcelain receptacles* are made (despite the fact that they are used for light bulbs and not attachment plugs). The 4-inch square box is also

Fig. 3-12. Several types of square junction boxes. (Courtesy Appleton Electric Company)

commonly used, and so is the 4 11/16-inch square box. All of these junction boxes are equipped with knockouts of assorted sizes, ranging from small cable size to 1/2- and 3/4-inch trade-size pipe openings. Extension rings that increase the depth of the boxes are also available; these attach with the cover mounting screws.

The so-called *handy-box* or *utility box* is quite popular for surface-mounting jobs. It can be used with either cable or conduit. The height and width are about the same as those of a wall box, and it will accept one device frame. These boxes are not gangable.

Boxes that are intended for mounting in a floor, flush with the surface, are of a special design. Generally they are cast in one solid piece and made dust-proof. They are mortised into the floor so that nothing protrudes above the surface. The receptacle is sunk deep into the box, and the cover is gasketed to keep out dust and moisture. The main cover is provided with a pair of smaller covers that thread into it. One is solid, to be used when there is no attachment plug in the receptacle, and the other has a hole through which the attachment cord runs, sealing the box when the attachment plug is in place.

Then there are *weatherproof boxes* designed to be used outdoors, with no protection from the weather. These are cast in one piece and have special gasketed covers or doors over the devices or equipment mounted in them. While they are not water tight, they will hold out rain and snow, as well as dust and dirt and harmful rays from the sun. Threaded openings are provided, which can be used either with pipe, EMT, or weatherproof cable and connectors as the circumstances dictate. There are various sizes, but the most common accept either one or two devices on standard frames. Junction boxes are also available, as well as specially designed enclosures for disconnect switches, circuit breakers, lighting fixtures, and other equipment.

Boxes of larger size than the above are called *pull boxes*. Their use is for making connections in large numbers of conductors, or for convenience in pulling long spans of conductors in from one raceway and rerouting them out through others. Unless the system is especially complex, the use of pull boxes in residential work would probably be confined to pulling in or splicing a service lateral. There is a great range of sizes and shapes available, some with screw-attached covers and others with hinged covers, some with various sizes of knockouts and others blank. Whatever the problem, there is a pull box to solve it.

DEVICES

The term *devices* covers a lot of area. By definition, a device, as it relates to the field of electrical wiring, is any unit of an electrical system that is intended to carry current but which does not consume any power; it is a carrier, but not a worker in itself.

Receptacles

The most common device is the *receptacle* (Fig. 3-13), found by the dozens in every home. Strictly speaking, a receptacle is a contact device installed at an outlet for the connection of a single attachment plug. Most of them are *double receptacles*, also called *duplexes*, or *duplex convenience outlets*, or maybe (and incorrectly) "double

Fig. 3-13. These are but a few of the many types of receptacles available for general wiring purposes. (A) 20A-125V single. (B) 15A-125V duplex. (C) 20A-125V duplex. (D) 15A-125V triplex. (E) 2-wire 15A-125V single. (F) 2-pole 3-wire 15A-125V locking. (G) 2-wire 15A-125V duplex on 4-inch cover. (H) 15A-125V clockhanger. (I) 15A-125V surface duplex. (J) combination switch and single receptacle. (K) 2-pole 3-wire 30A-125V surface power. (L) 3-pole 4-wire 30A-125V/250V panel power. (M) 15A-125V duplex. (N) combination 15A-125V single and 15A-250V single. (O) 15A-125V single weatherproof. (P) 2-pole 3-wire 30A-125V flush power. (Courtesy Leviton Mfg. Co.)

plugs." There are many types of receptacles made for a great range of purposes, but in the home you normally would have use for only a few.

Duplexes are designed to be mounted in standard wall boxes and will also fit handy-boxes and some junction boxes. They are rated for 120 volts AC, and those commonly used in the home have either 15- or 20-ampere capacity. There are two types, the *nongrounding* that takes a two-prong attachment plug, and the *grounding* that takes a three-prong attachment plug. Single 120-volt receptacles are also available, but seldom used nowadays except in floor outlets.

Specially designed receptacles are often used with heavy appliances or equipment such as ranges, driers, and welders. Different prong configurations are used for different uses and sizes (Fig. 3-14), and the ratings are 240 volts AC, at 30, 40, 50, or more amperes. Specially designed molded plug-and-cord sets called *pigtails* are used with them.

Some receptacles are made to be weatherproof and dustproof, and have a built-in seal at all openings. And *safety outlets* incorporate a plastic disc that covers the plug openings and must be rotated before the plug can be inserted. You can also buy a device with a single receptacle on one end of a frame and a wall switch on the other (Fig. 3-13J), which will fit into the same space as an ordinary duplex. Another variation on this theme is a row of three receptacles on one frame (Fig. 3-13D).

There are also several types of receptacles that are meant to be surface mounted and need no enclosure (Fig. 3-13I and K). Some have space for two or three plugs to be inserted, others are continuous strips three feet or more in length which will accept a plug at any point, or at certain spaced intervals. In short, manufacturers produce receptacles in a variety great enough to cover almost any possible situation.

Switches

Another device available in incredible array is the switch. All the dozens of types (Fig. 3-15) have the same function though—to turn the power off and on. The "wall switches" common to all home wiring are called *toggle switches*. The least expensive is called a *snap switch* because of the noise it makes; *quiet* or *no-click* switches operate silently, but on the same principle. There are also *pushbutton* switches that push on and push off. And *mercury* switches do away with the springs and blades and mechanical motion of ordinary toggle

3W 15A-125V	3W 20A-125V	3W 15A-250V	3W 20A-250V
3W LOCKING 15A-125V	3W LOCKING 15A-277V	2W T-SLOT 15A-125V	2W 15A-125V
3W 15A-277V	2W LOCKING 20A-250V	3W LOCKING 15A-125/250V	2W LOCKING 15A-125V
2p-3W POWER 30A-250V	2p-3W POWER 50A-250V	2p-3W POWER 30A-125V	2p-3W POWER 50A-125V
3p-3W POWER 30A-125/250V	3p-3W POWER 50A-125/250V	3p-4W POWER 30A-125/250V	3p-4W POWER 50A-125/250V

Fig. 3-14. Receptacle slot configurations, along with the matching attachment plug-prong configurations, are designed for specific purposes and have only limited interchangeability. Some of the more common types are shown here.

switches, using instead a small glass tube containing a globule of mercury. The conductor leads of a mercury switch pass through the glass to form open contacts within the tube. When the tube is tilted in one direction, the mercury falls to the bottom of the tube and immerses both contacts, completing the circuit. When the tube is tilted the other way, the mercury rolls back away from the contacts and the circuit is opened.

Some toggle switches are equipped with a tiny light in the handle that remains on when the switch is in the *off* position, so that the switch can be readily located in the dark. Others are joined with a red indicator light (part D of Fig. 3-15) that goes on only when the switch is in the *on* position, showing that the load is activated. Another type remains in the *on* position for about a minute after the handle is flipped to *off*, giving the user a chance to move away before being surrounded by blackness—especially useful at stairs and entries, garages and bedrooms. And, you can also get two or three relatively smaller toggle switches mounted in a single standard frame to fit a regular wall box (Fig. 3-15N).

Most toggles are designed for use indoors, or outside in special weatherproof housings. All are rated at 120 or 125 volts AC, and most have a current-carrying capacity of 10 amperes, though a few will only handle 5 amperes.

There are three control variations. The first is called a *one-pole, single-pole*, or sometimes a *one-way* switch, and is used to control one or more loads from a single location. There are two terminals on the switch, one for the hot line coming from the supply and the other for the hot line, electrically the same, going out to the load. Incidentally, the input side of many electrical items, where the supply is attached, is called the *line side*. The output, where conductors are headed out to the load, is called the *load side*.

The second type of control is called a *three-way* system. A three-way switch is used to control one or more loads from two different locations, but must be used in conjunction with another three-way switch. A typical use is as at the top and bottom of a stairway. There are three terminals on the switch, one for the line or load and two for cross-connections running to another switch.

The third type of control is a *four-way* system, which is used for controlling one or more loads from three or more locations. The four-way switch has four terminals, one pair on either side of the four-way. The four-way switch is wired between two three-way switches. For each additional switching location needed, another four-way switch is inserted between the three-ways.

Toggle switches appear in many other forms, too, usually quite a bit smaller and of a somewhat different design than those we have just mentioned. The most familiar types are the *bat, ball*, or *lever-handled* ones that are widely used wherever electric or electronic control is needed. You may find them in

Fig. 3-15. A few of the more common types of AC switches for general wiring purposes. (A) 20A-125/277V locking toggle. (B) 15A-120/277V rocker. (C) double single-pole 15A-120/277V toggle. (D) 15A-120V single-pole toggle. (E) 15A-120/277V quiet push. (F) illuminated single-pole 15A-120/277V toggle. (G) weatherproof single-pole 10A-125V toggle. (H) mercury single-pole 15A-120V toggle. (I) surface 10A-125V tumbler. (J) ball-handle 6A-125V toggle. (K) 1000-watt 120V photocell switch. (L) single-pole pull-chain 3A-125V toggle. (M) surface single-pole 10A-125V T toggle. (N) double single-pole 15A-120/277V toggle. (O) single-pole 10A-125V T toggle with neon light. (P) single-pole 10A-125V on 4-inch oil-burner cover. (Courtesy Leviton Mfg. Co.)

your automobile, for instance, or on radio or testing equipment, or on your table saw. Provided that switches with the right ratings are chosen, there is theoretically no reason why they could not be used in place of standard wall switches, but they seldom are.

Pushbutton switches are sometimes used in homes. The function is the same, except the switching action is caused by pushing a button instead of flipping a handle. A doorbell button is one example, and this is known as a *normally-open momentary contact* switch. An automatic door switch is similar, but is rated for a higher voltage and current-carrying capacity. This type of switch is contained in a special box, mortised into the door jamb, and when the door is opened the switch button pops out and makes the circuit. This is a *normally-closed momentary contact* arrangement.

Several types of *rocker* switches are made for residential use. In one type (Fig. 3-15B), the decorative plate that covers the wall box also serves as the switch operating lever. The plate is mounted on swivel pins at the center, so that by pushing in slightly at top or bottom, the circuit is opened or closed. Another type, much smaller, is used for low-voltage switching systems. As many as three can be mounted on a single frame to fit into a standard wall box. The operation is the same; push the top and it rocks back to complete the load circuit, push the bottom and the circuit breaks.

In addition, there are *rotary switches*, like the mode selector on your stereo set, or like the three-way switch in a big floor lamp that activates one or a combination of loads. *Slide switches* are found on some appliances and often on power tools. *Trigger switches* are also common to power tools. *Pull-chain switches* are used on some lamp sockets, and the *pull canopy switch* is often used on overhead fixtures, especially fluorescents.

THERMOSTATS

Thermostats are switches, too, which are tripped by temperature changes rather than manually. In home use they most commonly control the central heating and air conditioning. But in addition, you can use them on heat cables and tapes, vent fans, water circulating systems, over- or under-temperature protection systems, and the like. There are dozens of different designs, both high- and low-voltage types, and any number of current-carrying capacities and contact arrangements. Some are adjustable through a certain

Fig. 3-16. Attachment plugs are made in great variety and should be matched to the specific requirements of each job. (A) armored dead front. (B) rubber. (C) high-abuse nylon. (D) power. (E) armored locking. (F) phenolic locking. (G) molded extension cord. (H) molded heavy-duty air conditioner. (I) molded heavy-duty power supply. (Courtesy Leviton Mfg. Co.)

temperature range, others have a manual override, and many are preset and designed for one specific application such as thermal overload.

PLUGS

Attachment plugs (Fig. 3-16), as used on the end of a lamp cord or appliance cord or range pigtail, constitute another class of devices. These, however, are obviously not a part of the built-in wiring system, even though they are a part of the residential system as a whole.

BOXES AND DEVICE ACCESSORIES

There are a number of accessory items that go along with boxes and devices. Covers are used to close up junction boxes, and they are bought separately from the boxes. Some are blank, and they are so named. Others have a knockout of one size or another centered in them. There are also *duplex* covers, in which one or two double convenience outlets can be mounted; others are made for one or two toggle switches. *Finish plates* are used with wall boxes.

Covers come in brown or ivory plastic, brushed aluminum, chrome-plated steel, stainless steel, wrought iron, copper, porcelain, wood, and heaven knows what else, in various degrees of plainness or ornamentation. They can be had blank, with single switch or receptacle openings, or in a multitude of switch and receptacle combinations (Fig. 3-17).

Also available are *plaster rings* for junction and wall boxes in a number of sizes and shapes. In use, they are attached to a box recessed into a cavity. The cavity is then plastered over, leaving a smooth finish with a metal-rimmed opening of the proper size and shape to accept the intended devices. Then there are *canopies*, which have finish trim rings or covers used to hide the inner parts of lighting fixtures, as well as the box that supports them. Usually they are supplied with the fixture, but some are bought separately.

There are also a number of strange bits and pieces used to support fixtures, called *crowsfeet, hickeys, fixture studs, tripods*, and so forth. These are not supplied with fixtures, but are resorted to in different mounting problems. The usual routine is to match the fixture with whatever items are needed to overcome the particular mounting problem. There is always a way.

RACEWAYS

A raceway is any channel for holding wires or cables, and it is used and designed for this purpose alone. There are a number of different kinds, but those which are most often used in residential work are *rigid metal conduit, electrical metallic tubing* (EMT), *rigid nonmetallic conduit,* and *flexible metal conduit*.

Rigid metal conduit, called *rigid, conduit,* or just plain *pipe*, looks similar to galvanized water pipe. The two must never be interchanged, however. Rigid is usually corrosion-proofed on the inside and outside, and is made so that there are no burrs or rough spots on the inside to harm the conductor insulation. Steel is the usual material, though aluminum and other metals are used as well. Conduit has fairly thick walls, comes in 10-foot lengths, and is threaded with standard pipe threads. Standard pipe-trade sizes are used, ranging from a minimum of 1/2 inch, through 3/4, 1, 1 1/4, 1 1/2 inch, and on up through 5 inches in 1/2 inch increments, with 6 inches being the largest.

Electrical metallic tubing is called *tubing, thinwall,* or *EMT*. It is lighter and less expensive than rigid, is not threaded, and is designed especially for electrical work. It too comes in standard pipe-trade sizes, from 1/2-inch to 4-inch, and in 10-foot lengths or "sticks," 10 lengths to the bundle. The

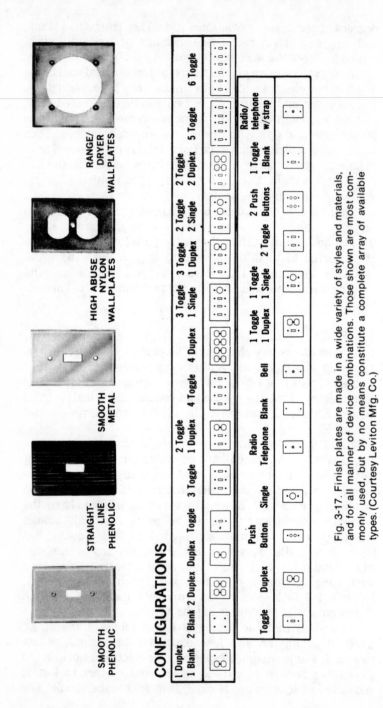

CONFIGURATIONS

Fig. 3-17. Finish plates are made in a wide variety of styles and materials, and for all manner of device combinations. Those shown are most commonly used, but by no means constitute a complete array of available types. (Courtesy Leviton Mfg. Co.)

walls are rather thin and not as rugged as rigid, but it is generally easier to handle and install. Steel is the preferred material, and like rigid, EMT is corrosion-proofed, seamless, and burr-free on the inside.

Plastic pipe is roughly equivalent to rigid nonmetallic conduit. This has recently become popular in some areas and probably will be extensively used in years to come. It is smooth, tough, light, will not rust or corrode, not terribly expensive, and is easy to work with. Though not the answer to all raceway problems, there are times when plastic pipe will perform better than any other. High-density polyethylene is one type of plastic pipe that uses threaded fittings. Another is made from polyvinylchloride (PVC) and is unthreaded. In addition, fibrous and asbestos-cement types can be used for some purposes.

Flexible metal conduit has a spiral metal jacket and can be easily routed around obstacles and through cavities because of its flexibility. Called *flex* or sometimes *greenfield*, it must be well supported. Flex is bought by the foot or by the coil, with the length of the coil depending upon the diameter of the flex. Liquid-tight flex is the same thing with a liquid-tight coating over the metal coil, usually a plastic of some sort. Both types are available in trade sizes of 3/8 to 4 inches.

There is a wide selection of fittings made to go with these raceways, with the greatest range of choice applicable to rigid and EMT. All rigid fittings are threaded and include a wide range of couplings, connectors, and offsets, plus a host of items now called *conduit bodies*, but also referred to as *Condulets, LB's, elbows*, and so forth. These conduit bodies come in offsets, 90° bends, 45° bends, and other configurations, often with at least one removable plate to facilitate the pulling of wires through the raceway.

About the same selection of fittings is made for EMT, but none of these are threaded except for the outside thread on the connector, where it is secured to a junction or pull box by a locknut. Instead, the fittings slip over the pipe and are secured with either one or two setscrews in the body of the fitting, which jam against the wall of the EMT. These, neatly enough, are actually called *setscrew fittings*. Sometimes threaded conduit bodies can be used by screwing an EMT connector (minus its locknut) into the body and then slipping this assembly onto the EMT. There is another type of fitting also in common use called *raintight*. These also slip over the pipe, but are secured by tightening a collar down upon a compression ring that squeezes against the EMT wall.

Fittings for rigid nonmetallic conduit are made of the same material as each particular type of pipe and are not interchangeable. In addition, the variety of fittings currently available is less than for rigid or EMT, which reduces the versatility considerably. The threaded sort work in the same manner as rigid fittings. Those used with PVC pipe, however, slip on over the pipe. They are cemented in place with a specially prepared glue that reacts with the plastic on both mating surfaces and actually welds them together. Once in place, they can only be removed by cutting them out.

Flex does not have fittings of the same kinds as the other raceways because of its different usage, but there are connectors that slip on over the end of the flex and clamp in place. The outside-threaded end of the connector is then slipped into an appropriate knockout hole in a box and secured with a locknut. There is no need for openable conduit bodies, since flex is designed for relatively short runs from junction box to junction box. Because of its inherent characteristics, little help is needed from elbows and such when changes of direction are necessary.

RACEWAY ACCESSORIES

There are also a few accessory items used with raceway assemblies. *Bushings* are rings with rounded outer surfaces that screw onto the end of connectors after the locknut is in place. Their purpose is to cover the sharp edge of the connector to protect the conductor insulation where the conductor bends into the junction box.

Insulated bushings are for the same purpose, but are made of plastic and insulate as well as protect. And then there are various types of *bonding clamps* and *grounding clamps*, which are used to preserve the ground integrity between lengths of raceway or raceway and fittings, or to attach ground conductors from other points in the system. And of course there are assorted *clips, drive-clips, clamps, straps,* and *hangers* used for mounting or suspending the raceways. The type you use will depend upon a combination of specific circumstances and personal preference tempered by what items your supplier may have on hand.

WIRE AND CABLE

Wire is what ties the whole system together, and we may as well get the definitions straight at the start. A *conductor* is a substance or body capable of transmitting electricity. A *wire* is a slender metal rod of indefinite length. In practical electricity, a conductor is one or more wires which transmit the electricity. The term *pair* refers to two conductors, usually working in concert. A *bare conductor* is just that; it has no covering over the raw metal. An *insulated conductor* is covered with an *insulator*—a substance that transmits electricity in a poor or negligible fashion under normal operating conditions. An *electrical cable* consists of two or more conductors that are usually insulated but may also include bare conductors, wrapped in an all-encompassing outer covering, or *jacket*. Unfortunately, the word *wire* is often used to loosely name what should be called *cable*, and it is also incorrectly applied to conductors at times.

Electrical conductors are made from two metals, as far as residential wiring is concerned: *aluminum* and *copper*. In some cases, copper-clad aluminum is used. A *solid conductor* is just that; one continuous strand of metal—a wire just as it comes from the forming die. A *stranded conductor* consists of multiple wires laid up in a particular pattern; they may be twisted around one another in a concentric spiral, or in a double spiral with one going clockwise and another layer going counter-clockwise, or whatever other arrangement is suitable. Most electrical conductors use the concentric lay.

Size

The size of an electrical conductor is determined by its diameter in *mils* (thousandths of an inch) measured at 68°F. In this country there is no official standard for trade sizes, but the one in common use is usually called the *American wire gauge* (AWG) system, or sometimes the *Browne & Sharpe* (B&S) system.

Since the AWG system does not cover large cable sizes, we use an additional method of sizing by the cross-sectional area of the conductor, measured in *circular mils* (CM) or thousand circular mils (MCM, where the first M is the Roman symbol for 1000.)

AWG wire sizes start at 4/0 (called *four-ought*), which measures 0.522 inches in diameter, and goes down to #50, which is barely visible and looks like a cobweb. The next largest step beyond 4/0 AWG is 250 MCM, and this system proceeds in steps up to 5000 MCM, which is as big around as your wrist (Table 3-1).

In general residential work, the largest size you might encounter could be the 4/0, but more likely 3/0. The smallest would probably be #18 as far as 120-volt wiring is concerned,

Size AWG MCM	Area Cir. Mils	Concentric Lay Stranded Conductors		Bare Conductors	
		No. Wires	Diam. Each Wire Inches	Diam. Inches	*Area Sq. Inches
18	1620	Solid	.0403	.0403	.0013
16	2580	Solid	.0508	.0508	.0020
14	4110	Solid	.0641	.0641	.0032
12	6530	Solid	.0808	.0808	.0051
10	10380	Solid	.1019	.1019	.0081
8	16510	Solid	.1285	.1285	.0130
6	26240	7	.0612	.184	.027
4	41740	7	.0772	.232	.042
3	52620	7	.0867	.260	.053
2	66360	7	.0974	.292	.067
1	83690	19	.0664	.332	.087
0	105600	19	.0745	.372	.109
00	133100	19	.0837	.418	.137
000	167800	19	.0940	.470	.173
0000	211600	19	.1055	.528	.219
250	250000	37	.0822	.575	.260
300	300000	37	.0900	.630	.312
350	350000	37	.0973	.681	.364
400	400000	37	.1040	.728	.416
500	500000	37	.1162	.813	.519
600	600000	61	.0992	.893	.626
700	700000	61	.1071	.964	.730
750	750000	61	.1109	.998	.782
800	800000	61	.1145	1.030	.833
900	900000	61	.1215	1.090	.933
1000	1000000	61	.1280	1.150	1.039
1250	1250000	91	.1172	1.289	1.305
1500	1500000	91	.1284	1.410	1.561
1750	1750000	127	.1174	1.526	1.829
2000	2000000	127	.1255	1.630	2.087

though sound system loudspeakers and some aerials occasionally use smaller sizes. By far the largest amount of your work will undoubtedly be done with #12 AWG solid conductors. Incidentally, solid conductors are generally available only from #8 down through the smaller sizes;

anything larger than #8 would be difficult to work with in solid form. Stranded conductors are made throughout the entire range.

Wiring Standards

Standards governing the use of various conductors in general wiring work, as well as in many specialized areas, are set forth in the National Electric Code. With respect to our field of interest (600 volts and below) the standards are quite specific. This information has been gathered and evaluated for decades under the auspices of the National Fire Protection Association with input from a number of organizations in the electrical field; it is constantly upgraded, updated, and revised as the state of the art advances. Final recommendations are based on research, laboratory tests and evaluations, and practical experience in the field, and these constitute the best and most practical information the electrician has to work with. In addition, most of the wire and cable for general wiring purposes that is on the market today conform to these various specifications and each type is designated and approved for certain uses.

Insulation plays a most important part in the practical function of a conductor, and as you can see from Table 3-2 there are a great many approved types that can be used for various purposes. In addition to size, a conductor is known by the type of insulation that covers it. One of the most widely used is called *TW*. As noted in the table, this insulation is made from a moisture-resistant thermoplastic, and it has a maximum operating temperature of 140°F. That is, at temperatures beyond this point, the insulation will start to degrade, then rapidly deteriorate. Unfortunately, the workable low temperatures are not given, but note that many insulations will become so brittle at sub-freezing levels that they will split open or break off when the conductor is bent. You can also see that TW can be used in either dry or wet locations. Other tables not included here give further specifications; TW is also flame-retardant, and in conductor sizes of #14 to #10, the insulation thickness is 30 mils, so there is no additional outer covering. A conductor with an AWG size of #12 with a TW insulation covering would be called #12 TW. All told, there are some fifty types of insulative coverings in the general wiring catagory alone, but under normal circumstances you will encounter only a few of them.

Table. 3-2. Conductor Types as Listed in the 1975 NEC

TRADE NAME	TYPE	MAXIMUM OPERATING TEMPERATURE	LOCATIONS PERMITTED
Heat-Resistant Rubber	RH	75°C 167°F	Dry
Heat-Resistant Rubber	RHH	90°C 194°F	Dry
Moisture- and Heat-Resistant Rubber	RHW	75°C 167°F	Dry Wet
Heat-Resistant Latex Rubber	RUH	75°C °F	Dry
Moisture-Resistant Latex Rubber	RUW	60°C 140°F	Wet Dry
Thermoplastic	T	60°C 140°F	Dry
Moisture-Resistant Thermoplastic	TW	60°C 140°F	Dry Wet
Heat-Resistant Thermoplastic	THHN	90°C 194°F	Dry
Moisture- and Heat-Resistant Thermoplastic	THW	75°C 167°F	Dry Wet
Moisture- and Heat-Resistant Thermoplastic	THWN	75°C 167°F	Dry Wet
Moisture- and Heat-Resistant Cross-Linked Synthetic Polymer	XHHW	90°C 194°F 75°C 167°F	Dry Wet
Mineral Insulation (Metal Sheathed)	MI	85°C 185°F	Dry Wet
Underground Feeder and Branch Circuit	UF	60°C 140°F	Dry Wet Corrosive Buried
Underground Service Entrance	USE	75°C 167°F	Dry Wet Buried
Silicone-Asbestos	SA	90°C 194°F	Dry
Asbestos and Varnished Cambric	AVA	110°C 230°F	Dry
Asbestos and Varnished Cambric	AVB	90°C 194°F	Dry

Cables are designated in a somewhat similar fashion by capital letters, which can be a bit confusing until you become familiar with them. Table 3-3 will give you the basic information for the types in common use; further details are contained in the NEC. To take an example, NM stands for nonmetallic and refers to the composition of the outer jacket. Inside the jacket there are two or more conductors. If they are of #12 AWG size, then the cable is called #12/2 NM (this particular cable also is called *Romex*). If there are three conductors, two insulated and one bare, then the cable is referred to as #12/2wg NM—the *wg* means *with ground* (the bare conductor). The two insulated conductors are covered with type TW insulation or something similar, and were they not encabled, they would just be called TW conductors, as before. There are not nearly as many types of cables as there are single conductors, nor are cables made in very many combinations using the different insulations. Sometimes cable must be discarded in favor of using a raceway with conductors covered by an insulation that will fulfill the requirements of some special conditions.

There are two further groupings of conductors used in electrical wiring, though to a much lesser degree than those for general wiring just discussed. One group of about four dozen types comes under the heading of *flexible cords and cables*. The group called *fixture wires* contains another twenty or so types. Sizes vary upward from #10 AWG *tinsel cord*; specific uses range from extremely high-temperature wiring for lighting fixtures to elevator control wiring. Each type in both groups has its own particular uses, characteristics, and specifications that may be found in the NEC.

As you can see, there are literally hundreds of possibilities open to you in choosing the various component parts of your wiring system. Some of your choices will be governed by necessity; you must have certain articles. Availability is another big factor, since a particular box or device or trade name in common use in one part of the country may well be completely unknown in another. And, of course, local codes and rules may play a part in what you can or cannot use.

One of the best possible ways for you to add to your knowledge, so that you can pick and choose both intelligently and economically, is to collect a pile of catalogs. Borrow them from your supplier or a local contractor or request them by direct mail to the manufacturers. And pore over them, taking note of uses and specifications of the various items to

Table. 3-3. Cable Types and Uses

TRADE NAME	TYPE	LOCATIONS PERMITTED	LOCATIONS NOT PERMITTED	SUPPORT INTERVALS	BEND RADIUS
Mineral-insulated Metal-sheathed	MI	Dry Wet Indoors Outdoors Exposed Concealed Masonry embedded Hazardous Corrosive Underground (protected)	Destructive corrosive	6 feet	5 × dia.
Aluminum-sheathed	ALS	Dry Wet Exposed Concealed Underground (protected)	Direct burial Corrosive Masonry Hazardous vapors	6 feet	3/4-inch: 10 × dia. 1 1/2-inch: 12 × dia.
Copper-sheathed	CS	Wet Dry Exposed Concealed Corrosive (protected)		6 feet	3/4-inch: 10 × dia. 1 1/2-inch: 12 × dia.
Metal-clad	MC	Exposed Concealed Dry Wet under some conditions	Where physical damage might occur	6 feet 2 feet from every box and fitting	7 × dia.
Metal-clad	AC	Exposed Concealed Underplaster extensions Masonry air voids	Damp Wet Corrosive Below grade	4 1/2 feet 12 inches from every box and fitting	5 × dia.

Type	Abbr.	Uses	Additional uses	Support	Bend radius
Metal-clad	ACL	Exposed Concealed Outdoors Masonry Wet Underground (protected)	Direct burial	4 1/2 feet 12 inches from every box and fitting	5×dia.
Nonmetallic-sheathed	NM	Dwellings Exposed Concealed Dry Masonry voids	Corrosive Embedded in masonry Embedded in plaster Hazardous Service entrance	4 1/2 feet 12 inches from every box and fitting	5×dia.
Nonmetallic-sheathed	NMC	Dwellings Exposed Dry Damp Corrosive Masonry block	Embedded in concrete Embedded in plaster Hazardous Service entrance	4 1/2 feet 12 inches from every box and fitting	5×dia.
Service-entrance	SE	Service entrance Interior wiring in some cases	Underground		
Service-entrance	USE	Service entrance Underground Interior wiring in some cases			
Underground feeder and branch circuit	UF	Underground Direct burial Wet Dry Corrosive	Service entrance Hazardous Embedded in masonry Exposed to sunlight		

determine which might be best for your purposes. Then you can balance that knowledge against price and availability as you make the final selections. And by the way, you can find names and addresses of all the electrical equipment manufacturers in the country at your local library in a set of volumes called *The Thomas Register*. The librarian will be glad to show you how to use them.

Chapter 4
Electric Heating

Over the past twenty years or so, electric space heating has come into widespread use in homes throughout the country. Until just recently, it was touted by the utility companies as being the *ne plus ultra* of all heating systems, without which any self-respecting homeowner should never be. The furor has died down somewhat since the energy crisis, higher electric bills, and potential power shortages took over the headlines. Then too, some of the glow has faded from the bright picture of electric heat because of occasional misuse and inappropriate installation, leading to high operating costs and unsatisfactory heating. But the fact remains that comfort heating by electricity, especially with the recent advances in the state of the art, is undoubtedly well worth considering for any new home in any location, and may indeed be the best system in many cases. It may well be the best bet for the future, especially as a back-up to solar heating.

There are a number of fine advantages to this system. For instance, there is no open flame, anywhere. In most equipment, outside shell temperatures are too low to cause combustion or any serious human injury. No on-premises fuel storage is necessary and there is no possibility of running out of fuel at awkward moments, except in the case of a power outage when most other heating systems would also be out of commission. The equipment takes up little space, and with only a couple of exceptions, no special area need be allocated within the building for the machinery. Electric heat is the only type that is 100% efficient, converting all of the electrical energy consumed to usable heat in the home. It is as clean as the air in which it operates, takes no oxygen from the air for combustion, and there is no possibility of fuel odors or lethal combustion gases or explosion in the fuel supply. Most types of electric heating units are unobstrusive and quite easy to install, and some provide a tremendous flexibility in decentralization and heat zoning that can only be achieved at considerable difficulty with other types of heating. Maintenance for most types of equipment is minimal, and cleaning or servicing often is not required at all.

ECONOMICS

The economics of electric heating vary widely, but on the whole are favorable. Though each installation should be analyzed separately, more often than not the initial cost of the system plus professional installation is considerably lower than that for the more traditional oil- or gas-fired hot-air or hot-water systems, and the job is both quicker and easier. This can mean an immediate savings as well as a long-term savings in the total cost of a new dwelling.

For instance, let's suppose that the choice is between a gas hot-water heating system and a baseboard-unit electrical system. The hot-water system might cost $5000, the electrical system $2500, both professionally installed. So there is an initial savings of $2500 by putting in the electrical system. In addition, assuming a 20-year 80% mortgage on the home at a rate of 10%, the total cost of the hot water system would be the initial outlay of $1000 in cash plus $4000 at 10% interest for 20 years, for a total of $9264. The electrical system would cost an initial sum of $500, plus $2000 at 10% for 20 years, for a total of $4632. Thus, the total savings would amount to $5132 in the long run, not to mention the need for less capital at the outset.

Operating costs for an electric heating system depend upon a number of factors that are specific for each individual case. The principal factors are:

- the total heat loss of the house itself
- the cost of electricity per kilowatt-hour (kWh) at the site
- the number of degree days of heating needed during the year
- the design temperature difference at the site
- and a highly variable item known as the experience factor.

All of these factors we will go into in some detail so that you will be able to make your own particular determinations.

In general, however, the cost of electric heating compares favorably with other types and probably will continue to as all fuels keep rising in price. And with our natural gas supplies dwindling and the oil situation becoming more critical, electricity may eventually become the most reasonable and dependable basic source of home heat.

Remember too that the savings afforded by low initial installation costs may provide a substantial portion of the money needed to cover any monthly difference in the operation cost between electric and some other type of heating. The savings in interest and principle payments shown in the preceding example, for instance, would provide $19.80 a month toward the fuel bill every month for 20 years. This would be sufficient to entirely pay for the heating bills of an average 1500-square-foot home, designed for a temperature differential of 90°, with 6000 degree-days of heat, at 2¢ per kWh power cost.

DETERMINING THE HEATING CAPACITY OF A ROOM

The electrical calculations, wiring, controls, and installation of the heating units in an electrical heating system are simple enough, and they follow much the same basic requirements and practices as any other part of the residential electrical system. The problem comes in determining what the heating capacity of those units should be in the first place. Until you know what you need, you can't figure out what electrical provisions to make for them, or how they will be fitted into the building. The first step, then, in designing an electric heating system is to calculate the *total heat loss* of the structure. This, incidentally, can be done before the building plans are actually complete, provided that you know or can find out as you go along the dimensions of rooms, doors, and windows, and the building materials that will be used.

Heat is always in a continual process of transference from warm areas to cold areas, much like water running downhill to seek its own level. If the air outside a house is warmer than inside, heat will pass through the elements of the structure into the interior, and if the air outside is colder, heat will move outward to the exterior, always in an attempt at equalization. The heating system must be capable of replacing this escaping heat with a rate at least equal to the rate of loss, in order to maintain a reasonable level of comfort for the inhabitants. In addition, the homeowner must take steps to insure that the heat loss is minimized as much as is both possible and practical in order to keep costs down and comfort up. There is no way, however, to stop heat loss entirely.

111

Proper insulation is critical to an electric heating system. The usual recommended minimums are the equivalent of 3 1/2 inches of either fibrous batt or roll thermal insulation in all exterior wall spaces, 6 inches in ceilings or roofs, and 2 inches in floors over unheated areas. Glazing on windows and doors should be double, with weatherstripping used around all openings, cracks well caulked everywhere, fireplace dampers tight, and vent fans equipped with automatic dampers or louvers. Perimeter insulation must be used all around poured concrete slabs. Also, full vapor barriers must be used on all building sections that are exposed to the colder temperatures. Unheated crawl spaces and attics must be properly ventilated to carry moisture away and maintain the integrity of the thermal insulation (moisture lowers its effectiveness). In short, the heated areas of the house must be absolutely "tight" to the weather.

Heat Transfer Factors

In calculating total heat loss of a structure, the first determination involves *transmission heat losses*. Every element of the building, carpeting, wall board, plaster, and shingles, has a certain resistance to the passage of heat. This resistance is symbolized by the letter *R* and is called the *R value*. Most common building materials and combinations thereof have been tested extensively and a specific numerical value assigned to each. You will see 6-inch thick fiberglass roll insulation, for instance, designated as having a value of *R-19*. By contrast, an 8-inch concrete block is rated *R-1.04*, and 1/2-inch plywood is rated at R-*0.62*. It would take 31 layers of 1/2-inch plywood to create the same total resistance to the passage of heat as one layer of 6-inch fiberglass insulation.

From these resistance values, we can obtain two other values. The *U factor* represents the coefficient of heat transfer in Btu per hour per square foot per degree Fahrenheit of temperature differential between inside and outside. The *W factor* is the same, but has been converted directly into watts of electrical power instead of Btu of heat. This gives rise to two similar formulas:

$$HT = U \times A \times TD$$
$$\text{and } HT = W \times A \times TD$$

HT is the transmission heat loss, *U* and *W* are the heat transmission coefficients, and *TD* is the difference in Fahrenheit degrees between the air on the inside and the air on the outside of the house. *A*, of course, is area. The calculation is not as difficult as you might expect, and we will use the second formula for an answer directly in watts.

Calculating Areas

To begin with, make up a sheet of paper for every room in the house which will be heated, like the one in Figure 4-1. Then measure the length and the height of each wall that is exposed to the outside; multiply one by the other to get the square footage that will be exposed. If there is a wall facing an area that will be heated, but to a lesser temperature than the living quarters, such as a shop or a garage, jot this down as a separate figure. Enter these figures as "wall gross."

Next, do the same for each floor or part of a floor that is over an unheated area or an area heated to a lesser temperature. Keep the two figures separate and enter these square footages as "floor."

Now do the same thing with the ceilings, remembering that if the floor above is heated, that part of the ceiling is not included. Enter this as "ceiling."

Now measure the square footage of exposed glass area and exposed door area in each room. Keep separate figures for single glass, double glass, triple glass, single 1 3/4-inch door, and double door, and list them all as such, totaling each category.

HEAT LOSS BY CONDUCTION

Living Room: $12' \times 18' \times 8'$

SECTION	AREA $\times$	W $\times$	TD $=$	WATTAGE
Wall gross, cold	312			
Wall gross, cool	72			
Floor, cold	108	0.012	80	104
Ceiling	216	0.013	80	225
Glass, double	40	0.17	80	544
Door, 2″ with storm, cold	21	0.10	80	168
Door, 2″, cool	16 1/2	0.15	50	124
Wall net, cold	251	0.02	80	402
Wall net, cool	55 1/2	0.02	50	56

TOTAL $H_T = 1623$ watts

HEAT LOSS BY INFILTRATION

	V $\times$	F $\times$	N $\times$	TD	$+(CF \times TD)=$	WATTAGE
	1728	0.0053	0.75	80	— —	550
Fireplace, average Damper					14.6 80	1168

Total $H_I = 1718$ watts

TOTALS

$H_R = H_T + H_I = 1623 + 1718 = 3341$ watts

Use 3500 installed watts

Fig. 4-1. Sample heat-loss/heat-required calculation sheet for one room or area.

Last, subtract the total square footage of doors and windows form the "wall gross." Enter this as "wall net."

W Factors

The next step is to assign W factors to each of the exposed square footages that you have listed. Studies have shown that while all materials are slightly different and that installation practices and procedures do have some bearing on the final results, the differences in general construction are really rather minimal. You can make specific calculations for every bit and piece in the building, but it is much easier and almost as effective to use generalized tables made up for the purpose.

In practice, any errors are usually too small to be noticeable in the actual operation of the system under field conditions. In any event, these calculations can only provide us with approximate answers since we have no control whatsoever over the wildest variable that gives rise to the need for heat in the first place—the weather.

From Table 4-1, then, pick the appropriate construction combinations and assign them as necessary. The walls may all be standard wood frame, for instance, with 3 1/2-inch insulation, for a factor of 0.02. All single-glass windows and doors will be 0.33, double 0.20, and so forth. By the way, if the doors have glass in them, list them as being *all glass*!

Temperatures

Now you must determine what temperatures to use. The inside temperature, the one to which you will heat the living quarters, can be whatever you are used to or think is adequate, such as 68° or 70° or 72°. If elderly people will be occupying the premises, perhaps 75° or 78° might be better. But the higher the temperature, the higher the operating costs.

Choosing the outside temperature is a matter of judgement, too. It is not necessary to use the *record low figure* for your area, but on the other hand you will want enough capacity to remain reasonably comfortable during a cold snap of several days' duration. If −10° occurs perhaps once every five years and −15° each winter is unusual, the −10° would be a good choice. Probably zero would not. You may be able to get a good idea of exactly what to expect from local weather bureau, fuel supply company, or utility company records covering previous years.

To arrive at the TD (design temperature), just subtract the low outside temperature T_0 from the high inside temperature T_1. If you have settled upon 70° and −10°, your TD will be 80. If one wall faces a garage that will be heated to

Table. 4-1. W Factors for Various House Constructions

CONSTRUCTION	FIBROUS INSULATION				
	None	1 1/2"	3 1/2"	6 1/2"	13"
Frame wall	0.10	0.04	0.02		
Brick veneer wall	0.10	0.04	0.02		
8" masonry wall	0.16	0.04	0.02		
12" masonry wall	0.11	0.04	0.02		
8" log wall	0.29				
Ceiling, normal	0.18	0.06	0.02	0.013	0.007
Floor, normal	0.10	0.05	0.02	0.012	
Floor, 4" concrete	0.20	0.06	0.02		
Single glass	0.33				
Double glass	0.17				
Triple glass	0.12				
Storm sash	0.20				
Door, 1" wood	0.30				
Door, 2" wood	0.15				
Door with storm door	0.10				

50°, this TD will be $70° - 20° = 50°$. The TD for the garage itself will be $50° - (-10°) = 60°$.

Enter the appropriate figure beside each listed category on your room sheets, and then multiply each area (except wall gross) times the W factor times the TD. Total all of the category answers, and you will have the sum wattage of installed heat necessary to replace the heat which escapes by transmission loss in each room.

Infiltration Heat Loss

The second element of heat loss which must be calculated is called infiltration heat loss. This formula is expressed as

$$HI = V \times F \times N \times TD + (CF \times TD)$$

where HI is the total heat loss from infiltration factors, V is the volume of the area to be heated, F is the air heating coefficient, N is the number of air changes in the heated area per hour, *CF* is the loss from construction factors, and *TD* is the design temperature.

These calculations recognize the fact that air is always moving through a home for one reason or another, completely replacing itself periodically. The reasons may include strong winds, doors opening and closing, people entering and bringing capsules of cold fresh air with them, leakage through cracks and around structural elements, normal air movement through flues and vents, the relative quality of materials used in the construction of the building and the expertise with which they were installed, and so on. While such factors differ widely, certain average figures can be used for nearly all installations with a surprising degree of accuracy. As with the previous calculations, any errors are practically inconsequential.

Go back to your room sheets again (Fig. 4-1) and determine the volume of each room by multiplying the height times the length times the width. Where the shapes are odd or irregular, break the room up into imaginary regular sections, calculate, and add the results together. Enter the figures under a column headed *V*. Under the heading *F*, enter the figure 0.0053, which is a generally used figure to express the number of watts per cubic foot per degree Fahrenheit, the temperature difference needed for heat replacement in moving air in residences.

The value for *N* is a variable and will require a bit of thought on your part. The rate of air change in the house is something over which you have little practical control, assuming that the structure is well caulked and sealed and all vapor barriers are intact, windows and doors well fitted, effective weatherstripping applied, and so forth. The relatively high values of *N* suggested here are not intended to imply that your house is extremely drafty, but are in fact realistic values intended to permit your heating units to respond quickly to an influx of cold outside air, as from a door opening. The suggested value for the average home, whatever that may be, is somewhere between 0.5 and 0.75 air changes per hour. If you are in an exposed location where the wind blows frequently, or where periodic windstorms can be expected, it might be wise to use one change per hour, or even a bit more, so that *N* might equal 1.0 or 1.1 or whatever seems reasonable. If a number of active folks live in the house and there is constant door opening, a value of 1.0 would be a good choice. On the other hand, a working couple with quiet living habits who leave the house shut up most of the time and live in an area of quiet or infrequent winds might expect only half an air change, or $N = 0.5$. Choose a suitable figure of no less than 0.5 but no more than 1.25 and enter it under N in each room formula.

To find the values for CF, refer to Table 4-2, pick the appropriate numbers, and enter them on the room sheets as necessary. As you can see, all of the categories are quite specific. If you have a construction feature that is similar to those listed but not quite the same, use the charted figures as a comparative guide and interpolate some figures of your own.

The one remaining symbol, TD, is the design temperature and will be the same as the one you used in the previous formula.

When you have all of the figures jotted down, work out each room formula by adding all the CF figures for each given room and multiplying by TD, then adding that figure to the product of V times F times N times TD. Each completed room sheet should look something like the one in Figure 4-1. Your answer, HI, will be the total infiltration heat loss in watts, indicating the amount of installed heat wattage you will need for replacement.

The total HT and HI, which can be written as HR (heat required), equals the minimum amount of installed heat-unit wattage or capacity that you will need in each room to cover all heat losses based upon the particular set of figures you have used (Fig. 4-1). This figure may or may not be the one which you will ultimately use, for reasons we will go into a bit later.

TOTAL HEATING REQUIREMENTS AND COSTS

The next chore is to determine what the total heating requirements for all rooms will be for the entire year and what the operating cost is likely to be. This uses the formula

$$kWh_T = \frac{HR \times DD \times C}{TD}$$

Table. 4-2. CF Factors for Various House Constructions

CONSTRUCTION FEATURE	WATTS/SQ. FT.
Fireplace, tight damper	5.9
Fireplace, average damper	14.6
Fireplace, no damper	43.9
Exhaust fan, no louver or damper	11.7
Exhaust fan, louver or damper	0

TOTAL HEAT LOSS

ROOM	HR WATTS
Living room	3341
Master bedroom	2142
Guest bedroom	1978
Kitchen	1361
Bath	338
Dining room	2039
Basement	3846
Garage	2980
Hall	376

Total = 18401 watts

TOTAL COST

$$kWh_T = \frac{HR \times DD \times C}{TD} = \frac{18.4 \times 5150 \times 15}{80} = 17,768$$

17,768 kWh times the local power rate of $0.0248 per kWh
is $17768 \times 0.0248 = \$440.65$ per year.

Fig. 4-2. Sample total heat loss (top) and electric space heating operating cost per year calculations (bottom).

where kWh_T is the total number of kilowatt-hours or power required. HR is the total heat requirement, not in watts this time, but in kilowatts (1 kW = 1000 W). DD is the number of degree days of heating required in your particular location, C is the experience factor, and TD is the design temperature.

First, tally up all of the individual room HR figures into one combined wattage requirement for the entire building, as in Fig. 4-2. Then find out what the number of degree-days is per heating season, or calendar year, in your area. This, by the way, is a hypothetical measure of the amount of heating needed on the average in a given location during each 24-hour period of the heating season. It is usually found by subtracting the mean of the high temperature and the low temperature for the day from 65°. Thus, if the high for January 14 was 26° and the low was −10°, the mean would be the sum of 26° and −10°, or 16°, which when divided by 2 equals a mean of 8°. And this subtracted from 65° equals 57 degree-days of heating needed during that 24-hour period.

Degree-day figures are compiled as year round totals and then averaged out over periods of many years to arrive at a reasonably constant figure. Chances are that you will have to contact either an area fuel supplier or the nearest office of your serving utility company to get a close average number of degree-days for your particular spot. If this is not possible, your only recourse is to find the figure for the area closest to you, perhaps from a large weather station, and adjust upward or downward to suit local conditions as closely as possible. It is safer to use too many degree-days rather than too few.

The Experience Factor

The experience factor, C, is a great variable. This considers living habits, such as turning heat off in areas of the house that are not in use, closing the house and turning the heat down substantially for extended periods of time, the number of occupants and how much time they spend in the house, how often heat-producing appliances are used, how often fireplaces are used and for how long, how often outside doors are used, how high a temperature is maintained inside the house and whether some sections are warmer than others, and whether any windows may be opened. Another consideration is the siting of the house; how much sun does it receive, how sheltered is it from the wind, how much roof overhang is there to block sun, is the position of the house relative to prevailing wind direction, and what is the position of glass exposure. The general quality of workmanship in the construction of the house and the effectiveness of caulking and weatherstripping play a part; so does the quality, type, and condition of the building materials. Obviously all of these details cannot be figured exactly, and they will probably change from time to time anyway throughout the occupancy life of the dwelling.

Studies of experiences with many thousands of homes, however, have led to the use of certain figures as a generality. The continued use of these figures has proved them to be sufficiently accurate to do the job. In some areas, the power supplier may have an applicable set of experience factors based upon electric heating installations in its district, so check with them first. If no figures are available from this source, you can use the precalculated figure of $C = 17$ if your chosen value for $N = 0.50$ changes of air per hour, or $C = 15$ if the value for $N = 0.75$ changes of air per hour. A reasonable figure for 1.0 air change per hour would be 13, and you can figure other approximate numbers for C by relating it to your chosen value of N; the relationship is direct, for all practical purposes.

The value for TD is the same as in the previous formulas. If you are using two or more TD values, use a separate kWh_T formula for each. Multiply all of the factors as shown in the formula in Fig. 4-2, divide by TD, and the answer will give you the total projected power consumption of your heating system for one calendar year.

Heating Costs

Obtain a rate schedule from your serving utility company, or better yet, ask them what the average cost per kilowatt-hour (kWh) of power is in your area for electric heat customers, and multiply this times your projected kWh_T figure to determine approximate operating costs. The problem with using a rate schedule is that most of them are figured on a sliding scale. The more power you use, the lower the cost per kWh. This means that each month the first few hundred kWh will cost X cents per kWh, the next few hundred will cost Y cents per kWh, and so forth. This makes accurate calculations most difficult, especially since your other figures are based on a span of a year instead of a month. Also, the costs of the rest of the electrical system are lumped in with the heat and cannot be specifically separated, unless the heat system is separately metered.

With an average cost figure, however, you can still come up with a relatively accurate total. An old rule of thumb that electricians have sometimes used in the past says that if the cost of "bulk" power—that is, more than 1500−2000 kWh per month—averages about 2¢ per kWh or less, electric heating is a feasible system. Today, however, with fuel and power prices increasing so rapidly and with the shifting price differentials between fuel types, each installation should be individually calculated and compared for the particular locality.

INSULATION

Insulation is really the only readily controllable major factor in the total heat loss of a structure. But as far as cost goes, there is first a point of diminishing returns and then a practical cutoff point in the amount that can be used. Suppose, for instance, that your first heat-loss calculations were based upon 2 inches of fibrous thermal roll insulation in the walls and 4 inches in the ceilings (or roof). By recalculating on the basis of 3 1/2 inches in the walls and 6 inches in the ceiling, you might find the total heat loss lowered considerably, with consequent savings in kWh_T costs.

But you must balance the expense of the extra insulation against the savings in operating costs. Those savings might pay for the extra insulation in only a few months. But if you were to make a third set of calculations based upon 10 inches of thermal insulation in the ceilings, you might find the kWh_T savings to be so small that it would take 15 years to regain the added cost of the insulation; the returns have diminished.

There are other factors to consider too. For example, 6 inches of thermal insulation in the walls might give you substantial savings, but if the building is already constructed with 4-inch studs, you have reached the practical cutoff point at 3 1/2-inch insulation. It is a good idea, especially in the colder climates, to run through several sets of calculations using different values in the equations in order to see which works the best in your particular case.

COMPARING HEATING SYSTEMS

In making comparisons between electric heat and other types of fuels, you can use the same methods and formulas described earlier for total heat loss. Other heating systems are based on a loss stated in terms of *Btu per hour* rather than watts, so you must convert. One watt is equal to 3.413 Btu per hour, so multiply your *HR* times that figure to arrive at the total heat loss in Btus. You can then correlate this figure to other fuel system costs by simple multiplication.

Once you know what the heating capacity of the system must be, you can then set about choosing the kind of equipment that best suits your needs. There are several possibilities. Where the heating system and central air conditioning are to be used together, the electric *heat pump* is sometimes used to provide both heating and cooling. Central heating only requires an *electric furnace* to provide heat for a forced hot air or hot water network. Another method uses *unit heaters*, usually freestanding or freehung and equipped with blowers. *Radiant panels* can be mounted in or on ceilings or walls. *Resistance heating cables* are made for embedding in plaster ceilings or concrete floors, or sandwiching between layers of gypsum board on ceilings. Probably the most commonly used type consists of a series of *baseboard heaters* that circulate warm air by convection rather than with blowers.

Heat Pumps

The heat pump (Fig. 4-3) is actually an air-conditioning unit working in the reverse-refrigeration cycle. This system is finding increasing use in the more temperate areas of the country, especially where summer air conditioning is a desirable if not essential feature of the home. In such cases, special central air-conditioning units are made to supply all or at least a part of the winter heating loads. Using essentially the same equipment, these heat pumps make both practical and economic sense. However, this type of electric heating has not enjoyed widespread use because as the outside temperatures drop in severe winter weather areas, so does the heating capacity of the equipment. New designs and capabilities are slowly overcoming this problem, so you may soon see practical heat-pump designs used in more northern states.

WINTER

1. Cool outdoor air in.
2. Cold air out.
3. Refrigerant pumps heat indoors.
4. House air in.
5. Warmed house air to rooms.

Heated Air

■ High-Pressure Hot Gas
■ High-Pressure Warm Liquid
▨ Low-Pressure Cold Liquid/Gas
▨ Low-Pressure Cool Gas

SUMMER

1. Warm house air in.
2. Cooled air out.
3. Refrigerant pumps heat outdoors.
4. Warm outdoor air in.
5. Hot air out.

Cooled Air

■ High-Pressure Hot Gas
■ High-Pressure Warm Liquid
▨ Low-Pressure Cold Liquid/Gas
▨ Low-Pressure Cool Gas

Fig. 4-3. Typical heat pump unit. (Courtesy General Electric Co.)

Electric Furnaces

The electric furnace (Fig. 4-4) is a relatively small, self-contained, resistance-heating unit. A wide range of capacities is now available. The medium of heat exchange can be forced hot air, blown through ductwork and out through registers in the usual fashion, or it can be forced hot water pumped through baseboard or other types of radiation units.

1. 100% of the electrical power is converted into heat.

2. Complete controls and cool operating furnace jacket UL listed.

3. Fan and limit control prevent sudden cold drafts and hot blasts.

4. NICHROME wire heating elements — 5 year warranty.

5. Sequenced controls (5 KW steps) circuit breakers on LE051, 68, 85 and 102 D1B.

6. Multi circuit electrical hook-up for low installation costs.

7. Tufskin cabinet insulation for efficient operation.

8. Direct drive blower for smooth starting, quiet operation.

9. Filter-Cleans the air.

10. All welded steel framework and cabinet for long life.

Fig. 4-4. Typical electric furnace unit. (Courtesy General Electric Co.)

The forced-air type can also be used in conjunction with a central air-conditioning system using the same ductwork.

The typical advantage of electric heat in requiring no flue or chimney for operation is particularly useful here, since the electric furnace is small and light enough to be tucked away virtually anywhere in the house. The cost of an electric furnace is low, perhaps 25% less than a comparable gas- or oil-fired furnace, and there is no chimney expense involved. The installation of necessary ductwork or piping and radiation units, however, places total costs close to those of other types of systems. There is also some loss of efficiency as the heat is transferred from the source to points around the house. And as with heat pumps, maintenance is likely to be more of a factor than with some other types of electric heat.

Zoning is also something of a problem with electric furnaces. They are usually operated by a single standard wall thermostat, although two or more thermostats could be wired to operate appropriate zone equipment in a complex system.

Unit Heaters

Unit heaters (Fig. 4-5) are small individual resistance heaters which are available in a variety of sizes and designs. Some are equipped with built-in thermostats, while others can be operated from wall thermostats. Many of the smaller capacity units are designed to be mounted either in or on a wall, and some are suspended from the ceiling. They are equipped with fans or blowers for either low volume air movement or high volume, depending upon the application.

Common ratings for units used in domestic service are 1250, 1500, 2000, 4000, and 5000 watts. Though they could be used to make up a complete home heating system, more often they are installed as supplementary heat or located in areas other than the living quarters. One particular exception to this is the type made for installation in bathroom ceilings for quick additional heat at the flip of a switch.

Radiant Panels

Radiant panels are factory-made units that come in certain shapes and sizes and heat outputs for installation on or flush with the finished surface of walls or ceilings. They are relatively expensive and do not readily lend themselves to some types of construction or decor, but are effective.

Correct panel sizing and placement is important; they cannot be altered or "rebuilt" on the job site. Each unit is complete within itself and must be correctly installed. Wiring connections are made to leads built into the panel. Control is

Fig. 4-5. Electric unit heaters for residential ceiling installation. (Courtesy NuTone Division of Scovill)

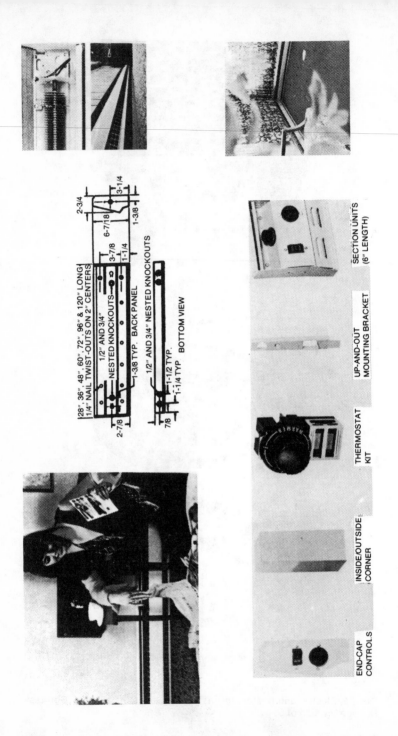

28", 36", 48", 60", 72", 96" & 120" LONG
1/4" NAIL TWIST-OUTS ON 2" CENTERS

1/2" AND 3/4"
NESTED KNOCKOUTS

1-3/8 TYP. BACK PANEL

1/2" AND 3/4" NESTED KNOCKOUTS

BOTTOM VIEW

END-CAP
CONTROLS

INSIDE/OUTSIDE
CORNER

THERMOSTAT
KIT

UP-AND-OUT
MOUNTING BRACKET

SECTION UNITS
(6" LENGTH)

Rebel Baseboard Heaters

Medium Watt Density

Volts	Lgth.	Watts	BTU/hr.	Cat. No.
120	28"	500	1707	BB-H-21
	36"	750	2360	BB-H-31
	48"	1000	3413	BB-H-41
	60"	1250	4266	BB-H-51
	72"	1500	5118	BB-H-61
208	28"	500	1707	BB-H-28
	36"	750	2360	BB-H-38
	48"	1000	3413	BB-H-48
	60"	1250	4266	BB-H-58
	72"	1500	5118	BB-H-68
	96"	2000	6824	BB-H-88
	120"	2500	8547	BB-H-108

Medium/Low Watt Density

Volts	Lgth.	Watts	BTU/hr.	Cat. No.
240/208	28"	500/375	1707/1280	BB-HI-24
	36"	750/565	2360/1927	BB-HI-34
	48"	1000/750	3413/2360	BB-HI-44
	60"	1250/940	4266/3108	BB-HI-54
	72"	1500/1125	5118/3840	BB-HI-64
	96"	2000/1500	6824/5118	BB-HI-84
	120"	2500/1875	8547/6399	BB-HI-104

Medium/Low Watt Density

Volts	Lgth.	Watts	BTU/hr.	Cat. No.
277/240	28"	500/375	1707/1280	BB-HI-27
	36"	750/565	2360/1927	BB-HI-37
	48"	1000/750	3413/2360	BB-HI-47
	60"	1250/940	4266/3108	BB-HI-57
	72"	1500/1125	5118/3840	BB-HI-67
	96"	2000/1500	6824/5118	BB-HI-87
	120"	2500/1875	8547/6399	BB-HI-107

Low Watt Density

Volts	Lgth.	Watts	BTU/hr.	Cat. No.
120	28"	375	1280	BB-I-21
	36"	565	1927	BB-I-31
	48"	750	2560	BB-I-41
	60"	940	3108	BB-I-51
	72"	1125	3840	BB-I-61
277	28"	375	1280	BB-I-27
	36"	565	1927	BB-I-37
	48"	750	2560	BB-I-47
	60"	940	3108	BB-I-57
	72"	1125	3840	BB-I-67
	96"	1500	5118	BB-I-87
	120"	1875	6399	BB-I-107

Fig. 4-6. Typical baseboard heating units and accessories. Note that none in this particular line are of the high-density type. (Courtesy Emerson-Chromalox)

accomplished through line or low-voltage thermostats, either in zones or larger sections of the home, using relays as necessary when combined unit loads are too great for the thermostats to handle. Design and installation of this type of system is most often done by professionals.

Heating Cables

Heating by cable is an effective system, though installation costs can run high when professionally done. Cables can be easily installed by the do-it-yourselfer, in which case the costs are low, but some skills are required, and the energy output (human) runs high.

One type of cable is made for use in ceilings, where it is embedded in the fresh plaster after being stapled to a gypsum lath or plaster subsurface base. The ceiling type can also be used with dry ceiling construction by first stapling the cable to gypsum wallboard sheathing, then buttering the back of the finish layer of gypsum wallboard with a quarter of an inch of wet plaster and immediately nailing the sheet into place. A second type is for use in poured concrete floors, but it is not the same as that used for melting snow in drives and walks.

Cables are bought in certain standard wattages, each of which is a different length. Spliced to each end of the cable is a short lead for making connections to the serving branch circuit. The entire cable must be laid in even spacings above (or below) the entire area to be heated, avoiding lighting fixtures or other metallic objects. The two leads travel either to a junction box or thermostat.

When power is applied, radiant energy is produced evenly over the entire area, changing to heat as it strikes the objects in the room. In effect the whole ceiling (or floor) becomes a radiant heat panel.

This system provides comfortable, even, and effective heat, takes up none of the living space of the home, and is completely unobstrusive. Zoning is simple, even to the extent of providing a separate thermostat for each room. One disadvantage is that heating from a cold start—in order to bring the temperature up to 70° from 40°, for instance—requires quite a bit more time than other types of equipment. Once the proper level is reached, however, cycling is short and the temperature remains remarkably even provided that no great quantities of cold air are introduced at any one time.

Baseboard Heaters

Heating by standardized baseboard units, which look much like conventional hot-water radiation sections (Fig. 4-6) is probably the most common method in dwellings. Each unit consists of a long resistance element equipped with fins to aid convection, all enclosed in a metal cabinet. A second type makes use of a fluid-filled core that is heated by the resistance element. Both styles are in common use, with the dry type being less expensive and the wet claimed to have somewhat better and more even heating capabilities.

Baseboard units come in ratings of 500, 1000, 1500, 2000, and 3000 watts, with roughly standardized lengths of 2, 4, 6, 8, and 10 feet. They are designed to be surface mounted at floor level against the walls or baseboards. Surface temperatures are sufficiently low that they can be attached to combustible materials with no danger. One type of unit is made to set into a hole in the floor with a flush register grille over it, to provide heat in areas where floor-to-ceiling glass walls are used and baseboard mounting is not possible. Most models are also equipped with thermal safety switches so that if air circulation is blocked, the units will shut off when they become too hot.

Some units are equipped with individual built-in thermostats, while others are used with wall thermostats, either line- or low-voltage. The latter employs a transformer and relay in a small enclosure (Fig. 4-7) that is added to the first

Fig. 4-7. High-density baseboard heating unit, opened, with low-voltage thermostat and controlling relay unit at left.

heating unit in a circuit of several heaters, or can be separately mounted. Zoning can be set up with each individual unit, but more often is done room by room, floor by floor, or section by section. Installation is very simple and the equipment and wiring costs are low.

Summary

Once you have chosen the type of equipment, all that is left is to pick the appropriate capacity to cover your heat losses. With a central type of heating, such as the heat pump or furnace, simply select a standard unit tha delivers the closest amount of wattage (or Btus) above you *HR* or total heat loss for the structure. Where individual units are to be used, go back to the room calculation sheets and pick one or a combination of units whose total output in watts is approximately equal to or greater than each room's heat-loss figure. The combinations can also be juggled around to provide the best physical sizes of units for convenient installation.

Chapter 5
Lighting Design

Illumination by means of the light bulb was the first practical application on a broad scale that resulted from the harnessing of electrical power. Thomas Alva Edison started the process when on October 19, 1879 he sucessfully coaxed a feeble and shortlived glow from a filament made of fine cotton sewing thread enclosed in a hand-blown glass envelope. "After this, we will make electric light so cheap," Edison said, "that only the rich will be able to burn candles." On September 4, 1882 he backed up his words by opening a power station in New York City. There was one dynamo and a mere 85 customers, but the Electrical Age was under way.

Today, electrical illumination still remains cheaper than candles—not to mention being a little more effective. Of all the appliances, tools, devices, gadgets, and toys that surround us in the home, few are more practical, more widely used, or more convenient than the electric light bulb. The returns, actual, psychological, and esthetic, that we get back from the expenditure of the small amount of money required to operate a light is phenomenal, though most of us take it for granted.

And yet, good lighting is probably the most disregarded aspect of a residential electrical installation. More often than not, lighting is only an afterthought, a minor item that is stuck into the plans after everything else has been taken care of. Very few homes have good lighting, most have barely adequate systems, and many are horribly substandard. Tract houses and those built on speculation are among the worst offenders. The builder has no knowledge of the specific ways in which the houses will be used after they are occupied, so he cannot plan a proper lighting system. And anyway, good lighting raises the cost a few dollars. So why bother? Most old houses have never been updated, except perhaps for replacing old, worn out lighting fixtures with new ones which hardly do a better job. Even in many new and expensive homes, architects, designers, and interior decorators select lighting fixtures and their locations more on a basis of decorative effect and physical compatibility, and often price, than how well they will fulfill their intended function in lighting an area.

Most home dwellers are so inured to lighting that is minimally functional or below par that it seldom occurs to them that the situation could be vastly improved. Then too, most people do not have the knowledge of what constitutes an excellent lighting system, nor do they realize the benefits that can be derived from such a system. They do not know how to go about upgrading the lighting in their homes or planning a complete system for their dream home soon to be built.

This is a rather sad situation, because the difference in cost (in a new home) between a substandard lighting arrangement and one that is at least adequate and properly functional is a matter of only a few dollars. And the difference between an adequate system and a luxurious one—a showcase installation—is a bit more than just a few dollars but still not a great amount. To illustrate, let's take a house that costs $45,000, about average in today's market, that has an average (barely adequate) lighting system. To convert this to an excellent or even outstanding system might cost an additional $1000. Put another way, this is just a bit over 2% of the total cost of the house. Even if the differential cost were higher, the percentage of the total cost of the house remains quite low.

If you are about to invest that much money in a home of your own, perhaps even custom-built and of your own design, doesn't it make sense to invest an extra two or three percent in such an important feature as excellent lighting, rather than settling for just a passable system? It does. And bear in mind, too, that the extra amount spent is not just an outright cost—money down the drain and not recoverable. That additional outlay not only buys extra enjoyment and convenience during your occupancy, but also enhances the dollar value of your property. At least a part of the investment will return to you upon resale of the home.

GOOD LIGHTING HAS ADVANTAGES

There are quite a few advantages to a well-designed, complete, and extensive lighting system. We have already noted the increased property value in a new house; the same can be even more true of an older house that is extensively remodeled with good lighting in mind. Sometimes the increase in value will amount to more than the cash outlay for lighting renovations, particularly when done by the owner-occupant.

Increased safety for occupants and guests, visitors, tradespeople, or anyone who might have occasion to be on the premises, is another plus factor. This is an especially important point when there are small children, elderly persons, or semi-invalids present. Properly lighted passageways, stairs, stairwells, outside steps, walkways, and entries, as well as the rooms themselves, are a must.

Efficiency can be increased through the use of good lighting, too. Granted that this might not be as important to the home dweller who is on his own time as it is to the factory owner who wants to maximize production. Nevertheless, most people prefer not to waste their time in needless monkey-motion simply because they cannot see well enough to do a good job, especially if they want to get on to something else. And there are many tasks, such as intricate sewing, model building, or handloading firearms ammunition, that plainly cannot be accomplished readily or with any degree of accuracy and proficiency unless the lighting is top-notch.

Closely tied to efficiency is *convenience*—an abstraction that we enlightened citizens of a technological age set great store by. The more convenient things are, the better we like it! But if a kitchen counter is so dim and shadowed that you cannot see to slice vegetables for a salad, or you have to turn sideways to admit enough light to the cutting board, that is hardly convenient. If the kids set up the cardtable to do a picture puzzle and have to swipe all the lamps in the room so that they can enjoy their pastime, that isn't so convenient either—for them or for the rest of the household. And if Dad takes the screens into the basement to paint them and they turn out to be maroon instead of black when brought back into the light of day, a family crisis occurs. A well-designed lighting system takes into account the life style, interests, and daily activities of the residents, and provides for all possibilities within reason.

Odd though it may sound, health is another consideration. This is tied to safety, of course, since inadequate lighting can lead to bumps and bruises, broken bones, even cuts and burns. But you must also protect your eyesight, especially if you are a heavy reader. Children need good study centers, the cook needs well lighted working space, and so forth. This involves not only the intensity of illumination, but proper angling, contrast, reflection, and other factors. Eyestrain, which can lead to headaches and upset stomachs, are all too often caused by poorly designed lighting arrangements. And then there is the mental health factor since inconvenience, or lack of efficiency, or both, can quickly lead to annoyance. Annoyance can become irritation. Irritation can then lead to family squabbles, tension, nervousness, and a host of unsavory problems. That trail can lead off in many directions, or in none at all—no problem if those involved simply shrug off the effects, which inevitably exist, of poor lighting. But it is safe to say that under those conditions, life in the home is not as happy and as satisfying as it could be.

There are other intangibles, too, such as enjoyment. The total enjoyment of a home is the sum total of a lot of small things. For instance, if you have a corner cabinet loaded with a collection of fine antique cut glass, but you can't properly view them without a flashlight, that isn't much fun. Part of the enjoyment in making such a collection is gone. Or if Junior wants to take you on in a few games of Ping-Pong after supper, you may refuse if that means fussing around with portable lamps and dangling extension cords. You have lost an enjoyable session with you son. The examples are endless, but what it all boils down to is that for maximum enjoyment of life in and around the home, you need a good lighting system.

Pride also plays a part—pride in ownership and display of your home. You may wish to highlight certain architectural niceties, especially if you have designed and built them yourself. Or you may want to show off, to one degree or another, your collections, your treasures, your gardens, or maybe the house itself. Some homeowners feel a certain distinct satisfaction in knowing that they have a fine lighting system. Others feel that they must have the best, the biggest, the most complex, the most expensive, whether it be lighting or automobiles, and this too is a valid reason. After all, the keeping-up-with-the-Joneses syndrome has always been a boon to American manufacturers. To some, the challenge of designing and installing a complicated and complete system could well be a motivating factor.

And there is a social aspect to good lighting. Everyone wants his friends and visitors to feel comfortable and at home, and to have an enjoyable time. Lighting plays a role in this situation, just as it does for the occupants themselves. And from an opposite standpoint—anti-social if you will—consider the recent rapid rise in crime in residential neighborhoods. A well lighted home, within and without, is a definite deterrent to wandering vandals. No burglar is going to hang around for long when an aroused inhabitant flips a bedside switch that turns on lights all through the house and grounds in a matter of seconds.

These are a few of the attributes of a superior lighting system, and doubtless there are more. Condensed, what it all means is that in the course of modern living to which today's homeowner is accustomed, excellent lighting should no longer be considered a luxury, but a necessity. It has often been said that a man's home is his castle—his presumably applies equally to his family—so why should he not strive for that completeness and fulfillment that is part and parcel of a true home?

And all of this can be accomplished with light? You bet. With a knowledge of the principles of light and lighting, the material available to work with, and some simple instructions and recommendations, you can design and put together a lighting system tailored exactly to your needs and superior to virtually all the systems you will see.

LIGHT

The physics of light is an extremely complicated field, one that would take volumes to explain and extensive study to understand. Obviously we can do no more here than to look at a few of the most general aspects that affect home lighting. In any case, more than that would be impractical and unnecessary, not to mention confusing, since most of the material would be of little direct use in planning a home lighting system anyway. But if you have some idea of what light is and how it works, then you will be better able to understand more readily what is involved in home lighting and why some ideas are effective and useful and others are not.

Light is actually a form of energy. It is a part of the electromagnetic spectrum, a wide range of radiant energy that includes radio waves, X-rays, gamma rays, and many others. Light is generally referred to as *visually evaluated energy*—in other words, you can see it. Light occupies that part of the spectrum between about 3900 angstroms to about 7500 angstroms. And an angstrom, should you be interested, is equal to one ten-millionth of a millimeter.

Color

So-called *white light* is a combination of the principal colors: red, orange, yellow, blue, green, and violet. All of these colors, any combination thereof, are actually sensations that our eyes preceive, excited by certain wavelengths of energy and assigned an arbitrary name. Thus, when we see wavelengths of about 4000 angstroms, we see *violet*. Wavelengths of around 6200 angstroms would produce *orange*, and so forth. There is no sharp dividing line between colors, of course, but a gradual phasing from one to another which gives us tones and shades. *Ultraviolet light*, some of which is visible, is at the bottom or short end of the scale. Infrared is at the long end and gradually fades into invisibility as the wavelengths grow longer.

As you well know, if you have ever pored through a book of paint color chips, there are literally thousands of gradations of color, and there is no way we can put a name to each particular shade that will have any real meaning. But color, and thus light, can be measured with great precision. Every gradation of light is composed of a mix of color, and each has a certain intensity and temperature, measured in degrees *kelvin* (K), which can be determined by a process called spectrophotometry. Using this information, it is possible then to match or contrast light to objects being lighted, or to make compensation for unwanted or inadequate effects.

Furthermore, different colors have different energy levels and different levels of sensitivity to the eye. Blue-green, through yellow, to yellow-orange is the most visible colors to us and brings about the strongest impressions. This sensitivity tapers off on each side of the scale; blue through violet on the short-wavelength side, and orange through red on the long. As the light dims, the colors of lesser visibility diminish rapidly, to the point where they become gray and then black, disappearing while yellow still remains.

In working with the effects of light and color, we consider three colors to be the primaries—blue, green, and red. The combination of these three will make white light. Now, as you probably know, any one of several things can happen to a ray of light when it strikes an object. It can *refract*, or bend, and in this case the ray spreads out into its various color components, as when passing through a prism. The ray can also *reflect* as a sunbeam does from the chrome on your car. Or, the ray can be *absorbed*, being no longer visible. It may also be *transmitted*, by passing through a transparent object and out again with no change. In actuality, what generally occurs is a combination of any or all of these circumstances.

It is primarily absorption that permits us to perceive color. All the objects around us, whatever they may be, will absorb a certain amount of the light that strikes them, reflecting most of the remainder. The reflected portion of the light excites a color sensation in our eyes. Thus, when light shines on what we see as a bright red pillow, the red component is reflected to our eyes while the remaining color components of the light, blue and green, are absorbed by the fabric. If you were able to shine a pure green light on the same pillow, it would appear black (the absence of color) because there would be no red component in the light that could be reflected back to your eyes. In practice, of course, none of the colors are pure, but subtle blends of various component shades.

These, then, are some of the elementary principles of light. As we proceed, you will see the practical applications of these principles as they affect home lighting.

LIGHT SOURCES

The main source of our light is a natural one, the sun. Consciously or not, we tend to judge our surroundings on a basis of that familiar natural light. One problem with this is that, though we rarely realize the fact, natural light is continuously changing in color. This fact is illustrated in Fig. 5-1. Direct sunlight in the early morning starts at a color

Fig. 5-1. Chart of light temperature and color.

temperature of about 1900°K, slowly rises to a noontime peak of 5300—5400°K, and then begins to cool again. In the meantime, reflected and diffused natural light may range around 7000°K under a bright but evenly overcast sky to as high as 25,000°K in a clear, brilliant blue northwest sky. So we live with an ever-changing mishmash of light that affects our perception of color, tone, shade, and texture. Some standard of comparison with artificial light was necessary, so an arbitrary figure of 5400°K, which is called *mean noon sunlight* as computed at Washington, D.C., was finally selected. The composition of this light is a balance of 33% red, 34% green, and 33% blue.

There are many artificial sources of light, but only a few are of interest to the home lighting designer. The most widely used are *incandescent* lamps and *fluorescent* lamps. *Mercury, sodium*, and *quartz-iodide* lamps sometimes have application for residences or farms. *Neon, neon glow*, and *ultraviolet* lamps sometimes have home uses too, as do *sunlamps* and some *infrared* lamps. *Black light* lamps may be used for decorative or novelty effects, and *grow lights* are becoming increasingly popular as an aid to raising indoor plants. Figure 5-1 will also give you an idea of the temperature and color relationships of various light sources.

Tungsten Lamps

The tungsten filament, incandescent bulb was introduced back in 1907, and it still remains the most widely used light source in homes today. Literally thousands of types of these bulbs have been designed over the years for hundreds of applications, but we need be concerned with only a few. All of them are designated by code numbers that indicate their shape and dimension. The *A* bulb is standard and probably the most familiar to you, along with the *PS* bulb (pear-shaped). An A-19 bulb would be a standard bulb with a maximum dimension (height) of 19 increments of 1/8 inch, making a total of 2 3/8 inches high. The *T* bulb (tubular, or sometimes called showcase) has a more limited use. The *G* (globular) bulb has enjoyed a recent upswing in popularity, and so have various kinds of *F* (flame) bulbs, which are used in many decorative fixtures. There are also a number of small bulbs used decoratively, Christmas lights being the most well known, and some special-purpose bulbs for ovens, refrigerators, sewing machines, and such.

Those general-purpose bulbs, though once all made with clear glass envelopes, now are manufactured with several features that serve to diffuse and soften the light. The most common, and the most efficient, is the inside-frosted style. These have a hard, smooth outer surface that makes them easy to clean, and the frosting on the inside is thin enough to allow most of the light produced by the filament to shine through. If you look carefully, you can see a slight shadow of the filament where it is close to the top of the bulb.

White bulbs, made of a translucent white glass, are more expensive and provide somewhat less light, perhaps as much as 20% less than clear glass, but there is a more even diffusion and less glare. For some applications this can be an important factor.

There are also a few delicately tinted bulbs. Green, pink, and blue bulbs, for instance, are mainly used for decorative purposes. They appear rather strange to the eye in most circumstances and have small practical value except to warm or cool a particular decor as background lighting.

Flame and smaller decorative bulbs, sign bulbs, and the like may be clear, or toned in amber or smoke, or done in brilliant colors. The purpose is for effect rather than just light.

A somewhat different sort of incandescent lamp is made in the shape of a cone or modified cone. The sides of the bulb are opaque, and all of the light output is directed through the relatively flat front, or top, surface. The type *R* (reflector) is designed for indoor use, while the *PAR* (parabolic reflector) is made from special glass to withstand the rigors of weather in outdoor applications, without need for any further protection. Both of these types come in either *flood* or *spot* form. Flood lights have a coating of silver on the inside surface and use a frosted front surface to diffuse and spread the light. Spots, on the other hand, have little or no frosting and the light becomes a directed beam, much like a flashlight, that is concentrated on a small area. The size of the area depends to some degree upon the size of the bulb, but more so on the proximity of the bulb to the object being lighted. The closer the bulb, the smaller the lighted area, with a correspondingly limited peripheral light spill.

There is another type of incandescent bulb that is perhaps not as widely used in residences as it might be, but under the right circumstances can be very effective. This is the *Lumiline*, a slender glass tube of about an inch in diameter and of varying lengths up to several feet. Figure 5-2 shows some of the incandescent bulb types generally available.

There are some drawbacks and limitations to incandescent bulbs. The standard for artificial lighting is based upon light from a tungsten filament at 3200°K, which consists of color components in the amounts of 49% red, 34%

Fig. 5-2. Various types of incandescent bulbs. The size and shape of a bulb is designated by a letter or letters followed by a number. The letter indicates the shape of the bulb while the number indicates the diameter of the bulb in eighths of an inch. For example, "T-10" indicates a tubular-shaped bulb having a diameter of 10/8 or 1 1/4 inches. The illustrations show some of the more popular bulb shapes and sizes. (Courtesy Westinghouse Electric Corporation)

141

green, and only 17% blue. Nearly all incandescent bulbs produce light at temperatures within a few hundred degrees kelvin of this standard because higher temperatures greatly reduce lamp life. So incandescent light is of necessity quite warm in tone, which in turn has an effect on decorating schemes and visual perception. Also, because of their inherent shape and size, these bulbs provide only small sources or points of light and are more suited to small areas. Several used together will cover a larger area, but must be handled properly to avoid hot spots and shadows. The one exception to this is the Lumiline, which will provide a broad source or wash light and can cover larger areas.

Incandescent bulbs also generate a great deal of heat, and the bigger the bulb the more heat is produced, a factor which must always be considered during lighting design and installation. In addition, though reasonably long-lived when installed in the proper position and not subjected to vibration and hard knocks, they are relatively expensive to operate. On the other hand, they are effective when properly used, convenient, easy to replace, inexpensive to buy, available in wide variety, and affort great flexibility of lighting. When desirable, the level of illumination can be readily controlled by the use of dimmers.

One thing which seems confusing to incandescent lamp buyers, but need not be, is the base size (Fig. 5-3). Amost all lamps used in the home are of the screw-base type, and there are only five sizes. The smallest is the *miniature* base, as used on most Christmas tree lights and some nightlights. The next step up is the *candelabra*, the principle use of which is implied in the name. The next is the *intermediate* size, which is used in a number of lighting fixtures. The *medium* size is by far the most prevalent and is found in the majority of lamps and fixtures. The *mogul* base is the largest, and in the home it is usually found only in a limited number of big floor lamps or overhead fixtures.

Incandescent bulbs are rated both by voltage and by wattage. Though the general range is from 1 1/2 volts to 300 volts, they are designed to operate most efficiently at certain particular voltage levels, such as 12 volts, 115 volts, or 230 volts. General-use bulbs come in wattage ratings of 7 1/2 to 300 (to 500 watts in the mogul base size). There are also three-way bulbs, which must be used with a special switch and socket, available in combinations of 30−70−100 watts, 50−100−150 watts, and 50−200−250 watts in the medium base size. Mogul base three-way bulbs have 50−100−150 and 100−200−300 watt ratings.

Fig. 5-3. Comparison of incandescent bulb bases. From the left, mogul, standard, intermediate, candelabra, and miniature.

Wattage ranges of single-filament bulbs vary with the specific type of bulb, but there generally is some degree of choice with almost all of them. The wattage rating is related to the power consumption and roughly indicates the amount of light you can expect to get from a given bulb. The output from a 100-watt bulb is more than a 50-watt bulb, usually being more than twice as much.

In operation, electrical current passes through the filament of an incandescent bulb, where it encounters resistance and causes the filament to heat and glow at a certain level of intensity. As time passes the light output decreases because the filament, which in today's bulbs is tungsten operating at around 4800°F, slowly evaporates and condenses on the glass envelope. The result is that slowly enlarging black spot you will sometimes see inside the older bulbs around your house. In addition, as the filament operates at a lower temperature and so produces less light. Other factors are at work that affect the length of life of a bulb, too, such as shock and vibration, voltage fluctuation in the power line, or operation at a higher than rated voltage. Incandescent bulbs should ideally be burned in a vertical position, with the base up, for longest life.

Fluorescent Lamps

Fluorescent lighting is relatively new, being introduced in 1938 and coming into fairly widespread use by the late 1940s. Today it is widely used in the home and serves well in certain capacities, though it does not have the flexibility that incandescent lighting has.

All fluorescent lights require a *starting device* to get them going, and there are several main types of lights, according to the methods used. The *pre-heat* lamp requires a separate starter—a small replaceable cannister that is inserted into a special socket from the outside of the fixture. This is the kind of fluorescent light that is usually found only in the smaller sizes today. It requires several seconds to warm up and finally come to life, usually blinking several times in the process. Other types have built-in starting circuits that require no attention or consideration. One is the *rapid-start* type, the other is the *instant-start*; both of these will light up almost immediately when the switch is flipped. The point here is that not all fluorescent light tubes are interchangeable, but are dependent to a degree upon the starting system used.

Fluorescent lights also have internal regulating devices called *ballasts*. These may consist of series resistors, capacitors, reactors, or transformers, depending upon the design of the fixture. Ballasts do generate a certain amount of heat, and some of them produce a hum that is clearly audible and can be annoying in a quiet room. Though the ballasts are usually mounted inside the fixture case itself, they can also be established in special cabinets in a remote location, thus doing away with both noise and heat generation problems. Remote ballasts help to minimize another common fluorescent light characteristic—radio interference. This problem, however, can also be eliminated by keeping radios and lights well apart, or if this is not possible, by installing interference filters on the radios.

Fluorescent light bulbs are long, skinny tubes that vary slightly in diameter and greatly in length. They are designated by a *T* (tubular), followed by a number that indicates the diameter in eighths of an inch. Thus, a T12 bulb would be 1 1/2 inches in diameter. If another number is present, this indicates the length of the lamp in inches; a 96-T8 lamp would be eight feet (96 inches) long and an inch in diameter. Overall lengths range from six inches to eight feet. The fixtures themselves are made to accept one, two, or four lamps.

Table. 5-1. Properties of Fluorescent Lamps

Lamp Description		Atmosphere	Light Output	Color Rendering Ability	Color Flattery Ability	Coordinated Color Temperature	Lighted Appearance ICI Color Coord.	
							x	y
Cool White	CW	Cool	100%	100%	100%	4100°K	.372	.375
Cool White Deluxe	CWX	Cool	71	126	103	4200	.369	.363
White	W	Warm	104	91	99	3500	.409	.394
Warm White	WW	Warm	104	73	96	3000	.435	.402
Warm White Deluxe	WWX	Warm	71	101	101	3000	.430	.389
Daylight	D	Cool	85	93	102	6500	.313	.337
Living White	LW	Cool	75	124	105	4300	.369	.363
Cool Green	CG	Cool	92	—	—	6100	.315	.366
Sign White	SGN	Cool	75	—	—	5300	.332	.350
Natural	N	Warm	70	—	—	3650	.388	.361
Supermarket White	SMW	Cool	74	—	—	4500	.362	.375
Merchandising White	MWX	Warm	78	—	—	3450	.409	.396
Red	R	—	06	—	—	—	.690	.295
Pink	PK	—	45	—	—	—	.526	.356
Gold	GO	—	60	—	—	—	.510	.469
Green	G	—	140	—	—	—	.264	.630
Blue	B	—	45	—	—	—	.205	.183

The "cool white" lamp is used as the basis for comparison for other lamp types. (Courtesy of Westinghouse Electric Corporation.)

Fluorescent bulbs are rated for operation at certain voltage levels, and also by wattage just as the incandescent variety is. The largest are only a bit over 200 watts, while small lamps may be rated as low as 6 watts. The amount of light produced, though, is three to four times as much per watt as incandescents. Most lamps are fitted with two contact pins at each end, referred to as a *bipin base*, either in the miniature size or medium size; these pins are inserted into sockets in the fixture. Some types of lamps, however, such as the slimline, have only one contact at each end.

Tubes are available in several different color values (Table 5-1). *Cool white* is most common, next is *warm white*. Both of these lamps have counterparts known as *deluxe* types, and these give somewhat better color renderings, especially of red and orange, and also blend rather well with incandescent light. This makes them particularly useful in residential applications where both light sources are likely to be used side by side. *White* is a compromise between *cool white* and *warm white*, and *warm white deluxe* gives off a pinkish glow that sometimes causes strange effects. Don't be misled by the so-called *daylight* type of tube, which reads out at about 6500°K; it emits a harsh, cold, blue-white light that makes everyone look gray and wan, as though just recovering from some exotic illness. Table 5-2 lists the generally available fluorescent tubes and their characteristics.

Though there are numerous small mechanical and electrical design differences in the fluorescent light family, most of them operate in the same general way. When the power is switched on, a small filament located at each end of the tube is caused to heat up to the point where an outward electron flow begins. A precalculated surge of voltage throws an arc from one end of the tube to the other through the mercury gas sealed into the glass envelope. This arc then continues until the power is turned off.

If the tube were of clear glass, you would see only a faint purple glow, since the light emitted from the arc is ultraviolet and close to the short end of the visible light spectrum. The white inside coating on every fluorescent tube, however, is not just a coloring but actually a coating of *phosphor*—a mixture of chemical powders that fluoresce when struck by the ultraviolet rays. The phosphor glows brightly to cause the highly visible light. Incidentally, though at one time the chemicals in fluorescent tubes were considered highly dangerous if ingested or inhaled from a broken tube, but this no longer holds true. Still, the tubes should be handled with care because they will shatter if struck just right.

Fluorescent lights have the advantage of producing a lot of light at low operating cost. They are cool running with only the ballasts generating any appreciable amount of heat. They can effectively provide a broad source, long line, or wash light. Shadow and glare is minimal, and general background illumination is easy to arrange. Except for a few specially made (and expensive) models not common to home use, their output level cannot be controlled, as with a dimmer, because after a slight drop in voltage they simply go out. Nor can tubes of different wattage ratings be substituted in any given fixture as they can with incandescent lights.

Often their size and shape or appearance gives rise to difficulties in installation. From practical and esthetic standpoints, there are many places in the home where they simply cannot be used for those reasons. And, too, the color temperature of the light emitted sometimes makes fluorescent lighting difficult to work under—even downright undesirable—depending upon the specific lighting design features needed. A six-foot strip dangling over the head of the bed, for instance, is somehow not very appealing, despite the fact that there would be plenty of light available. Cost is not necessarily a determining factor, because though the fixtures are initially expensive and tubes more costly to replace than incandescent bulbs, tubes and fixtures both have a remarkably long life, making the total expense per unit of effective light quite low.

Other Lights

The remaining types of light sources are of only limited interest to the average homeowner. Mercury and sodium lamps are primarily used for outside lighting—you have seen them along highways and in big parking lots. These are sometimes found in residential areas or on farms as yard lights. Often these lamps are supplied and installed by the serving utility company. Quartz iodide lamps are a special type of high-intensity incandescent lamp, but have much the same use outside.

Neon lamps are all too familiar to us, since we are surrounded by them in every business district and shopping center, buzzing away garishly on all manner of signs. Under certain circumstances there might be occasion to use this type of lighting in the home, but it would be unusual. Neon tubes can be formed in all conceivable shapes, though, so they can be used effectively in architectural lighting, as in formal gardens. Small neon glow lamps are mostly used as indicator lamps and nightlights.

Table 5-2. Some Available Fluorescent Lamps

Lamp	Ordering Abbreviation	Base	Approximate Ampere	Operating Volts	Minimum Starting Volts	Average Preheat Amperes	Cathode Heater Voltage Nominal
Preheat							
4W T-5 6"	F4T5/CW	Min. Bipin	0.135	32	110	0.15	—
6W T-5 9"	F6T5/CW	Min. Bipin	0.147	47	110	0.20	—
8W T-5 12"	F8T5/CW	Min. Bipin	0.170	56	110	0.22	—
13W T-5 21"	F13T5/CW	Min. Bipin	0.160	95	176	0.20	—
14W-T-12 15"	F14T12/CW	Med. Bipin	0.385	39	110	0.55	7.5
15W T-8 18"	F15T8/CW	Med. Bipin	0.300	55	110	0.55	7.5
15W T-12 18"	F15T12/CW	Med. Bipin	0.330	46	110	0.55	7.5
20W T-12 24"	F20T12/CW	Med. Bipin	0.380	56	110	0.55	7.5
25W T-12 33"	F25T12/CW	Med. Bipin	0.480	58	110	0.75	—
30W T-8 36"	F30T8/CW	Med. Bipin	0.355	98	176	0.53	—
90W T-17 60"	F90T17/CW	Mog. Bipin	1.530	63	132	1.80	—
Preheat—Rapid Start							
40W T-12 48"	F40CW	Med. Bipin	0.430	105	256	—	3.6
40W T-10 48"	F40T10/CW/99	Med. Bipin	0.420	108	256	—	3.6
40W T-12 22-7/16"	FB40CW/6	Med. Bipin	0.430	100	256	—	3.6
Rapid Start							
30W T-12 36"	F30T12/CW/RS	Med. Bipin	0.430	75	250	—	3.6
High Output							
24" T-12 38W	F24T12/CW/HO	Recessed D.C.	0.800	41	225	—	3.6
48" T-12 60W	F48T12/CW/HO	Recessed D.C.	0.800	75	256	—	3.6
72" T-12 87W	F72T12/CW/HO	Recessed D.C.	0.800	113	405	—	3.6
96" T-12 112W	F96T12/CW/HO	Recessed D.C.	0.800	165	465	—	3.6

Super-Hi							
48" T-12 110W	F48T12/CW/SHO	Recessed D.C.	1.500	86	250	—	3.6
72" T-12 160W	F72T12/CW/SHO	Recessed D.C.	1.500	128	350	—	3.6
96" T-12 215W	F96T12/CW/SHO	Recessed D.C.	1.500	172	470	—	3.6
96" T-12 219W	F96T12/CW SHO-II	Recessed D.C.	1.500	172	470	—	3.6
Cold Weather							
48" T-10/14 105W	F48T10J/CW	Recessed D.C.	1.500	—	—	—	3.6
72" T-10/14 150W	F72T10J/CW	Recessed D.C.	1.500	—	—	—	3.6
96" T-10/14 200W	F96T10J/CW	Recessed D.C.	1.500	—	—	—	3.6
Instant Start (Base Pins Shorted Inside Base)							
40W T-12 48"	F40T12/CW/IS	Med. Bipin	0.425	104	385	—	—
40W T-17 60"	F40T17/CW/IS	Mog. Bipin	0.425	107	385	—	—
Slimline							
42" T-6 25.5W	F42T6/CW	Single Pin	0.200	150	405	—	—
64" T-6 38.5W	F64T6/CW	Single Pin	0.200	233	540	—	—
72" T-8 37.5W	F72T8/CW	Single Pin	0.200	218	540	—	—
96" T-8 50W	F96T8/CW	Single Pin	0.200	290	675	—	—
48" T-12 39W	F48T12/CW	Single Pin	0.425	100	385	—	—
72" T-12 56W	F72T12/CW	Single Pin	0.425	149	475	—	—
96" T-12 73.5W	F96T12/CW	Single Pin	0.425	197	565	—	—
Circline							
22W T-9 8¼" Diam.	FC8T9/CW	Four Pin	0.380	60	185	—	3.6
32W T-10 12" Diam.	FC12T10/CW	Four Pin	0.430	80	205	—	3.6
40W T-10 16" Diam.	FC16T10/CW	Four Pin	0.415	108	205	—	3.6
Reflector							
40W T-12 48"	F40CW/RFL	Med. Bipin	0.430	105	256	—	3.6
90W T-17 60"	F90T17/CW/RFL	Mog. Bipin	0.530	63	132	1.80	—
48" T-12 39W	F48T12/CW/RFL	Single Pin	0.425	100	385	—	—
96" T-12 75W	F96T12/CW/RFL	Single Pin	0.425	197	565	—	—

Ultraviolet lamps sometimes find their way into homes. When used as germicidal lamps, they are usually bought as complete portable fixtures for use in sickrooms.

Grow lamps are used in greenhouses or over planters or potted groupings to keep plants healthy and growing well, especially during the long dark winter days. These are akin to small fluorescent units and may be portable or permanently installed, usually in an overhead position.

Black light sources, mostly tubular but now also available in screw-base bulb form, are used for decorative effect or just amusement. They can illuminate certain types of murals, paintings or posters, making them fluoresce. A mineral collector will tell you that many drab minerals displayed under black light show startling beauty. Small social gatherings in a room lit by black light can be fun, too, because of the eerie and unusual effects produced by fluorescing clothing, cosmetics, and surroundings.

Ultraviolet sunlamps are specially designed self-contained units, often with timers, used to acquire an artificial suntan. Most are portable, but they can be built in.

Infrared lamps transmit infrared rays from the high end of the visible spectrum and somewhat beyond. These are used primarily for heating. Here again, some are portable while others are designed, sometimes in combination with ventilation fans, to be permanently installed, as in a bathroom. Other types of infrared fixtures can be used to keep prepared foods warm or for fast-drying purposes.

LIGHTING FIXTURES

Lighting fixtures, as you are already aware, appear in an incredible array of sizes, shapes, decors, styles, motifs and prices, offered for what appears to be every conceivable need or taste. The simplest of all is the screwbase lampholder, which is merely a socket with a couple of contact screws and some means of attachment to its support, even if that support is only a dropcord. Fixture designers have taken this as a starting point and gone off in all directions. The same situation is true of fluorescent lights. In this case the simplest form is a plain steel case that holds the ballast, internal wiring, and the lamp sockets, but the fancier models are barely recognizable as such.

In designing your home lighting system, you will have to give consideration to a great many styles of fixtures. The best way to go about this is to pore through plenty of catalogs and visit some lighting showrooms to get a good idea of what is currently available. Some of your lighting units will have to be, and should be, floor and table lamps, and perhaps pin-up or bracket lamps as well with their own built-in switches. For the most part, these portable fixtures will be used at appropriate points for localized lighting.

General lighting, as well as broad-area high-level lighting, is usually accomplished with the use of permanently installed fixtures and their attendant remote switches. There are many such fixtures, both incandescent and fluorescent, which are designed to complement various decorating schemes and fulfill various lighting requirements. They may be surface mounted, or they may be of the recessed sort with only a trim ring and usually a diffusing panel visible. But in many cases you can use the simplest, most unadorned types of basic fixtures, hidden away behind decorative diffusers of your own design. Quite frequently this is the most satisfactory course to follow, especially in the case of a new house or a remodel job where the lighting can be designed right into the structure rather than tacked on as a secondary thought, as must often be done when relighting an existing structure.

When you choose lighting fixtures, there are a number of factors to keep in mind. Cost is important, naturally, and so is the physical appearance—the way the fixture will complement or blend in with your decorating scheme. You should also look for good quality and solid construction, and preferably stay with brand names of good repute. But above all, tailor your choice to the specific application. Know ahead of time, as closely as possible, what sort of performance will be necessary for the fixture to effectively do its particular job, and choose accordingly. A poorly chosen fixture that does not live up to expectations is not only frustrating, it is expensive as well.

TYPES OF LIGHTING

Artificial lighting is generally grouped into five categories, all of which overlap to one degree or another (Fig. 5-4). This refers not to the ultimate mix of lighting that occurs in a room, but to the way in which light is emitted from the lighting fixture after it has been installed in its permanent position.

Direct lighting occurs when almost the entire output of a light source is directed downward. At least 90% of the light remains below the level of the fixture, only 10% or less travels upward. An example would be an enclosed, recessed ceiling fixture with a flat diffuser or lens flush with the ceiling.

At the opposite end of the scale is *indirect* lighting. In this situation, nearly all of the light from the source is directed upward to reflect from the ceiling, only 10% or less of the rays go downward. Example—a cone-shaped fixture aimed directly at the ceiling.

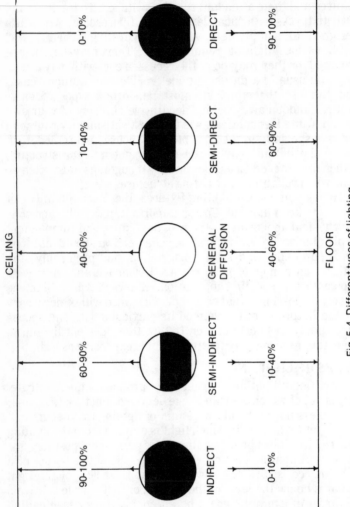

Fig. 5-4. Different types of lighting.

CEILING

DIRECT 0-10%

SEMI-DIRECT 10-40%

GENERAL DIFFUSION 40-60%

SEMI-INDIRECT 60-90%

INDIRECT 90-100%

FLOOR

DIRECT 90-100%

SEMI-DIRECT 60-90%

GENERAL DIFFUSION 40-60%

SEMI-INDIRECT 10-40%

INDIRECT 0-10%

A *general diffusion* system, sometimes difficult to achieve, spreads its light on a fifty-fifty basis, half going up and half going down. The maximum spread here is 40—60% in either direction. Example—a hanging ceiling fixture with a round translucent glass globe having the light source centered within.

The remaining two categories are in-betweens. *Semi-direct* lighting allows about 10—40% of the light to go up, as in a floorlamp partly shielded over the top, while about 60—90% of the light is directed down below the source. *Semi-indirect* lighting is just the opposite, 60—90% of the light travels upwards, while the remaining 10—40% goes down. An example of this latter case is a ceiling-mounted fixture that has open bulbs with a translucent diffuser suspended below.

These general catagories are based upon the way in which the light from a source or particular fixture is distributed, but there are other terms as well that are in common use and often somewhat general in their meaning. *Spot light*, for instance, is just that—a round pool of light, usually small but sharply brighter than its surroundings. The edge of the light pattern is well defined, and beyond that edge the brightness level falls off sharply to zero.

Diffused light actually means any light that is passed through a substance to spread the light. Most bulbs or tubes do this to begin with, and the light is then often further diffused by panels, lenses, or shades that are a part of the lighting fixture. The term implies softness and freedom from glare, but this is not always the case. Even well-diffused light can be irritating to the eye if contrasts are sufficiently high.

General light is the result of all the light sources in a room, including natural light. *General background light* is a term sometimes used to include all of the light in a room exclusive of lights used to illuminate specific small areas, like a reading lamp. Background light provides enough illumination to move around in and readily distinguish objects, but is generally insufficient for close or detailed work.

Work light, or *local* light, is just the opposite. This refers to small-area lighting primarily intended for use while undertaking some specific activity of a fairly concentrated nature, like cooking or playing cards. It does not necessarily need to be as highly localized as a reading lamp, though. In some kitchens, local light is actually widespread. But all work light is of a relatively high intensity. It also has a second function of increasing the general level of illumination by the very fact of its output. This is also called *supplementary* light and will comprise a large part of your overall lighting system.

PORTABLE LIGHTING

Portable lighting has some obvious advantages, the greatest of which is its portability. Floor and table lamps, and to a degree pin-up or bracket lamps, can be moved about here and there to satisfy the needs of the moment. They provide localized light, usually for specific purposes such as reading or knitting. When enough of them are used in concert, they also provide a low level of general illumination, but at the same time cause a lot of shadow and unevenness. To provide the best and most satisfying results, always use portable lighting along with general lighting.

Portable lamps that direct most of the light downward into the viewing area are hard on the eyes when there is no other general lighting because of the high contrasts between the lighted area and the rest of the surroundings. Used together with other lighting, though, they are excellent for concentrated, close work like model building. Many portables send a portion of the light upward by means of a translucent white bowl within the fixture, and these do a better job of providing some general light and minimizing high contrast areas. In general, both the size of the light pool and the intensity of the illumination should be scaled to the use of the lamp, the level of general light in the room or area, and the reflection-absorption characteristics of the surroundings. In other words, they should blend with, complement, or supplement the rest of the lighting system.

CEILING LIGHTING

Broad source lighting is one of the most effective ways to provide good illumination in a home, and *luminous ceilings* or *ceiling panels* do a remarkably good job along those lines. Furthermore, they are often simple to install, relatively inexpensive, and will fit in with most decors. They are particularly suited to the remodeling of older homes with high-ceilinged rooms.

Luminous Ceiling Units

Luminous ceiling units work on a principle of presenting a large area of light which illuminates a room or area fully, without shadow and glare, by wide-scale diffusion and reflection. At the same time, the lighted panels are of a sufficiently low level of brightness that they do not call too much attention to themselves; they can be looked at directly with no discomfort. Though luminous ceiling areas should be uniformly soft to look upon, their light output must be adjusted to suit the size of the room, the area lighted, and the use of the room.

There are several factors involved here. The number and light output of incandescent bulbs or fluorescent tubes located above the diffusing panel have a direct effect, of course, and so does the distance from the light source to the panel. The size, design, construction, material, and thickness of the panels can be varied to adjust the amount of light transmitted, and the ratio of lighted to unlighted ceiling in a given room can be arranged to suit the conditions. The color and reflectivity of walls, floors, counters, and furnishings also make a difference. In addition, if the light sources are mounted in a deep and large ceiling cavity above the panels, much light will be lost; conversely, it will be intensified in a small cavity or one in which reflective panels have been installed to direct light downward onto the diffusing panels. The light level can also sometimes be varied with the use of dimmers, and augmented if necessary with other types of lighting.

Luminous ceiling panels that are commercially available are little more than rather complicated lighting fixtures. These are designed so that the lighting units, which usually consist of two or four fluorescent tubes, are secured directly to the existing ceiling and wired directly into a lighting circuit. Then a diffusing panel is hung below the fixtures in a decorative frame, hiding them and providing a broad spread of light.

Another variety is actually a part of a suspended ceiling arrangement, available from most building supply houses. This consists of a system of prefinished steel or aluminum wall angles, main tees and cross tees that lock together to form an open gridwork. The spaces between the tees and angles, which can be fitted to suit any room, are than filled with drop-in fiberglass or wood-fiber ceiling tiles specially designed for the job. The whole affair is supported at an appropriate height by tiewires attached to the joists or rafters above.

Some of these ceiling systems offer specially designed fluorescent lighting fixtures that will drop right into a full-sized grid space, usually measuring two by four feet. The fixtures can be used end-to-end or side-by-side in whatever numbers are necessary, and these are supported by additional tiewires. These fixtures, called *troffer lights*, usually contain two or four four-foot fluorescent tubes spaced wide apart. You can buy them without any diffusing panels at all, but usually they are furnished with a built-in diffuser of smooth white, prismatic, crackled texture or eggcrate plastic.

Fig. 5-5. Custom-built ceiling lighting installation.

UPPER FLOOR

1/2" AIR SPACE

PAINT WHITE

2×10 JOIST

LAMP

4 1/2" APPROX.

9 1/2" APPROX.

6" APPROX.

STRAPPING

CEILING

TRIM

DIFFUSER PANEL

TRIM

CEILING

STRAPPING

2×10 JOIST

Other ceiling systems offer only the diffusing panels alone, which are inserted into the grid at appropriate spots in the place of the ceiling tiles. Then the light sources, which can be ordinary incandescent strips of whatever size works best, are mounted above the translucent panels. This system is the least expensive and offers the greatest amount of flexibility.

There is another possibility that will work in certain cases, but you will have to plan carefully and do the job well to get good results. This consists of mounting the light sources, and again they can be either fluorescent or incandescent, in the bays between joists or rafters (Fig. 5-5). The bays should be of a nominal depth of at least eight inches (a two-by-eight joist, for instance, actually measures only about 7 1/2 inches wide). If there are a lot of wires and pipes already in the bay, you may not get good results. The line of light that this method produces will be only the width of one bay; if you need more, the best method is to skip one or more bays before installing the next line. This does away with the narrow band of shadow caused by the adjacent joist or rafter.

The diffusing panel should be at least six inches away from the light sources, and preferably an inch or two more. If you use incandescent bulbs spaced a foot or so apart, the distance may have to be greater to prevent hot spots, depending upon the wattage of the bulbs. On the other hand, the farther beyond these minimum distances you place the sources from the diffusers, the less effectively the light will be transmitted into the room. Deep, cavernous spaces above the ceiling soak up tremendous amounts of light. If this is a problem, you may have to paint everything you can reach above the ceiling level a flat white. Or, you can install pieces of wall board or other material, painted white, or even strips of aluminum foil, in such a way as to direct as much of the available light as possible onto the diffusers. Conversely, if your luminous ceiling is too bright, the easiest way to dim it may be to paint or otherwise tone down the ceiling cavity with darker colors.

Cove Lighting

Cove lighting (Fig. 5-6) is another type of broad source illumination that works quite well, especially in large rooms, open-plan homes, or cathedral-ceilinged rooms. This consists of a line of light, usually fluorescent but it can be incandescent, which is mounted on the walls a minimum of one foot below the finished ceiling. The centerline of the light source should be about five inches out from the wall to prevent glare and harshness on that surface. The light fixtures and sources are hidden from view below by a wide trim piece that extends from

WALL

LIGHT THROW LINE

SIGHT LINE
(SOURCE HIDDEN)

6'
APPROX.

FLOOR LINE

CEILING

INTERSECTION OF TWO
LINES DETERMINES
POSITION OF FACEBOARD
LIP AND ANGLE.

12-16"

5"
APPROX.

WALL
LAMP

Fig. 5-6. Cove lighting.

the wall like a shelf and sweeps upward and outward at an angle, which further hides the fixtures and directs the light upward to shine on the ceiling. The angle of this faceboard depends upon the height of the cove above the floor and the dimensions of the room. A tall person should not be able to see any part of the light source itself from any point in the room.

Within reason, the further below the ceiling you can mount cove lighting the more effective it will be, because a greater wash of light will spread across the ceiling and reflect back into the room. For best results, the wall, ceiling, and the inside of the cove itself should be finished in a light, flat color with high reflectivity. No gloss finish should ever be used, however, as this causes glare. Textured ceilings are not a good choice to use with cove lighting, either. The texture traps the light and reduces the amount that is effectively thrown into the room, and because of the hundreds of tiny shadows cast on the ceiling surface, it looks dirty. Almost always, cove lighting should be used together with other lighting since it has a flat, neutral effect with a fairly low illumination level. Its main purpose is for general background light, with other sources for localized and decorative use.

A somewhat similar technique is to use one but preferably a series of wall-mounted incandescent fixtures that shine up and outward on the ceiling. Such fixtures should have opaque and rather deep hoods or reflectors so that the light source cannot be seen directly, and they must be mounted at eye level or above. On the other hand, if they are too close to the ceiling, the light pool will not spread sufficiently. You can use standard bulbs, but floods work better. This type of lighting works best with a series of fixtures positioned so that the light pools overlap somewhat. The trick is to arrange them so that the light is as even as possible and not distracting, and so that there is no glare or harshness.

Skylights

Skylights are a much neglected double-duty source of light in the home. Properly set up, they will admit natural light during the daytime and then provide excellent ceiling light after dark. Of course, skylights have some drawbacks—heat loss in the cold parts of the country, potential trouble spots for water leakage, introduction of too much heat or radiation in the hot climates, and so on. But today's modern designs have at least minimized most of the problems to the point where skylights are worthy of serious consideration for any homeowner.

Fig. 5-7. Artificial lighting added to a skylight.

ROOF

FLOOD BULB

STRAPPING

CEILING

TRIM

DIFFUSER PANEL

SKYLIGHT

FLOOD BULB

ROOF

STRAPPING

CEILING

TRIM

It is a relatively simple matter to mount one or more light sources high inside a skylight well (Fig. 5-7). These can be floodlamps in simple holders, standard incandescent bulbs, perhaps even used in a recycled cast-off fixture without its exterior trappings. Or, in the case of larger skylights, you can use fluorescent strips. If space is tight or mounting a problem, the fixtures can be suspended from a piece of conduit to get them as close to the top of the skylight as possible. Then, all that is left is to mount a diffusing panel in a removable frame and attach it to the bottom of the skylight well.

Sometimes the fixtures can be recessed into the sides of the well, which serves to hide and shield them completely. You might also be able to arrange a luminous ceiling panel directly below a skylight. The light fixtures would have to be mounted in such a way that no shadows are cast, and the luminous ceiling would then transmit a broad, soft field of natural light during the daytime and be electrically lit at night.

Soffit Lighting

Soffit lighting is a common method, especially in new construction where it can be planned for and designed into the structure (Fig. 5-8). Usually the fixtures are recessed into an opening in a dropped ceiling section, which often as not is

Fig. 5-8. Soffit lighting.

WALL
LINE

LAMP   FACEBOARD

SOFFIT   DIFFUSER PANEL

Fig. 5-9. Cove-soffit lighting.

specifically built to contain them. Specially made soffit-and-fixture combinations can be used, or they can be made up from lampholders and diffusing panels as with luminous ceilings. The most common application of soffit lighting is to illuminate a built-in section such as a reading nook, breakfast nook, hobby center, or over lavatory basins and dressing tables.

Another method that can be quite successful is a *combination* of cove and soffit lighting (Fig. 5-9). The bottom of the shelf that projects from the wall under the line of cove lights is the soffit, which by definition is the underneath surface, usually exposed, of an architectural feature. But instead of being opaque, all or part of the shelf is translucent, made of a diffusing material that lets part of the light fall below the cove. Because of the lack of a reflecting surface directly above the light line, however, the amount of light diffused downward is small and of a low brightness level.

Open-plan architecture and vaulted or cathedral ceilings can be enhanced by other variations of soffit lighting. A floating or canopy soffit (Fig. 5-10) is used to direct light downward onto an interest center. By leaving the top off the assembly, or using a narrow diffuser panel set into the top, you can construct a floating cove-soffit unit that will project a portion of the light upward along the ceiling, and a correspondingly lesser proportion downward.

Fig. 5-10. Lighting by means of a floating soffit.

163

Sometimes there is little distinction between soffit and troffer lighting, except that troffers are usually commercially manufactured, complete fixtures that are recessed into the main ceiling portions of rooms to give a fairly high level of illumination to a particular area. But these too can be made up from diffusers and lampholders, and generally use fluorescent tubes. There also is a gray area of definition where troffer and even some soffit lighting leaves off and simple ceiling fixtures take over. Ceiling fixtures are usually small, using one to four incandescent bulbs, and may be either recessed or enclosed for surface mounting or open surface mounting. Recessed fixtures direct nearly all of their light downward. Enclosed surface-mounting fixtures may beam their light down or may have translucent side or dropped panels that throw some of the light outward, making a bigger light pool. The open surface-mount types diffuse part of their light down, while the remainder is thrown up against the ceiling to reflect out into the room as general illumination.

WALL LIGHTING

In wall-lighting techniques, the principal object is to bounce light off the walls, rather than the ceilings, to provide a general illumination level throughout the room or area. This technique is particularly effective in homes that have high or cathedral ceilings, and can be readily used in the remodeling of older homes as well. For the most part, wall lighting also tends to be a less expensive method than ceiling lighting, as well as being easier to install, maintain, and design. Wall lighting is also quite versatile and can be arranged to suit nearly any decor.

Valance Lighting

Valance lighting is probably the best known and most widely used system. This is installed directly above windows, shining down upon and reflecting from the closed drapes. To build a valance unit (Fig. 5-11), first mount a series of light fixtures—fluorescent or lumiline work best—to spacer blocks on the wall so that the centerline of the tubes is at least two inches out beyond the face of the drapes, but not more than about four inches. The light line should be at least ten inches below the finished ceiling to avoid excessive glare.

The next step is to mount a faceboard on brackets to hide the light line. The width of the facer should be about the same in inches as the distance from the bottom of the light line to the floor is in feet. Mount the facer so that its top is about even with the top of the light line. If you prefer fancy scrollwork on

Fig. 5-11. Valance lighting.

the bottom of the facer, figure the material needed for this in addition to the width of the board, not as a part of it. Depending upon the characteristics of the room and the placement of furniture, you may have to add a narrow lip on the inside bottom of the faceboard as a further light shield, so that no glare will be visible. Center the facer lengthwise over the light line and try to keep the overhang at the ends as short as possible, preferably about six inches. You can make the facers yourself out of 3/4-inch wood stock or buy them ready-made from plastic or metal.

WALL LINE

FACEBOARD

LAMP

5″ OR MORE

4″ APPROX.

2−4″ APPROX.

Fig. 5-12. Wall bracket lighting.

Wall Bracket Lighting

Wall bracket lighting is pretty much the same thing, except that no draperies are involved. The principles of construction are similar (Fig. 5-12), and the system can be varied to suit many purposes. You can mount the light line at virtually any height on the wall, close to the ceiling to wash down across a book wall, tapestries, or a collection of paintings, or lower down on the wall, over the head of a bed or a desk top.

In most cases you can attach the line of light directly to the wall, so that the centerline of the source is about four inches out away from the surface. Again, six inches is usually a sufficient width for the faceboard, though it can be more if you wish. Its interior surface should be about six inches or a bit more from the wall. In any case, be sure to leave room enough to maneuver replacement tubes or bulbs into the fixture sockets. The faceboard can be slanted in either direction to let more or less light spill from top or bottom as the occasion demands. Whether or not the facer should be centered directly over the line of light, or should be a bit up or down from center, depends upon the height of the light line and its particular purpose. You may have to experiment a bit to get the best placement and angle.

A modification of this method that works nicely over a desk top or breakfront is shown in Figs. 5-13 and 5-14. These use a shelf of whatever width is necessary or desirable on the wall. The line of light is fastened close to the outer edge of the shelf so that the tubes extend beyond the edge. A faceboard is then attached, usually tilted in a few degrees at the top as in Fig. 5-13, covering the line of light but leaving a two- or three-inch crack between the shelf and faceboard to allow a portion of the light to shine upward onto objects on the shelf. This is particularly effective in doing away with a contrast dark spot directly above a work area that is principally lit by a low line of light. A further variation is to mount the line of light

Fig. 5-13. A wall bracket light to fit over a serving counter.

Fig. 5-14. A wall bracket light designed to fit over a desk.

well out from the wall with a solid shelf-and-faceboard assembly (Fig. 5-14) containing an inset diffusing panel. The result is a lighted shelf that diffuses a certain amount of soft light upward.

Cornice Lighting

There are times when valance lighting is impractical because of low ceilings or insufficient space between the window tops and the ceilings. One effective solution is to install cornice lighting (Fig. 5-15). In this instance, you mount the line of light directly to the ceiling so the centerline of the tubes is four or five inches from the wall and at least a couple of inches clear of any drapes. Then, mount a faceboard tightly to the ceiling about an inch beyond the outer edge of the light fixtures.

The width of the faceboard must be determined by the dimensions of the room and arranged so that the line of light is invisible. Six inches would be a minimum. If the faceboard has to be so wide as to be out of proportion to its surroundings, build a small slanted lip along the inside bottom edge of a proportional faceboard to hide the light line.

You can also vary this system by installing a plastic diffuser panel along the bottom of the opening between the wall and the faceboard. This, of course, lowers the amount of light considerably, but also makes it softer. A further possibility is to mount the line of light either on the upper wall or on the ceiling, then box it in with diffuser panels in frames on both bottom and side. This allows light to bounce from both wall and ceiling, as well as to diffuse directly into the room. In a large room, a double row of tubes could be used, giving sufficient light for general low-level illumination. The decorative effects can be interesting.

Luminous Wall Panels

Another idea with a lot of possibilities is to take a luminous ceiling panel and stand it on end in a wall. The luminous wall can be of any size and shape that is architecturally feasible and handily installed. All such panels must be of a low brightness level, hardly more than a glow, since it is impossible to escape looking directly at them.

Luminous wall panels can run from floor to ceiling, or be of any lesser height, positioned at any level, set into the bays

Fig. 5-15. Cornice lighting.

WALL STUD

HEADER

LAMP

MOLDING

DIFFUSER PANEL

LAMP

HEADER

WALL COVERING

Fig. 5-16. Custom-built luminous wall panel.

between wall studs, or built into specially constructed wall openings just like a window. A double-faced panel mounted in an interior wall will serve two rooms. This gives the additional effect of extra spaciousness and depth, especially in a small house. In climates where heat loss is not a concern, they can also be placed in outside walls, either with one interior panel or with an exterior one to light an entry, patio, or porch. In locating the panels, though, make sure that they will not interfere with placement of large furnishings, wall hangings, concealed pipes, ducts, and wires.

Constructing luminous wall panels is no more difficult than ceiling panels, and it follows approximately the same principles (Fig. 5-16). Since most wall bays are apt to be only about 3 1/2 inches deep, however, the light sources should be mounted below and above, or to the sides of the diffusing panels, so that their shadows cannot be seen and the panels present an even glow. Narrow panels generally work better than broad ones since it is easier to balance out the light. Sometimes, however, properly mounted spotlights will light a broad panel nicely, depending upon the specific design details of the installation. Once the lampholders are located in their

170

best positions, all that remains is to mount the diffusing screens (which may have decorative designs on or in them) in frames and attach them to the walls over the openings. For best results the interior surfaces should be painted a flat white to reflect the maximum amount of light and help with even distribution.

Wall And Ceiling Lamps

Finally, we come to the most versatile and probably the most widely used method of wall lighting—that of small fixtures containing either flood or spot incandescent wall bulbs. These come in the form of pin-up lamps, bracket lamps, and wall lamps, as well as some types designed to be ceiling mounted such as cone lamps, recessed "eyeball" lamps, and track lamps. The general idea is the same for all—to project or cast a wash of light onto the wall surfaces that will reflect back into the room for low level illumination (Fig. 5-17). Some fixtures serve a decorative purpose as well, others to highlight specific objects or features. They can be used in hundreds of different ways and work most effectively.

Cone and bullet lamps can be ceiling mounted to throw a bright wash of light down or angularly across a wall surface. Eyeball lights do the same and are also adjustable to a degree. Some cones, if the hood is properly formed, can be mounted directly upon the wall with the light spilling downward. There are other styles meant to be positioned on the wall at about eye level. A small amount of light is directed upwards, the greater part down, with the source well shielded from the eye.

PLANNING THE LIGHTING SYSTEM

There are a number of factors that you should keep in mind along with the elementary principles of light discussed at the beginning of this chapter. In effect, they comprise the practical application of those principles.

The reflection and absorption of the light coming from the fixtures you install play a great part in the overall efficiency and satisfactoriness of your lighting system. You depend primarily upon reflected light to define all the objects that surround you. When no light is reflected, you no longer see the objects, but your shins can find them readily enough. You also depend upon reflected light for the various color sensations. In order to get the final effects that you hope for, you will have to coordinate your light sources and their placement carefully with your decor, architecture, furnishings, appliances, work areas, traffic patterns, and so forth.

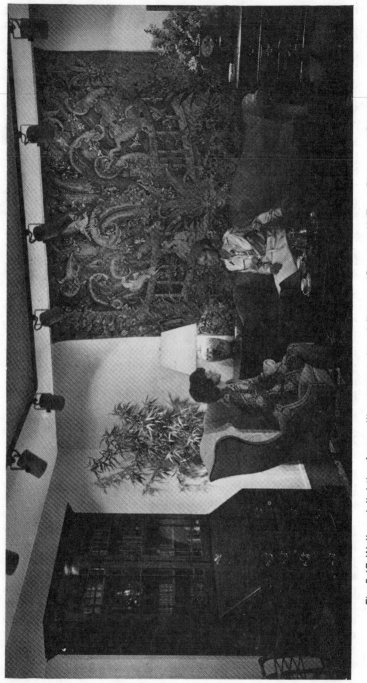

Fig. 5-17. Wall wash lighting from ceiling-mounted track fixtures. (Courtesy NuTone Division of Scovill)

Remember that light reflection and color are tied tight together. Regardless of whether the finish of the reflecting surface is hard or soft, matte or glossy, textured or smooth, more or less light will be reflected or conversely, absorbed, depending upon the color. The lighter the color, the more light will be reflected (Table 5-3). White has the highest reflectance and bounces back around 80% of the light that strikes it. Some of the very pale tints, like eggshell or ecru, also approach this figure. Pale blue and green reflect about 70%. But the deeper and darker the shade becomes, the less light is reflected and the more absorbed. A dark blue might reflect only 25%, chocolate somewhat less. Black, as you might expect, is at the bottom of the scale at about 10% or less.

The precise reflectance value depends upon the precise shade and tone. Most paint manufacturers can supply you with the reflectance values of their products. Furnishings, fabrics, and finishing materials such as wall paneling, ceiling tile, and flooring, however, are another matter. Most likely you will have to make educated gusses as to what the final results of any given combination of specific elements will be.

Lighting engineers use certain value ranges of reflectance in their calculations that they have found to be the best and

Table. 5-3. Reflectance Percentages for Various Colors

COLOR	REFLECTANCE
White	80 – 100
Pastel rose	78 – 80
Pastel yellow	78 – 80
Ivory	75 – 80
Pastel blue	72 – 75
Pastel green	72 – 75
Cream	70 – 75
Pastel beige	65 – 70
Pastel lilac	65 – 70
Buff	60 – 65
Light green	60 – 65
Light gray	55 – 60
Medium gray	50 – 55
Tan	45 – 50
Mustard	30 – 35
Medium blue	20 – 30
Dark gray	20 – 20
Natural concrete	20 – 30
Medium brown	20 – 25
Dark green	15 – 20
Dark brown	10 – 15
Red	10 – 15
Black	0 – 10
Dark woods	0 – 10

easiest to work with. Floors should reflect about 20—30% of the light reaching them, while walls may range from 30—60%. Ceilings should have the highest reflectance for good lighting, 60—80% or better. These reflectances provide a well-balanced and effective overall lighting scheme. The trick is to maintain the relative balance of the three elements. Thus, if a floor is exceptionally dark, say 15%, the ceiling should not be 80% because the balance is off and the result will probably be unsatisfactory. The ceiling should be toned down to 55—60% and the walls kept at a mid-point of 45—50% to maintain the correct ratio. If one element is high or low, the other should be modified to suit. Higher overall reflectances in a room or area will require less installed lighting wattage to achieve a given level of general illumination than low reflectances.

Lighting Problems

Not all reflected light, however, is useful. That which is not can be distracting and perhaps irritating as well; this is called *glare*. In its most obvious form, glare is light that reflects brilliantly—and sometimes painfully—from a shiny surface. This is *specular* light, and it is usually easy to correct by moving either the reflecting object or the light source, or interposing something between the two.

Direct glare is a common problem and frequently comes from the light source itself. An example would be a luminous wall panel that has too high a brightness level; it hurts the eyes to look at it. Or the problem may be a floor lamp with too small a faceboard, allowing the light line to be seen. Glare could also come from some object in the room that is too brightly lighted, like a white refrigerator in the wash of a recessed kitchen floodlight.

Stray light can be annoying, too. This does not occur often in small conventional homes because the spaces are relatively confined and usually well boxed in. But in large, open-design houses with multiple levels, lofts, cathedral ceilings, few solid walls, and a lot of off-shaped divider partitions, light sources installed for specific use in one area may spill unwanted and sometimes bothersome light into another. For instance, a person in a sleeping loft might be able to look down directly onto the unshielded bulbs of a pendant fixture in the dining area.

Contrast is another factor that must be considered. In lighting this is simply the degree of difference between light and dark. A certain amount of difference is always necessary for proper perception, but extremes are to be avoided. A lack of contrast, where all the elements in a scene are of similar value, appears washed out and flat, almost two dimensional. This could happen, for instance, in a white kitchen lighted too brilliantly by cold-tone fluorescent tubes.

On the other hand, too much contrast is equally bad. This can lead to a mix of visual highs and lows, hot spots and darkness, that is confusing and distracting and at worst can wash out form and color in the scene. Consider, for instance, a single white-shaded table lamp sitting next to a mahogany wall.

Both extremes of contrast inevitably lead right back to glare problems. The higher the contrast between two objects, the more difficult they become to view comfortably.

The situation is much the same with *shadow*. Shadow gives the dimension and depth and gradation of tone that is essential to visual perception. But again, it is the middle ground that is most important as a rule. Too much shadow leads to dullness and finally invisibility, light and dark pockets that have limited tonal gradations, and a certain amount of danger to those who try to navigate or work under such conditions. Too little shadow, on the other hand, washes out the scene, reduces depth and three-dimensionality, and is generally rather dreary.

As you can see, all of the above factors are interrelated and virtually inseparable. As you select the type and size of your lighting fixtures and sources, determine their locations, and work out the techniques that you want to use. Keep all of them in mind. The final effectiveness of your lighting layout will depend to a great degree on these factors, and the planning will involve a lot of balancing, compromise juggling, and thought.

LIGHTING CALCULATIONS

The usual method of lighting most homes is to proceed by guess, backing that up with previous experience on what seems to work pretty well and what doesn't. Lighting engineers, however, go about the job a bit differently. They have formulas, calculations, charts, and tables for just about every conceivable situation. And while it is neither necessary nor possible to go into this field in any depth here, there are a few points that may be of help to you during the planning process.

First, the intensity of light is measured in *candles*—a term originally meaning the strength of the flame of a standard candle. The present definition, however, is much more involved and precise, but from it we derive the basic unit of illumination, the *footcandle*. This is the amount of illumination that falls on a surface that is everywhere one foot distant from a uniform point light source of one candle intensity. To bring this a bit closer to home, a standard 60-watt inside-frost bulb provides about 60 footcandles of light.

Another unit of light measurement is the *lumen*. This is a unit of *luminous flux*, and one lumen per square foot of surface area is the equivalant of one candlepower. It is a measure of brightness, and most light sources are rated in lumens.

Now light, as you probably realize, diminishes as you move farther away from the source. It does so mathematically, falling off as the square of the distance from the source. Let's say that the illumination on a one-square-foot piece of cardboard one foot away from a light source is 60 footcandles. At a point one foot further back, or two feet from the source, the same amount of light that covered the piece of cardboard now covers an area of four square feet instead of one. So, the illumination is now only one-fourth, or 15 footcandles. At 4 feet from the source, the area covered is 16 square feet, and the illumination has dropped to 3 3/4 footcandles.

This means a couple of things in particular to the person designing his own lighting system. First, since light falls off so rapidly from the source, relative distances between source and lighted object or surface are important. Distance and intensity of source have to be carefully coordinated. And second, the candlepower distribution curve of any given lighting fixture can be computed. This in turn gives the illumination in footcandles at any given point for light cast by any given source in any given sort of fixture. This can be determined by various formulas for light on a perpendicular, vertical, or horizontal plane. These computations can be fairly complex and involve the use of sine, cosine, and tangent tables. But by using such figures, taken either from manufacturers' precomputed distribution curves or from spot measurements by a meter designed for the purpose, you can then position and adjust fixtures and light output to arrive at an optimum illumination level.

General Guidelines

Fortunately, there are also general guidelines that you can use for levels of illumination, as illustrated in Table 5-4. For close work, especially of an intense nature such as making tiny models, you should have a level of at least 100 footcandles spread over the immediate work area, perhaps a bit more at the point of concentration. Immediately outside the work area, you should have 80 – 90 footcandles, tapering away gradually.

Table. 5-4. Illumination Levels for Various Activities

ACTIVITY	FOOTCANDLES
Concentrated tasks: sewing, model making, fly-tying, etc.	100 – 125
Less concentrated tasks: reading, studying, workbench, cooking.	50 – 75
Familiar tasks and activities: household chores, dining, recreation.	25 – 50
Low-level activity: conversation, watching TV, casual work, and chores.	10 – 25
General basic mobility.	2 – 10

If concentration, speed, or efficiency is not necessary to a task, the lighting level can be reduced to about 50 – 60 footcandles. This also makes for good general illumination of a fairly high level, such as is often used in kitchens.

An average general level that might be found in a living room or family room could be anywhere form 20 – 50 footcandles. Much would depend upon the mood desired, the activities taking place, and so forth. Supplementary lighting is generally used along with this.

General activity that requires no concentration or high degree of visibility, like watching TV, after-dinner drinks, and conversation, could be well served by a general level of 10 to 20 footcandles.

An even lower level of 5 to 10 footcandles is sufficient for some areas. Under normal circumstances this would be plenty for an attic, basement, crawl space, garage, or perhaps even a laundry room. Again, supplementary local lighting should be available for use when needed. Halls, passages, stairways, and entries that have no obstructions or dangerous spots are often lighted at less than five footcandles, though frequently the level is much higher for increased mood and decorative effect. Outdoor walkways and steps are usually at a very low level, using two or three footcandles to prevent glare, excessive contrast, and deep shadow.

The Lumen Method

Another way of calculating your lighting needs is called the lumen method. Here, you first determine what the *average* level of illumination is to be in a given area, or upon a given surface. For every footcandle of light level you want, you will need one lumen of light per square foot of area. Let's say you want to light a drafting table measuring 2 by 4 feet in size, you want a level of 100 footcandles. The surface is 8 square feet, multiplied by 100 equals 800 lumens. You then need to provide a fixture and bulb combination that will supply that amount of light. This is quite easy since you can find the lumen rating of bulbs whether printed upon the package or in the manufacturer's specification sheets.

You can figure large areas or whole rooms in the same manner, but here other factors come into play. First, you must consider light losses because of absorption. For instance, you will lose much more light to a dark blue wall-to-wall carpet than to a beige one. Note also the general size and shape of the room. A lot of odd angles will cut up the light flow and a high ceiling will soak up a lot of light. A large room can be lighted more efficiently than a small one because the light can be distributed more evenly and less is absorbed by the walls.

The desired spacing of the fixtures can make a difference, especially if there are some restrictions on placement. The type of fixtures you want to use, their height from the ceiling, and the particular way in which they emit their light must be considered. Once you make allowances for all these points, you can arrive at some final conclusions. Unfortunately, there is no handy rule of thumb that you can use for this type of calculation. Engineers in the field base their final decisions upon prepared charts for certain types of lighting fixtures that take all of these points into consideration.

The Wattage Method

There is one simple guide that you might wish to follow, however, which in most instances works fairly well. This is the wattage method (Table 5-5), which is predicated upon the use of incandescent bulbs. The tabulated figures reflect only an *average* amount of light that may be needed for general illumination; they do not take into account any extreme conditions. All the points mentioned before still come into play, but at least the figures give you a preliminary base from which to work.

Note that this system is *not* intended for fluorescent lights, which may emit three to four times as much light per wall as incandescent lamps. But if you want to use fluorescents in your

Table. 5-5. General Illumination Levels Based on Incandescent Lighting Wattage per Each Square Foot of Lighted Area

AREA	WATTS/SQ. FT.
Living room	10
Kitchen	10
Hobby and crafts	10
Laundry	8.5
Bedroom	8.5
Family and recreation	8.5
Workshop	8.5
Bathroom	7
Library and den	7
Dining room	5.5
Hallway	5.5
Stairs	5.5
Porch and deck	4
Walkway	2
Garage	2
Attic	1
Crawl space	1

figuring, you can adjust the values by simply dividing by an average factor of 3.5 or 4. Thus, a kitchen requiring 1000 watts of incandescent lighting might require only 250 watts of fluorescent lighting.

Part 2:
Planning

Chapter 6
Making a System Layout

As you have gathered by now, the residential electrical system is a fairly complicated creature. This could easily lead to a lot of mistakes, omissions, fumbling around, and scratching of heads. But if you are smart, you will see to it that a complete layout of the system is made *before* the job is begun, and this is true whether you or someone else is going to do the actual work. After all, errors on paper are easy to erase. And besides that, a well-drawn and engineered layout makes the system much easier to visualize and keep track of while the job is in progress. Even experienced electricians usually prefer to go by a set of plans and specifications, though they might be able to see and understand all the complexities of a job before they start and could probably carry through without any undue difficulties.

At the least, a large sketch and a few notes will serve you. They do not need be particularly fancy so long as they are reasonably complete and definitive, done in a manner that at least you can understand them. Far better, though, is a clean and detailed layout with all necessary specifications spelled out clearly so that no mistakes can be made later on during the flurry of installation activities.

Aside from making your own job easier, there may be some necessity for you to work up a good set of plans. For instance, your lending institution may require them as a part of the document package before releasing any funds to you. In some areas, building departments or the electrical inspector may need a set on permanent file before a permit can be issued.

If you plan to have any or all of the installation done by an electrical contractor, it is wise to have every detail down in black and white—ahead of time to avoid possible misunderstandings or problems. Good plans are necessary if you want to put the job out to bid, or if you must make accurate cost estimates, or if you will need to provide material takeoffs for your supplier. Whatever the circumstances, there is no question that a good layout will probably save time, money, and mental anguish, resulting in a faster, safer, and more satisfactory installation.

PRELIMINARY WORK

Working out an electrical layout is not difficult.In fact, it can be a rather enjoyable and challenging project. Even a first class job requires only a modest amount of equipment (Fig. 6-1). First, you will need a flat, smooth surface to work on. A drafting board is ideal, of course, but a table will serve if it is absolutely smooth and relatively hard surfaced. A chunk or 1/8- or 1/4-inch tempered Masonite is ideal, and you can attach it to a piece of plywood backing and prop it at a comfortable angle just like a drafting table.

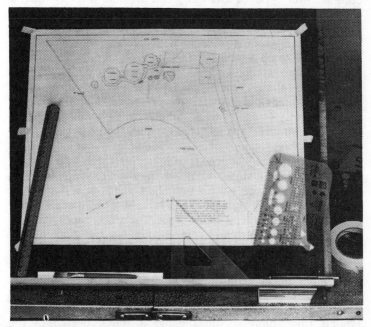

Fig. 6-1. An elaborate set-up is not necessary to turn out creditable drawings and plans. You can build your own drafting board and add a few inexpensive pieces of drafting equipment as necessary.

You will also need a pad of technical paper for tracing and drawing, such as draftsmen use. A common size that works well is 17 × 22 inches, but the larger the drawings, the easier they are to work with.

Next, arm yourself with an architect's concave rule, the three-sided rule shown on the left with foot scales. You will also need one or two transparent plastic triangles for drawing right angles. If you wish, you can also purchase special plastic drafting templates that outline electrical symbols, architectural details, appliances and furnishings, and even lettering. You can use ordinary pencils, but drafting pencils are more satisfactory. Either the standard wood-cased type or a mechanical lead holder will do the job. Hardness of the lead is a matter of preference largely, but number 2H gives fine results in general work. A sharpener or lead pointer, a soft rubber or art gum eraser, and perhaps a large sheet of graph paper ruled with 8, 10, or 12 squares to the inch.

This combination of materials will allow you to work up your own original drawings, which provides a couple of worthwhile advantages. First, you can easily make corrections, redraw, and change things around as you go along. Second, this type of drawing can be fed directly into a blueprint machine, so that you can have as many copies as you want. But if you don't want or need to go to all of this apparent trouble, you have at least a couple of reasonable alternatives. You can make up a series of freehand sketches, even doing it room by room on small sheets, though these are a bit difficult to read and follow correctly. Or, you can obtain copies of the floor plan blueprints of the structure and draw or sketch your layout directly on them. Bear in mind, though, that blueprints won't take many erasures before they become practically indecipherable.

Assuming that you are starting from scratch with original drawings, the first step is to attach the sheet of graph paper to the drawing surface with a few short strips of ordinary (or draftsman's) masking tape. Then lay a sheet of tracing paper on top, evenly, and tape it down taut. The graph paper will give you a continuous series of reference points and guidelines to work from. Next, lay out a complete scale drawing of the floor plan of the house, using a separate sheet for each floor, or each level or section for multilevel open designs or modular pod designs. If any outbuildings are to be wired, include them as well, perhaps to a smaller scale if the circuitry is simple. The ground floor plan should include a rough design of the immediate outside surroundings of the house if there is much outside wiring. Also, you may be required to provide a plot

Fig. 6-2. The plot plan shows the relationships of the major elements on the site, usually only to rough scale.

185

plan (Fig. 6-2) of either the entire lot or the part of the land close by the house, showing the location of the serving power pole or other supply source, any poles which must be added, and the service drop or lateral. If you already have blueprints of the floor plans and plot plan, you can save a little time by taping your tracing paper over them and tracing the plans out.

As you work, be sure to include all window and door openings, stairwells, and other major architectural features. Scale in the doors themselves in full open position. Add any of the fixed features that may not be on the usual type of floor plan, such as planters, half-walls, shoji screens, cornices, and built-in cabinetry. Then you will have to exercise your imagination a little bit. Drawing in block symbolic form and to rough scale, add all of the major pieces of furniture and large appliances at the points where you feel they will most likely be placed in the finished house. At the same time, try to think ahead to the items that you may acquire at some future date and indicate them as well, perhaps with dotted lines. The finished drawing should look something like the one in Fig. 6-3.

Once the floor plans are complete, you are ready to start locating all of the various devices and load items. But before you begin, go to the blackboard and write one hundred times, "I will plan for the future." For in nearly every case, difficulties that arise some years after an electrical system is installed are a direct result of not planning ahead, of not providing for future needs, expansion of family and home, upgrading of standard of living, natural increases of interest in hobbies or social entertaining; in short, a lack of foresight. And while you're at the blackboard, also write down "I will not use cube taps" (Fig. 6-4). For this is the specific point in most residential wiring systems where lack of planning ahead shows up first, and where trouble arises.

TEMPORARY SERVICE

Perhaps the first consideration should be the temporary service, or *construction loop* (refer back to Fig. 6-2). Even though probably not required on the plans that you may have to present to someone when you are ready to start work, no harm can come from including this detail. Temporary services must nowadays be installed almost as carefully as the permanent service and pass the necessary inspection under applicable codes. This means that even though temporary, the location should be chosen with care. As construction proceeds, the loop should not have to be moved around and perhaps reinstalled. It must feed the temporary power into the site with the fewest problems and the greatest efficiency.

186

FIRST FLOOR PLAN
JOHNSON RESIDENCE
SCALE—3/16" = 1'
DO NOT SCALE

Fig. 6-3. Floor plans, done for each floor or level of the building, show major dimensions and overall layout of the house to scale, but not in detail.

Fig. 6-4. Overloaded receptacles and masses of cords and plugs are one of the leading causes of electrical mishaps in the home. This situation should be avoided at all costs.

Check with your local power company to see if they have any special requirements or wishes. Then spot the temporary service on your plans so that it will be easily accessible but not in the way of site preparation, cement trucks, building material delivery and storage, and so forth.

PERMANENT SERVICE ENTRANCE

While you are talking with the power company about the temporary service, investigate the requirements for installation of the permanent service. The NEC spells out a number of pertinent rules, and the serving utility may have a few more for you to contend with. Generally speaking, the service equipment should be located at some convenient point in the building that is closest to a utility pole; the shortest distance from power source to main disconnect is best. If there doesn't happen to be one particular course that seems obviously best, the engineering department of the power company will probably be glad to help you work out a suitable arrangement.

The main disconnect, which may or may not include the main distribution center, should be completely accessible and easily reached from any point in the house. Years back, the general practice was to hide everything in the darkest, remotest corner of the basement, but this practice is now discouraged. For one thing, this is most inconvenient for servicing and for replacement of fuses. For another, the basement is the last place anyone would want to go to shut off the power in case of an emergency.

On the other hand, esthetics and decor demand that the service equipment not be stuck on a living room wall. A back hall is a good spot, or any point handy to an outside exit. Inside an attached garage is another possibility, or an outside wall of the building. But don't tuck it in a small closet, behind a door, or anyplace where physical damage might occur; there might be large objects or stacks of material or belongings piled against it.

The Service Drop

The size and shape of the house and the contours of the lot, as well as the location of walks and driveways, may also present some problems. The NEC specifies certain minimum clearances that must be maintained, but often local codes or utility company requirements will override them. A service drop (Fig. 6-5) must be at least 18 feet above a street or alley way, and 15 feet above any area that might be subject to truck traffic, such as a parking lot. Private drives must be cleared by a minimum of 12 feet, and over the finish grade of the lot, walkways, or any platforms like an open deck, the clearance is 10 feet.

In addition to this, the point of attachment of the service drop on the building must be at least 10 feet above grade, but not more than 30 feet. This means that on a low-walled house the service drop, in order to gain clearance, might have to run to a service mast going up through the roof overhang on the outside of the house, or through the roof itself from the inside. If the mast is not further back from the edge of the roof than 4 feet, the point of attachment can be no less than 18 inches above the roof at the mast. It can be higher, of course, and often is. Also, a service drop must clear any roof over which it passes by at least 3 feet, provided that the roof has a pitch of at least 4 inches in every 12 inches of run. A lesser slope requires a greater clearance of 8 feet. In some cases where this roof clearance cannot be achieved and other satisfactory arrangements cannot be made, a special support tower can be installed as a permanent structure on the roof to boost the lines up.

To further confuse the issue, note that the point of attachment for the service drop must be kept at least three

Fig. 6-5. Service drops require planning for adequate clearance.

feet below or to one side of any opening in the building. Also, the weatherhead should be higher than the service drop attachment point. If all of these clearances add up to a problem in your case, remember that you can mount a main disconnect, or even a main entrance panel, at some convenient spot on the outside of the building and run feeders from there to one or more load centers at convenient points inside the house. Or, you may wish to come in via the underground route with a service lateral.

The Service Lateral

A service lateral (Fig. 6-6) can be run in any direction from the power source and into the building at any point, which is sometimes a great advantage. The only installation requirements usually have to do with the mechanics of burial and protection of the lateral itself, which we will cover in detail a bit later. Though the cost of an underground service is likely to be greater than an aerial service, it has much greater flexibility of location, is better protected from the elements and mechanical damage, and is quite unobtrusive.

Symbols Used on Layouts

Once you get all of the dimensional and clearance details sorted out, choose the best rough locations for the various parts of the service entrance system, including the meter box. Locate them symbolically on your layout. No great amount of drafting detail or precise dimensioning is necessary, since minor changes will probably be made at installation time.

As you can see from Table 6-1, each item of an electrical layout has its own symbol, and you can pick them from the chart as needed. A letter after the symbol means that there is something special about that particular item, and refers back to a separate specification sheet containing the details. These are called subscript letters and are always in the lower case, not to be confused with the capital letters within the symbols or following the switch symbol, S.

LOCATION OF OUTLETS

The object of the next exercise is to locate all of the outlets (power utilization points) for the entire system. The order in which you do this really makes no difference—just as long as you get them all. But for the sake of simplicity, we will follow a certain procedure that you can alter to suit yourself. All locations and information will follow the standards set forth in the NEC. You should check against any local code that might pertain in your area and make changes or additions as needed.

Fig. 6-6. Service lateral entrances are sometimes used, and sometimes required.

DIST. POLE

8' MIN. CONDUIT PROTECTION

METER

GRADE LEVEL

BUSHING

BUSHING

DRIVE

WALK

LATERAL—METALLIC CONDUIT—6" MIN.

LATERAL—NONMETALLIC CONDUIT—18" MIN.

LATERAL—DIRECT—BURIED CABLE—24" MIN.

SLEEVE UNDER CONCRETE AND PAVING

BUSHING

BUSHING

INSIDE SERVICE EQUIPMENT

SERVICE-ENTRANCE CONDUCTORS FROM POINT OF ENTRY ON.

Table. 6-1. Commonly Used Symbols For Electrical System Drawings

Symbol	Description
	Receptacle outlet, ungrounded
	Receptacle outlet, grounded
	Duplex receptacle, ungrounded
	Duplex receptacle, grounded
	Fourplex receptacle, ungrounded
	Fourplex receptacle, grounded
	Range outlet
	Duplex receptacle, split wired
	Duplex receptacle and switch
	Clock receptacle
	Special purpose outlet (designate with subscript)
	Floor special purpose outlet
	Floor receptacle
	Floor duplex receptacle
	Telephone outlet
	Floor telephone outlet
	Fixture outlet, surface incandescent
	Fixture outlet, recessed incandescent
	Blanked outlet
	Fan outlet
	Junction Box

193

Table 6-1. (cont.)

Symbol	Description
(L)	Lampholder outlet
(LV)	Low-voltage controlled outlet
(L)$_{PS}$	Lampholder with pull switch
[O]	Fixture outlet, single surface fluorescent
[O R]	Fixture outlet, single recessed fluorescent
[O]	Fixture outlet, multiple surface fluorescent
[O R]	Fixture outlet, multiple recessed fluorescent
(MO)	Motor outlet
(WH)	Watt-hour meter
(G)	Generator
(T)	Transformer
S	Single-pole switch
S_2	Two-pole switch
S_3	Three-way switch
S_4	Four-way switch
S_D	Door switch
S_K	Key switch
S_P	Switch and pilot lamp
S_{LV}	Low-voltage control switch
S_{LVM}	Low-voltage control master switch
S_T	Time switch

Table 6-1. (cont.)

S_{MC} Momentary or pushbutton switch

S_{WP} Weatherproof switch

S_F Fused switch

Ⓢ   Ceiling pull switch

Pushbutton

Buzzer

Bell

CH   Chime

R   Radio outlet

TV   Television outlet

Annunciator

FS   Fire alarm device

H   Horn

T   Thermostat

Surface-mounted panel

Flush-mounted panel

Battery

Branch circuit, concealed in wall or ceiling

Branch circuit, concealed in floor

Exposed wiring

Auxiliary wiring (or use color code)

Table 6-1. (cont.)

Heavy lines indicate feeders

Arrows indicate home run, circuit number

Hash marks indicate number of conductors (no marks means two conductors)

⭕ a,b,c   Special outlets (designate in specs)

▢ a,b,c   Special auxiliary outlets (designate in specs)

a, b, c, d, e, f   Subscripts may be used with any symbol to designate special variations, conditions or installation procedures, or to call attention to specs.

Dot means conductors connect

Loop means conductors bypass

Ground

Convenience Outlets

First, the convenience outlets for duplex receptacles. These can be placed at any suitable height on the walls, from a position in the mopboards just above the floor level, to almost ceiling height in the case of a clock receptacle. The standard height seems to be about 12 inches from the floor, or 42 inches above the floor when placed over counters. But you can position them anywhere that will be most useful for you. Duplexes are not expensive, so allow for plenty of them; within reason, the more the better, so that you can avoid the aforementioned cube tap calamity.

Living Areas. In all of the living quarters of the house, receptacles must be spaced so that no point along the floor line in any wall space is more than six feet from an outlet (Fig. 6-7). This sounds more complicated than it is. First, the living quarters include all the areas where *normal* household activity is carried on, excluding bathrooms. This means living and dining rooms, rumpus and rec rooms, bedrooms, kitchens, dens, libraries, and so forth. It does not include garages, storage areas, cellars and crawl spaces, attics, porches, and other areas that are ancillary to normal living activities.

Fig. 6-7. Minimum spacing requirements of receptacles in residential applications.

197

Wall space is taken to mean stretches of *unbroken* wall, including permanent half-walls, bookcases, dividers or bar-type counters, and also patio doors or exterior sliding panels. It does not include room doors, closet doors, fireplaces, fixed glass curtain walls starting at floor level, or similar items.

In summary then, there can be no more than 12 feet from one receptacle to another on an unbroken wall space, and there can be no more than 6 feet from any break in the wall space to a receptacle. In addition, any wall space two feet or more in width that stands by itself, such as between two doorways or between a fireplace and a window-wall, must have its own receptacle. The point of all this measuring is to eliminate—or at least minimize—the number of extension cords used with cords about six feet long.

Kitchens. A somewhat modified rule applies in kitchen and dining rooms and areas above countertops. Here, any countertop space wider than 12 inches must have a receptacle mounted in the wall behind and above it. For instance, if you have a 2-foot counter, then a refrigerator, then a 3-foot counter, a cooktop, and a 12-inch counter in a row, every counter segment would require a receptacle—irrespective of the 6-foot rule used in the rest of the living quarters. But note that the *remainder* of the kitchen must follow the basic 6-foot rule. Note too that floor receptacles do not count toward the 6-foot rule *unless* they are positioned quite close to the wall. The minimum number of convenience outlets for the kitchen is two.

Other Locations. There are several other areas in the home that are required to have at least one convenience outlet. One must be placed outdoors at any handy spot, and this must be a weatherproof receptacle. You will also need one outlet in the garage if attached to the house, one in the basement, one for a laundry area whether you plan to have a washer or not, and one located near the wash basin in every bath or lavatory. Though not required, attics and crawl spaces should have at least one receptacle. Some heavy 120-volt portable appliances also need outlets, and these should be placed no more than six feet from where the appliance will be used. All of the general-purpose and small-appliance outlets should be spaced out as evenly as is practical.

Placement. As you locate all of these outlets, try to forsee any interfering elements. A duplex behind a door or heavy drapes has little value, and arrangements of furnishings may play an important part in the location of some duplexes. Keep outlets away from heating units. Place them so that attachment cords are not apt to drape in inconvenient, undecorative, or unsafe places such as across a register, past the corner of a window, or directly behind a wash basin or sink. You might find it helpful, too, to code each outlet with a letter referring back to a spec sheet that contains the mounting heights or any other specific instructions or details. This helps when installation time rolls around.

Fixed and Stationary Appliances

The next items to consider are the fixed and stationary appliances that operate on 115 volts, along with any other equipment that might fall into the same category.

Stationary Appliances. Outlets for freezers and refrigerators are usually located directly behind the units themselves, and the same is true for an undercounter trash compactor. Some well pumps and most sump pumps are simply plugged in. Sort through your mental list of possibilities, and wherever you will have a plug-in load, mark in a duplex receptacle outlet on the layout.

The 230-volt pieces of equipment require different symbols for receptacles, and they are next on the list. These are usually few in number, and include such items as the range and clothes dryer. Spot the necessary symbols for each item you have or hope to add later on.

Fixed Appliances. Each appliance or piece of equipment that is to be wired directly into the system also requires an outlet; this may be either a junction box or a connection loop. This category would include such items as a built-in oven, counter cooktop, undercounter dishwasher, garbage disposal, water heater, possibly a well pump, a hood over the range, ventilating fans, and auxiliary heaters.

Comfort heating and cooling equipment will also require an outlet. Fuel-fired heating apparatus generally calls for one 115-volt outlet. Heat pumps, central air-conditioning units, and electric furnaces require 230-volt outlets.

Place the appropriate symbol in the immediate area where the fixed equipment will be located. Electric heat panels, cables, wall and ceiling or floor heaters, and baseboard units require an outlet symbol for each individual piece of equipment. Cable heat conductors can be installed so that the starting point and the connection leads fall at nearly any convenient point, and that is where that symbol should be drawn in. The junction box outlet placement for heat panels varies somewhat with the make and size of the units, so you will have to be guided here by the manufacturer's specifications. Try to spot the outlets in easily accessible places, though.

Outlets for wall, ceiling, and floor heaters are drawn in at the approximate point where the equipment will be. The location of baseboard heating units is often quite specific; they should not interfere with furnishings or construction details. Draw in each unit to scale in its final location. Then place the outlet symbol at one end or the other of the unit, since connections can be made at either end, whichever is the more convenient.

LOCATION OF LIGHTING FIXTURES

Locating the lighting fixtures requires some thought and planning (see Chapter 5). Portable lighting fixtures are of no concern here, only those that will be permanently installed as a part of the built-in wiring system.

First, note with a fixture outlet symbol all of the points where the fixtures will be installed. Then add to each symbol a subscript letter for each different type of fixture. In other words, all of the four-foot two-lamp fluorescent lights might be coded "a," the matching pair of fixtures on either side of a bathroom mirror might be "b," and so forth.

Then make up a specification sheet for the various fixtures and spell out the meaning of each subscript. Subscript "c," for instance, might read, "Acme Mfg. #3360 decorated glass, one-light wall bracket, gold-tone finish." Special installation notes may also be added. As you make your choices, both of fixtures and locations, make sure that everything fits together satisfactorily so that you won't have to make last minute changes.

A check with the NEC will show you that there are some restrictions placed upon lighting fixtures, and you must also take these into consideration. For instance, damp or wet locations require a special type of fixture intended for that particular use. A *damp location* is a roofed porch, shower, some barns, basements, and garages. A *wet location* is a place exposed to any kind of weather. A fixture approved for a wet location can also be used in a damp location, but not the other way around. Fixtures not specifically marked for one or the other must be used only in dry locations.

Any and all clothes closets or general-purpose closets that are apt to have clothing hung in them come under a special set of rules. For instance, no *pendant fixtures*—sockets with or without shades, dangling from a cord—are allowed. But *flush-mounted recessed fixtures* can be mounted anywhere in a closet, since the fixture is considered actually to be outside the closet. However, any other type of fixture can be wallmounted above the closet door with a clearance from it to any point where combustible material of any kind might possibly be stored of a minimum of 18 inches. The fixture may also be mounted at any point on the walls or ceiling of the closet where a similar minimum clearance can be maintained in any direction from the fixture. The object is to keep heat-producing bulbs away from anything that might catch fire. Some synthetic fabrics in clothes and blankets are highly flammable and can easily be set off by a 100-watt bulb.

LOCATION OF SWITCHES

Once all of the outlets are out of the way, you can decide where to locate the switches and note the appropriate symbols. Every load that is to be remotely turned on and off will need at least one switch, and each room or area in the house should be provided with at least one switched light that can be controlled from a convenient location. If you would prefer not to have a built-in fixture in some room, you can switch a duplex receptacle (or half of one) to control a portable fixture, meeting the requirements that way.

The best situation is to have one or more light sources controlled at each point of entry and exit in every room. This should include areas such as the garage, crawl space, attic, and any other little-used parts of the house. The best possible combination is to group switches on *both sides* of each doorway in every room, permitting the control of lights both ahead of and behind the user. You can thus turn on a light from outside of the room you are entering, and then turn off the light to the room you are leaving from within the room you have entered. You are never in darkness, and there is no insecurity or fumbling in the dark for a switch. This is especially helpful for elderly or infirm persons.

Switches should be located at a convenient height, usually anywhere from 42 to 54 inches. Those at doorways should be at the knob side of the door, while those at steps or stairways are best located at least a couple of feet back from the outer edge of the top step, eliminating the danger of a misstep while reaching for the switch.

In theory, any number of switches can be wired to any number of lights. But in practice, this can become quite involved. If there are to be many three- or four-way switching legs, it is much easier to go to a low-voltage switching system (more about this later). Low-voltage control (LVC) is a bit more expensive in a small installation than standard switches, but as the system becomes larger and more complicated, the proportionate cost decreases and the installation becomes easier.

The key word in switch location is *convenience*. Take care that they do not wind up behind doors or drapes, or in awkward places where you will have to lean over funiture or past large appliances to reach them. It is also a good idea, at least in the case of standard line-voltage switches, to keep them at a reasonable distance from wash basins, sinks, and showers. And whether line-voltage or LVC, the cost of a switch is only a small part of the completed electrical system, so you might just as well include as many as you want to.

LOCATION OF THERMOSTATS AND DIMMERS

The remaining miscellaneous parts of the system can now be located. Thermostats are generally placed about five feet from the floor on an open section of wall where they will get normal, unimpeded air flow from the room air circulation pattern. They should be placed away from any heat producing equipment like registers or table lamps and where direct sun could shine on them. Nor should they be placed on an outside wall, or on an inside corner near a window, or where they might be affected by cold air from an opening and closing outside door.

Dimmers are placed in any handy location, usually close to the loads that they will control. Timers and time clocks should be placed out of sight, as a rule, but not in an inaccessible location. A common choice is to put such equipment right beside the main distribution panel. Other odds and ends that will be a part of the built-in electrical system, such as photocells or other special controllers, clocks that are to be direct-wired, heating tapes or cables for melting snow, and other special gear should not be forgotten.

CIRCUITRY LAYOUT

With most of the elements of the system now drawn into place, the easy part is done. The remaining portion of the layout consists of entering the circuitry, again in symbolized fashion. The circuitry is the network of conductors and cables that ties the whole system together and makes it function. Without a working knowledge of those conductors and cables—what they are, how they work, and the limiting factors that govern their use—you won't know what symbolic circuitry to set down. So now we must move off onto a slightly different tack, into an area of combined layout and engineering in order to complete the drawings.

Conductor Ratings

One of the most important characteristics of a conductor to the electrician working in the general wiring field is its current-carrying capacity, or to use a modern term, its *ampacity*. The ampacity of a conductor is dependent upon several things such as size, resistance, and temperature. Obviously a conductor in service cannot be loaded to the point where it will finally become white-hot and then melt. This is in fact what happens when a short circuit occurs. From a practical standpoint, a conductor has to be limited to a capacity that is safe and efficient, both for the conductor and for the power user.

The rated ampacity of conductors used in general wiring is determined by trade size, ambient operating temperature, the type of insulation with which they are covered, whether the use is continuous or noncontinuous, the physical conditions of use, the number of conductors in the same raceway or cable, the metal that they are made of, and sometimes the specific load that will be attached. Much of this information is boiled down, fortunately, into a series of tables that you can readily match a particular conductor to a particular job. All commercially available conductors and cables for use in the general wiring field conform to the specifications used in making up the tables.

Table. 6-2. NEC Amperage Ratings of Single Copper Conductor in Free Air at 30°C

AWG MCM	60°C (140°F)	75°C (167°F)	85°C (185°F)	90°C (194°F)	110°C (230°F)	125°C (257°F)	200°C (392°F)	250°C (482°F)	
	TYPES RUW (14-2), T, TW	TYPES RH, RHW, RUH (14-2), THW, THWN, XHHW	TYPES V, MI	TYPES TA, TBS, SA, AVB, SIS, FEP, FEPB, RHH, THHN, XHHW*	TYPES AVA, AVL	TYPES AI (14-8), AIA	TYPES A (14-8), AA, FEP*, FEPB*	TYPE TFE (Nickel or nickel-coated copper only)	Bare and Covered Conductors
18	....	....	....	25	....	....	....	....	....
16	....	....	27	27	....	....	....	....	....
14	20	20	30	30†	40	40	45	60	30
12	25	25	40	40†	50	50	55	80	40
10	40	40	55	55†	65	70	75	110	55
8	55	65	70	70	85	90	100	145	70
6	80	95	100	100	120	125	135	210	100
4	105	125	135	135	160	170	180	285	130
3	120	145	155	155	180	195	210	335	150
2	140	170	180	180	210	225	240	390	175
1	165	195	210	210	245	265	280	450	205
1/0	195	230	245	245	285	305	325	545	235
2/0	225	265	285	285	330	355	370	605	275
3/0	260	310	330	330	385	410	430	725	320
4/0	300	360	385	385	445	475	510	850	370

Table. 6-3. NEC Amperage Ratings for Three Copper Conductors in Cable or Raceway at 30°C

AWG MCM	60°C (140°F)	75°C (167°F)	85°C (185°F)	90°C (194°F)	110°C (230°F)	125°C (257°F)	200°C (392°F)	250°C (482°F)
	TYPES RUW (14-2), T, TW, UF	TYPES RH, RHW, RUH (14-2), THW, THWN, XHHW, USE	TYPES V, MI	TYPES TA, TBS, SA, AVB, SIS, FEP, FEPB, RHH, THHN, XHHW**	TYPES AVA, AVL	TYPES AI (14-8), AIA	TYPES A (14-8), AA, FEP* FEPB*	TYPE TFE (Nickel or nickel-coated copper only)
18	....	....	....	21	....	....	·.....	....
16	....	....	22	22	....	....	·.....	....
14	15	15	25	25†	30	30	30	40
12	20	20	30	30†	35	40	40	55
10	30	30	40	40†	45	50	55	75
8	40	45	50	50	60	65	70	95
6	55	65	70	70	80	85	95	120
4	70	85	90	90	105	115	120	145
3	80	100	105	105	120	130	145	170
2	95	115	120	120	135	145	165	195
1	110	130	140	140	160	170	190	220
1/0	125	150	155	155	190	200	225	250
2/0	145	175	185	185	215	230	250	280
3/0	165	200	210	210	245	265	285	315
4/0	195	230	235	235	275	310	340	370

For example, Table 6-2 shows the ampacity of single insulated-copper conductors in free air at a temperature of 30°C. You can see that #14 wire with TW insulation will carry 20 amps. But #14 with type AVA insulation will carry 40 amps because of its higher temperature rating.

Placing the same conductors in a raceway or cable holding not more than three conductors changes the picture, however, as you can see from Table 6-3. In this case, the allowable ampacity for #14 TW drops to 15 amps, and for #14 AVA it drops to 30 amps.

Furthermore, another change occurs when the conductor material changes. Tables 6-4 and 6-5 show figures for aluminum conductors. You can see that #14 is not even listed. The smallest usable size is #12, and is the equivalent of #14 copper in ampacity. And aluminum #12 AVA will only carry 25 amps instead of 30 as its copper counterpart does.

As the ambient operating temperature rises, the allowable ampacity for any given conductor falls. The correction factors are given in Table 6-6. For instance, if you wanted to use a TW insulated #12 conductor at a higher temperature rating than given in the ampacity table, you would have to derate it. If the proposed temperature were 40°C (104°F), you would find the

Table. 6-4. NEC Amperage Ratings of Single Aluminum Conductor in Free Air at 30°C

AWG MCM	60°C (140°F)	75°C (167°F)	85°C (185°F)	90°C (194°F)	110°C (230°F)	125°C (257°F)	200°C (392°F)	
	TYPES RUW (12-2), T, TW	TYPES RH, RHW, RUH (12-2), THW THWN XHHW	TYPES V, MI	TYPES TA, TBS, SA, AVB, SIS, RHH, THHN, XHHW*	TYPES AVA, AVL	TYPES AI (12-8), AIA	TYPES A (12-8), AA	Bare and Covered Conductors
12	20	20	30	30 †	40	40	45	30
10	30	30	45	45 †	50	55	60	45
8	45	55	55	55	65	70	80	55
6	60	75	80	80	95	100	105	80
4	80	100	105	105	125	135	140	100
3	95	115	120	120	140	150	165	115
2	110	135	140	140	165	175	185	135
1	130	155	165	165	190	205	220	160
1/0	150	180	190	190	220	240	255	185
2/0	175	210	220	220	255	275	290	215
3/0	200	240	255	255	300	320	335	250
4/0	230	280	300	300	345	370	400	290

Table. 6-5. NEC Amperage Ratings of Three Aluminum Conductors in Cable or Raceway at 30°C

AWG MCM	60°C (140°F)	75°C (167°F)	85°C (185°F)	90°C (194°F)	110°C (230°F)	125°C (257°F)	200°C (392°F)
	TYPES RUW (12-2), T, TW, UF	TYPES RH, RHW, RUH (12-2), THW THWN XHHW, USE	TYPES V, MI	TYPES TA, TBS, SA, AVB, SIS, RHH THHN XHHW*	TYPES AVA, AVL	TYPES AI (12-8), AIA	TYPES A (12-8), AA
12	15	15	25	25 †	25	30	30
10	25	25	30	30 †	35	40	45
8	30	40	40	40	45	50	55
6	40	50	55	55	60	65	75
4	55	65	70	70	80	90	95
3	65	75	80	80	95	100	115
2	75	90	95	95	105	115	130
1	85	100	110	110	125	135	150
1/0	100	120	125	125	150	160	180
2/0	115	135	145	145	170	180	200
3/0	130	155	165	165	195	210	225
4/0	155	180	185	185	215	245	270

Table. 6-6. NEC Correction Factors for Conductors Used at Temperatures above 30°C(68°F)

C.	F.	60°C (140°F)	75°C (167°F)	85°C (185°F)	90°C (194°F)	110°C (230°F)	125°C (257°F)	200°C (392°F)	250°C (482°F)
40	104	.82	.88	.90	.91	.94	.95	....	....
45	113	.71	.82	.85	.87	.90	.92	....	....
50	122	.58	.75	.80	.82	.87	.89	....	....
55	131	.41	.67	.74	.76	.83	.86	....	....
60	140	....	.58	.67	.71	.79	.83	.91	.95
70	158	....	.35	.52	.58	.71	.76	.87	.91
75	167	....	....	.43	.50	.66	.72	.86	.89
80	176	....	....	.30	.41	.61	.69	.84	.87
90	194	....	....	....	....	.50	.61	.80	.83
100	212	....	....	....	....	....	.51	.77	.80
120	248	....	....	....	....	....	....	.69	.72
140	284	....	....	....	....	....	....	.59	.59
160	320	....	....	....	....	....	....	....	.54
180	356	....	....	....	....	....	....	....	.50
200	392	....	....	....	....	....	....	....	.43
225	437	....	....	....	....	....	....	....	.30

40°C line on the chart, then move across to the 60°C column since TW has a maximum rating of 60°C (140°F). The figure given is 0.82, which you would multiply by the 30°C rating of 25 amps (free-air single-insulated copper conductor) for an answer of 20.5 allowable amps. If the proposed temperature were 60°C, you would have to go to another type of conductor because that temperature is the maximum for TW. If type AVA is derated by 0.79, for instance, it would do the job.

These tables can be used in another way, too. Suppose that you want to use a #12 conductor, but the load will exceed the normal 20-amp capability of TW (three copper conductors or less together). If the current draw were 30 amps, for instance, then you would go to #12 AVA. This situation might occur when a new piece of equipment with a slightly higher current draw than the old is attached to a circuit in an existing line of conduit, where the conduit size restricts the pulling of larger conductor sizes. One further note; these tables include only those types that can be used for branch-circuit and feeder wiring, since a 15-amp capacity is the smallest allowable size.

To put this information into perspective, most residential wiring makes use of the 30°C temperature rating since in only a few places, such as within a fixture body, do the temperatures go higher than this. Also, single-pole branch circuits are almost always wired for 15- or 20-amp capacity and occasionally for 30 amps, using #14, #12, and #10 copper, or #12, #10, and #8 aluminum (respectively) either in conduit, armored cable, or NM cable. By the large, #12 copper is the

206

most popular wire size since most single-pole branch circuits are of the 20-amp size. The #10 size is more stiff and therefore difficult to handle. The #14 is a bit shy on ampacity, so you might just as well go to #12 for the few extra pennies and the little bit of additional labor involved. Use of the larger sizes and ampacities is determined by specific loads (more about that later).

Branch Circuits

The next step is to start laying out the branch circuits. First calculate either the power consumption or the current draw of each load item in your electrical system. Usually it is easier to convert everything to power figures to begin with. Go over your plans and jot down the wattage of every known item, from ovens to light bulbs. For those items you plan to add sometime in the future, you may have to make an educated guess, but make it on the high side rather than low so that you will have plenty of capacity when the time comes. Include all portable appliances and fixtures that normally remain plugged in or are frequently used, such as toasters, irons, floor lamps, and table lamps. In this way, you will cover all of the fixed and a good share of the variable load. In making the calculations, use 115 volts and 230 volts nominal. You can use the power formula equations found in Chapter 1 for loads related in amps rather than watts.

Required Branches. Let's start with the single-pole branches, some of which are already required by the NEC. You must have at least one 20-amp circuit feeding a receptacle in the laundry room or area. Find the proper receptacle and draw a short line away from it toward the main panel; label it #1. (A completed circuit layout is shown in Fig. 6-8.).

The kitchen must have two 20-amp circuits feed only small appliances and nothing else. These are usually the receptacles located above the countertops; each circuit can feed one or two duplexes. Actually, you could feed more than two, but since the object is to split the loads up as much as possible, the fewer outlets on these circuits the better.

Note that under the NEC, any branch circuit serving two or more receptacles (one or more duplexes) of a 20-amp rating *shall not* carry a load of more than 16 amps. If the receptacles are of 15-amp rating, the allowable load *shall not* be more than 12 amps.

For the sake of example, let's say that one branch circuit will serve one duplex receptacle where the toaster will normally be stationed, and the other will serve a pair of duplexes. From the toaster outlet, draw a short line with an

Fig. 6-8. Example of electrical layout drawing based upon floor plan. All receptacles are grounded.

arrowhead pointing back in the general direction of the distribution panel; label it #2. These short arrowheaded lines, by the way, are called *home runs*; they begin at the first (or only) point of power utilization on the circuit and extend back to the matching circuit number in the distribution panel. For the other branch circuit, draw a similar line from the duplex nearest the panel box to indicate circuit #3. Then draw a *loop* from that duplex location to the next handiest one that you feel should be on the same circuit. This could be the nearest duplex, or one across the room, or whatever you wish.

This same situation of two required 20-amp branches also applies equally to the pantry, dining room, breakfast room or alcove, or any other area where entertainment or family gathering is likely to take place, such as a rumpus room or family room. In other words, any spots that may see the use of small appliances on a reasonably regular basis must be served by these circuits. You can, if you wish, feed some of these receptacles with the two required 20-amp kitchen small appliance circuits, on the strength of the fact that if those small appliances are in use in one room, they probably will not be in another.

Optional Branches. With these first requirements out of the way, you can now go on to the remaining optional circuits. These are drawn out in the same manner, and the way in which they are arranged is a matter of judgement balanced against the load at each outlet, or what you feel the load might possibly be. There is one rule of thumb that states that there should be as many general-purpose branch circuits as there are rooms in the house, but most modern houses do in fact have considerably more.

The old spiderweb system of wiring so often seen in older houses is not a recommended practice, and for a number of technical reasons. Under this system, a wireman ran his *home run* from the panel to a centrally located junction box, often holding a ceiling lighting fixture. He then dropped a series of connecting loops out from the junction box to pick up about every outlet in sight. A far better way is to proceed in an orderly fashion from outlet to outlet so that each circuit represents a straight line through a series of utilization points.

Freezers and refrigerators are often put on separate circuits so that there is no chance that some other load might fault, thereby opening the circuit unnoticed and ruining a load of perishables. Some larger appliances, dishwashers for example, need circuits of their own because of their high current draw. In some cases, two or more appliances may be served by a single circuit, especially where one or two may not be operating when another is.

General-purpose circuits may serve both receptacles and lighting fixtures, or be confined to just one or the other, as you wish. Electricians will often split up circuits among various rooms, too. For instance, if two circuits are to serve two bedrooms, rather than running one circuit to each room, they will put half a room on one circuit and the other half on the second, and vice versa. Then, if a circuit should trip out, neither room will be completely without power.

LVC Systems. The situation may be somewhat altered if you have decided to use a low-voltage switching system. LVC uses one low-voltage relay for each load to be switched, but there are two methods of installing them. One is to place the relay directly in the outlet box that serves the load, and in this case the wiring is run directly to the load and relay location served by a branch circuit in the usual fashion. But LVC relays are also often mounted in banks inside special cabinets built for the purpose, and located either alongside the serving distribution panel or load center, or in some other convenient spot. In this second case, a circuit is usually taken directly from the distribution panel to feed several relays and loads, thus becoming solely a lighting branch circuit with receptacles and perhaps other loads placed on their own circuits.

Furnace Circuits. In many places there are special requirements for furnace circuits, and they vary widely under local codes. You may need a separate circuit, probably clearly marked, of 15- or 20-amp size even though the furnace may draw far less. This circuit may have to include a switch at the top of the basement stairs or some other handy but remote spot for emergency shut-down purposes.

In addition, you may also need a service or emergency switch located right beside the furnace burner, possibly with a special *lead-slug* switch, sometimes called a *fire-matic*, located directly over the front of the furnace. A local heating contractor or your inspecting authority can give you the necessary details.

Other Single-Pole Branches. To finish up with all of the 115-volt single-pole branch circuits, simply continue to join the various outlets in sequences that seem to make up reasonable combinations. Number the *home runs* for later reference. Keep each circuit as short as possible and use no more junction boxes or extra splices than is absolutely necessary. Every splice or extra connection is a potential trouble spot; try instead to jump from device to device or load with a minimum of conductors in each box. Sometimes this requires a bit of head scratching and possibly a little more cable, but the finished results are worth the effort.

This leaves only the conductors from the switches or other controllers to be accounted for. These are usually represented by dotted lines that run from the switch to the load being controlled. If there is more than one switch tied to a particular load, simply loop a dotted line from one switch to the next until all in that series are connected.

Two-Pole Branches. With all of the single-pole circuits and switch loops taken care of, the next step is to draw in the 230-volt two-pole circuits. These are almost always sized to serve one particular load and installed for that specific purpose. So, set up the necessary *home runs* from the outlets that will serve the range, wall-mounted ovens, water heater, well pump, or whatever other 230-volt loads you may have.

With electric space heating, you may have several units all on the same circuit. Or, you may have groups of units in zones controlled by separate thermostats for each group on an individual circuit. The latter case would be shown by a *home run* from the panel to the first unit of each group or to the controlling relay of the group, with connecting loops from there to each succeeding unit of each group.

Feeders

Now you can determine whether or not any feeders are needed. If all branch circuits are within the same building or confined to the immediate outside area, and all travel back to a single main entrance panel, then no feeders are used. But if there are outbuildings that contain more than one branch circuit, they will require suitable subpanels. A feeder from each must run back to the main entrance panel, or there could be a feeder link from a load center to subpanel. Distribution panels or load centers may also be advisable within the main building itself.

By looking at the layout you have just drawn, you can easily see where the concentrations of power utilization are. If the bulk of the circuits are in an area not too far from the main entrance panel, all is well and good. If they are strung out for long distances, however, it might be well to install load centers at strategic spots to save extra work and wire.

For example, take a house constructed in modular pod fashion. There might be three or four sections set out in a row a hundred feet or more long. If the service entrance is at one end, and if each pod requires a number of branch circuits or has fairly substantial loads, the best solution would be to install a distribution center in each pod, with feeders running from the service equipment to each panel. An alternative would be to mount a split-bus main-entrance panel in the first pod, with feeders running to load centers in the remaining three. Or, you might be better off to place a main panel at the first pod, with a subpanel in the third pod. The specific combination depends entirely upon particular electrical requirements, design of the structure, costs, and the personal preferences of the electrician. If you feel that load centers would be a good idea in your system, locate them now and draw in the necessary feeder lines. Figure 6-9 shows some of the possible distribution variations.

Special Circuits and Systems

If you plan to use any auxiliary systems (see Chapter 14), now is the time to sketch in the necessary major components at locations where there will be no interference with parts of the main wiring system. This is the time, too, to make any compromises and changes so that all elements of the system are compatible and to your satisfaction so far. Draw in the connecting cables for any auxiliary systems, perhaps color-coding them for easy identification and to avoid confusion. If any special subsystems are a part of your plans, check their respective requirements in Chapter 8 and enter these details onto the layout as well.

The circuitry for any outbuildings such as shops, garages, and barns should be drawn out in much the same fashion as the

Fig. 6-9. Variations in supply source distribution. There are several other possibilities as well.

main building. Make sure, too, that you have not forgotten any of the outside-lighting and small-appliance circuits for the patio, decks, gardens, and walkways. When all the entries are complete, you should have a layout that looks, in character at least, something like the one in Fig. 6-8.

Chapter 7

Sizing System Components

At this point, the layout you drew in Chapter 6 is complete. But as you can see, it is symbolic in nature and does not give a complete and detailed set of instructions and drawings on exactly how each item is to be installed, or even the specifications of each item. In fact, most of this detailing is often dispensed with, and the electrician simply uses the layout as a road map—choosing, sizing, and installing the proper items in the proper fashion (hopefully) as he goes along.

The more complete sets of plans, however, also include a series of specification sheets that call out details of every item or particular installation that the designer feels deserves special attention, or whenever he has specific wishes or instructions to be followed. And since these specifications are sometimes required, usually desirable, and often needed so you know how to compile them for your material take-off sheets, we will proceed with the aspect of the residential wiring system that could be called the engineering. You will probably find, too, that some of the circuits you have laid out will have to be rearranged as a result of the calculations that follow.

SIZING RECEPTACLES

Sizing of the devices is quite easy. Any receptacle connected alone to a branch circuit, such as the required laundry circuit, must be of the same rating as the circuit itself. In other words, a 15-amp circuit must have a 15-amp receptacle, a 20-amp circuit a 20-amp receptacle. But note, however, that any receptacles, regardless of number, connected to a 15-amp circuit must have *no more* than a 15-amp rating. Receptacles with 10-amp ratings are available, but their use is not recommended. Twenty-amp circuits with two or more convenience outlets may use either 15-amp or 20-amp receptacles. Beyond this point, the receptacle is sized exactly to the circuit; a 30-amp dryer circuit would have a 30-amp receptacle. Voltage ratings match those of the circuit itself and are most often classed as 125 volts and 250 volts, or occasionally both.

There is one point you should keep in mind regarding those receptacles that are a part of the general-purpose branch circuits, but not a part of any required circuits or circuits which feed specific computed loads. Each of these receptacles should be figured into the total circuit load on a basis of 180 volt-amperes each. With a supply voltage of 115 volts, this would mean a load for each receptacle of 1.57 amps (volt-amperes divided by volts). Thus, each duplex receptacle would be figured in as a 3.14 amp load. Turning this around, you can see that the maximum number of duplex receptacles allowable on any one 20-amp branch circuit would be six.

SIZING SWITCHES

The most commonly used switches are rated either 10 or 15 amps, which is generally quite adequate for household use. The better ones are also "T" rated, which means that their contacts will withstand the momentary surge of current encountered when tungsten lamp loads are turned on. Switches with higher ratings are also available for special purposes. In any such case where the load is unusual, match the capacity of the switch to about 125% of the load and you will have an ample safety factor. The same holds true of any other type of controller; match and include a safety factor. Voltage ratings are usually listed as 120V AC, 125V "T", 250V, or 120/277V AC. Some switches also carry a DC voltage rating.

SIZING CONDUCTORS FOR BRANCH CIRCUITS

The next consideration is the size and type of the conductors. Much of this, too, is quite simple. All of the required 20-amp circuits mentioned previously must obviously be served by 20-amp conductors. The type most commonly used is #12 TW or equivalent, since this is most easily obtainable and least expensive. Where nonmetallic sheathed cable is allowed for use, the recommended type is NM, or perhaps NMC. If armored cable is required, the type is AC. In some areas, the installation must be in conduit or tubing, and here single conductor #12 TW or something quite similar is used. In the case of aluminum conductor material, the comparable size would be #10 for all 20-amp circuits.

General-Purpose Circuits

All general-purpose circuits are usually wired with #12 conductors for a 20-amp capacity—for the sake of simplicity if nothing else. Wire is sold mostly by the 250- or 500-foot coil, so why fuss with two sizes when the one larger size will result in a better installation anyway? Many appliance circuits are similarly wired with #12 wire and a 20-amp capacity wherever possible. The reason is the same, plus the fact that #12 is much easier to handle than the bigger and much stiffer #10 with its 30-amp capacity. Most electricians recognize this fact and try to reduce everything possible to 20-amp circuits, simply because when all is said and done, the installation is simpler and usually better.

You will have a few loads, probably, that will exceed the 20-amp level, and so the conductors must be sized directly to them, using the next largest ampacity above the load requirement. If the two are quite close, as for instance a 29-amp draw on a prospective 30-amp conductor, then size the conductor up another step to 35 or 40 amps. Crowding a circuit capacity too close to the limit often results in problems of one sort or another, so it is not a recommended practice.

Special Circuits

Much the same can be said for two-pole 230-volt branch circuits. You will probably not have too many of these in your installation by comparison with the single-pole circuits. You can size each to its particular load by using the ampacity charts in Tables 6-3 or 6-5.

Appliances. Heavy appliances and equipment often come with detailed recommendations for wiring, and you can simply follow them. In many cases—electric spaceheating units, for example—the branch-circuit conductors must be sized to 125% of the rated nameplate current or wattage.

There are a few points to keep in mind as you size the circuits. Those which serve heavy appliances such as ranges, wall ovens, cook tops, and clothes driers are best sized at least one step higher than the minimum required by the specified current draw of the appliance. This will not only provide an additional safety factor, but will allow replacement of the appliance at a later date with one of greater size without replacing the wiring. At any point in the system where you think there may be added loads later, oversize the circuits and give yourself all the capacity that seems reasonable throughout the entire system. Erring on the long side is much better than having to do some rewiring five years hence, as is often the case.

Motors. Any and all motor loads in the system, including vent fans, blowers, furnaces, circulating pumps, well pumps, dishwashers, washing machines, or any other piece of equipment that contains a motor, should be figured differently. If there is one motor in a circuit, give it a rating of 125% of the nameplate running current draw. If a motor nameplate states that the draw is 10 amps, assess that motor at 12.5 amps for the purpose of circuit loading. If there are two or more motors in a single circuit, rate the largest one at 125%, and all of the remaining at 100% of their nameplate rating. This does not apply to the motors included in portable appliances, nor to any motor of less than one-eighth horsepower.

Inductive Loads. Inductive lighting loads—those which include transformers or ballasts such as fluorescent lights—should *never* be figured on a basis of the wattage of the lamp. Instead, use the current draw figure usually marked right on the fixture.

You might possibly encounter other types of inductive loads marked with a power factor, such as PF 90%, and then you may have to derate the circuit correspondingly.

Continuous loads must also be derated. These are defined as any loads that will be in operation continuously for three hours or more. Such loads must not exceed 80% of the ampacity of the conductors in the circuit. If, however, 67% or less of the load on a circuit is continuous, then the load is considered to be noncontinuous. If the circuit includes both types of loads, then the circuit load should be calculated as the entire amount of the noncontinuous load plus 125% of the continuous load.

There are other occasions when derating is necessary. For instance, a branch circuit supplying a single appliance must not carry a current load greater than 80% of its capacity. With a 20-amp circuit, this would limit the actual current capability to 16 amps. If a branch circuit supplies any combination of lighting with fixed or portable appliances, the total current draw of the appliances must be kept below 50% of the current rating of the circuit. The foregoing applies to 15- and 20-amp circuits. Any 25- and 30-amp circuits supplying a single stationary appliance must not be loaded to more than 80% of rated capacity.

216

Switches. The conductors used for switch and other controller loops must always match the size of the circuit conductors. Wall switches on a 20-amp general-lighting circuit would be wired with 20 amp conductors, usually #12 TW. Single-pole switches require two conductors, three-way switches need three, and so on. For a clearer picture of how this works, you may wish to refer to the connection diagrams in Chapter 11.

The situation is a bit different for low-voltage switches and controllers such as thermostats. The LVC system uses small three-wire cables, #18 in size, between all switches and between all switches and relays. Runs from any switch or relay to a master switch, or from the supply transformer to a switch, relay, or gang of relays, is two-conductor cable of the same size. Low-voltage thermostats, doorbell and chime systems, or other controllers or systems of a similar nature may use two, three, four or perhaps more conductors in a #18 or #20 size.

CONDUCTOR RESISTANCE AND VOLTAGE DROP

Now that you have taken all of these various factors and considerations into account, you should have all of your circuitry nicely sized, calculated, and arranged into a safe and workable system. The end of the process, however, has not yet arrived. There are still other details which can easily throw your designs into confusion.

Resistance

You'll recall from Chapter 1 that all conductors have some resistance. This is dependent upon four principal factors: cross-sectional area, length, ambient or operating temperature, and kind of metal. If the volume of a conductor is increased, the resistance will decrease, and if the temperature is increased, the resistance does likewise.

There are methods and formulas for computing any given conductor resistance, but the usual practice in the field is to consult tables compiled for the purpose. These tabulated figures are based upon a certain ambient temperature, usually 20°C (68°F) or 25°C (77°F), depending upon the authority conducting the tests. The resistance for each trade size of conductor is listed in terms of DC ohms per thousand feet (Table 7-1). Simple division will give you the resistance for shorter lengths.

You'll recall, too, that DC ohms are a purely resistive load, but AC ohms may include reactance factors. In the lighter wire sizes, from #3 AWG on up, DC ohms and AC ohms are

Table 7-1. NEC Values of Conductor Resistance

Size AWG MCM	D. C. Resistance Ohms/M Ft. At 25°C. 77°F.		
	Copper		Alumni-num
	Bare Cond.	Tin'd. Cond.	
18	6.51	6.79	10.7
16	4.10	4.26	6.72
14	2.57	2.68	4.22
12	1.62	1.68	2.66
10	1.018	1.06	1.67
8	.6404	.659	1.05
6	.410	.427	.674
4	.259	.269	.424
3	.205	.213	.336
2	.162	.169	.266
1	.129	.134	.211
0	.102	.106	.168
00	.0811	.0843	.133
000	.0642	.0668	.105
0000	.0509	.0525	.0836
250	.0431	.0449	.0708
300	.0360	.0374	.0590
350	.0308	.0320	.0505
400	.0270	.0278	.0442
500	.0216	.0222	.0354
600	.0180	.0187	.0295
700	.0154	.0159	.0253
750	.0144	.0148	.0236
800	.0135	.0139	.0221
900	.0120	.0123	.0197
1000	.0108	.0111	.0177
1250	.00863	.00888	.0142
1500	.00719	.00740	.0118
1750	.00616	.00634	.0101
2000	.00539	.00555	.00885

considered to be one and the same for general wiring purposes. Any difference that might occur is too small to worry about. But as the conductor size increases from #2 AWG up into the MCM sizes, that situation changes and the reactance factor becomes more pronounced. In addition to the factors listed above, we must consider also the type of cable jacketing, the type of raceway in which the conductor runs, and the frequency of the AC source voltage. Again, tables have been compiled that list multiplication factors to convert DC ohms to AC ohms (Table 7-2).

Table 7-2. NEC Factors for Converting DC to AC Ohms

Size		Multiplying Factor			
		For Nonmetallic Sheathed Cables in Air or Nonmetallic Conduit		For Metallic Sheathed Cables or all Cables in Metallic Raceways	
		Copper	Aluminum	Copper	Aluminum
Up to	3 AWG	1.	1.	1.	1.
	2	1.	1.	1.01	1.00
	1	1.	1.	1.01	1.00
	0	1.001	1.000	1.02	1.00
	00	1.001	1.001	1.03	1.00
	000	1.002	1.001	1.04	1.01
	0000	1.004	1.002	1.05	1.01

Voltage Drop

Voltage drop or *IR* drop has long been a consideration in commercial and industrial wiring. Within the past few years, this has been formally recognized as being of equal importance in residential installations. Even though the ill effects of severe voltage drop occur less frequently in household systems than in others, each circuit must be checked to insure that voltage drop does not fall below a certain level. That level is 3% of the supply voltage for branch circuits, or 3.45 volts for 115-volt circuits and 6.9 volts for 230-volt circuits. Your supply voltage may vary somewhat from 115/230, so in making your own calculations it would be best to use the actual value.

You might find the calculating task a bit easier if you work up a *circuit tree* such as the one shown in Fig. 7-1. This makes visualizing the circuits easier, and it will help you to keep the figures straight. The first step is to assign a current draw figure to each circuit, if you have not already done so. Next you must find the length of each circuit, which can be a tricky business. Scaling them off on your drawing is not a good idea, because the resulting figures will undoubtedly be far too low. Instead, you must try to visualize the actual run that each circuit will take in the building. A run going from the panel to an outlet will not travel 10 scale feet across a piece of paper, but go up the wall to the ceiling, around a corner, over the top of a window, back a ways across the wall, and drop down toward the floor for a total of perhaps 20 feet instead. Track each circuit to the very end, including any side branches; total up the footage as you go, trying for as accurate a figure as possible.

Once you have noted all of the circuit lengths on the tree, you can then compute the voltage drop for each branch. You

Fig. 7-1. A circuit "tree" sketch helps in visualizing the entire electrical system and in making the various calculations.

can do this in two ways, one quite accurate, the other less so but perfectly adequate for our purposes. The first method is to make an equivalent circuit drawing for each branch, showing the load resistance in series with the conductor resistance as in Fig. 7-2. You can find the load resistance from Ohm's law; $R = E/I$. Take the conductor resistance from Table 7-1. Then, find the IR drop just as was done in the examples in Chapter 1.

The second method is a bit simpler and uses the formula

$$E_{VD} = \frac{2L \times 10.7 \times I}{CM}$$

Where L is the length of the circuit in feet, I is the current in amperes drawn by the load (not the rating of the circuit itself), and CM is the area of the conductor in circular mils. The figure 10.7 is a "rule of thumb" assumption based on the fact that most commercial copper conductors one foot long with a cross-sectional area of 1 circular mil have a resistance of 10.6 to 10.8 ohms at 24°C. So you multiply the length of the circuit times 2 (for both conductors), times 10.7, times the current draw, and divide that answer by the area of the conductor, which you can find in Table 3-1.

230V   (A/C)   air conditioner
                 11,500 watts

Length of circuit: 200 feet.

#6 conductor: 0.410 ohms per 1000 feet
Total conductor resistance: $0.410 - (1000 \times 200 \times 2) = 0.164\ \Omega$
Air conditioner resistance: $E^2/P = 230^2/11500 = 52900/11500 = 4.6\ \Omega$

Equivalent circuit #1

$R_C = 0.164\ \Omega$

$R_{A/C} = 4.6\ \Omega$

Resistances in a series circuit are added, so:

Equivalent circuit #2

$R_T = 4.764\ \Omega$

Current flowing in equivalent circuit #2:

$I = E/R = 230/4.764 = 48.279$ amps

Therefore, since the same current flows in all parts of a series circuit, the current flow in equivalent circuit #1 is 48.279 amps.

So: Voltage drop across R_C: $E_{VD} = IR = 48.279 \times 0.164 = 7.917V$
Voltage drop across $R_{A/C}$: $E_{VD} = IR = 48.279 \times 4.6 = 222.083V$
Voltage drop R_C plus voltage drop $R_{A/C}$ is 230V

or, to simplify:

Current draw of 11,500-watt air conditioner on 230V circuit:
$I = P/E = 11500/230 = 50$ amps.

#6 conductor circular mil area: 26240 CM.

$$E_{VD} = \frac{2L \times 10.7 \times I}{CM} = \frac{2 \times 200 \times 10.7 \times 50}{26240} = 8.155V$$

Percentage voltage drop: $8.155/230 = 0.0354 = 3.5\%$

Fig. 7-2. Reduction of an actual circuit to a schematic equivalent to determine voltage drop.

Whichever method you use, the answer will be in volts. If you have already determined what the maximum allowable voltage drop can be for your power source, then all you need do is compare them to make sure that you fall within the specified limits. If you wish to find the percentage of the voltage drop, divide the supply voltage into the voltage drop. If the voltage drop exceeds 3% on any circuit, you have two alternatives. The first is to leave the circuit as is and go to the next size conductor, running through the formula again to make sure that this new size will fill the bill. The second alternative—often the more practical one as far as branch circuits with multiple loads are concerned—is to break the circuit up into two or more separate circuits until the voltage drop falls within the prescribed limits. If many of the circuits have a high voltage drop because of excessive lengths, you might be better off to shorten the circuits by means of load centers.

SIZING BOXES

Each connection point or device outlet must have a box, and this brings up another point which you must be aware of, a situation known as *box fill*. Each conductor in a box is required to have a certain amount of free air space in cubic inches. Turned around, this means that each particular size of box, regardless of type or use, can only have a certain number of conductors within it. Furthermore, each cable clamp, fixture hickey, or fixture stud within a box counts as one conductor. All of the ground conductors in the box together count as one conductor, too. Each strap or frame containing one or more devices is the equivalent of a conductor, but tap wires going out to a fixture do not count. A conductor that enters the box and passes on through with no break is considered to be one conductor. A conductor that enters the box and terminates there is also one conductor.

As an example, suppose you want to wire up a duplex receptacle midway in a circuit. The incoming line would count as two conductors, the outgoing line would count as two more. The ground wires together would make one, the receptacle itself would be another, and the cable clamp (one clamp will secure two cables) counts as one conductor. This makes a total of seven. You can see from Table 7-3 that if you are using #12 wire, the smallest wall box you can use measures 3 by 2 by 3 1/2 inches deep.

If the wire sizes in the box question are mixed, then Table 7-3 cannot be used. In this case, refer to Table 7-4 to find the amount of free air space required for each conductor size and

Table 7-3. NEC Values for Maximum Box Fill

Box Dimension, Inches Trade Size or Type	Min. Cu. In. Cap.	Maximum Number of Conductors				
		#14	#12	#10	#8	#6
4 x 1¼ Round or Octagonal	12.5	6	5	5	4	0
4 x 1½ Round or Octagonal	15.5	7	6	6	5	0
4 x 2⅛ Round or Octagonal	21.5	10	9	8	7	0
4 x 1¼ Square	18.0	9	8	7	6	0
4 x 1½ Square	21.0	10	9	8	7	0
4 x 2⅛ Square	30.3	15	13	12	10	6*
4 11/16 x 1¼ Square	25.5	12	11	10	8	0
4 11/16 x 1½ Square	29.5	14	13	11	9	0
4 11/16 x 2⅛ Square	42.0	21	18	16	14	6
3 x 2 x 1½ Device	7.5	3	3	3	2	0
3 x 2 x 2 Device	10.0	5	4	4	3	0
3 x 2 x 2¼ Device	10.5	5	4	4	3	0
3 x 2 x 2½ Device	12.5	6	5	5	4	0
3 x 2 x 2¾ Device	14.0	7	6	5	4	0
3 x 2 x 3½ Device	18.0	9	8	7	6	0
4 x 2⅛ x 1½ Device	10.3	5	4	4	3	0
4 x 2⅛ x 1⅞ Device	13.0	6	5	5	4	0
4 x 2⅛ x 2⅛ Device	14.5	7	6	5	4	0
3¾ x 2 x 2½ Masonry Box/gang	14.0	7	6	5	4	0
3¾ x 2 x 3½ Masonry Box/gang	21.0	10	9	8	7	0
FS — Minimum Internal Depth 1¾ Single Cover/Gang	13.5	6	6	5	4	0
FD — Minimum Internal Depth 2⅜ Single Cover/Gang	18.0	9	8	7	6	3
FS — Minimum Internal Depth 1¾ Multiple Cover/Gang	18.0	9	8	7	6	0
FD — Minimum Internal Depth 2⅜ Multiple Cover/Gang	24.0	12	10	9	8	4

* Not to be used as a pull box. For termination only.

add the amounts needed. Then find the volume of the box (the sizes listed in Table 7-3 are standard) by multiplying the width by the height by the depth in inches. If the box has insufficient capacity, you must go to a larger size to prevent jamming and damage to the conductors, as well as possible overheating. In addition, a box with plenty of room in it is much easier to work with from the wireman's standpoint.

SIZING PIPE OR CONDUIT

Other limitations in circuit layout arise when the installation must be made with pipe or conduit. This may occur only in certain sections of the system such as a masonry cellar or a furnace circuit, or it may include the entire system, depending upon the construction of the building and the local codes.

Neither conduit nor EMT is as easy to work with as cable wiring, but the smaller sizes are less expensive and easier to handle than the larger sizes. Most wiremen prefer to stay with 1/2- and 3/4-inch trade sizes as much as possible for branch

CONDUCTOR SIZE	FREE AIR SPACE IN BOX — EACH CONDUCTOR
14	2.0 cubic inches
12	2.25 cubic inches
10	2.5 cubic inches
8	3.0 cubic inches
6	5.0 cubic inches

circuits, though feeder circuits nearly always require the larger sizes. This means that the circuits must be carefully planned with an eye towards keeping the runs reasonably short and with a minimum of fittings and boxes. The pipeline routes have to be carefully worked out for the easiest installation, but at the same time, insure ample capacity and conductor space.

There are restrictions on the number of conductors that may be run in a given trade size of conduit or EMT. These relate to the percentage of the cross-sectional area of the conduit that may be filled. Table 7-5 shows those percentages. If there is more than one size of conductor in a conduit, then you will have to figure the cross-sectional area (Table 7-6) of each size with its insulative covering, add them all up, and check to see whether your answer is less than the appropriate percentage factor for that size conduit. If all of the conductors are the same size, then you can refer to Table 7-7 to find the maximum number of conductors for any given size of conduit. Note that ground conductors and equipment grounding conductors are to be included in the total number.

Flexible metal conduit in trade sizes of 1/2 inch or more follow the same fill requirements as rigid conduit and EMT. The one exception concerns the use of 3/8-inch trade size, which is often used in short spans of not more than six feet for taps to lighting fixtures or motors. These fill values are shown in Table 7-8.

NUMBER OF CIRCUITS

Assuming that you now have all of the branch circuitry well under control, the next step is to find out whether or not you have a sufficient number of circuits for the house under the recommendations of the NEC. This rule states that you must provide for a lighting and general-purpose load of three watts per square foot of living area in the dwelling. This does not include the required small appliance and laundry circuits,

Table 7-5. NEC Values for Maximum Conduit Fill by Percentage Cross Section

Area—Square Inches

Trade Size	Internal Diameter Inches	Total 100%	Not Lead Covered			Lead Covered				
			2 Cond. 31%	Over 2 Cond. 40%	1 Cond. 53%	1 Cond. 55%	2 Cond. 30%	3 Cond. 40%	4 Cond. 38%	Over 4 Cond. 35%
½	.622	.30	.09	.12	.16	.17	.09	.12	.11	.11
¾	.824	.53	.16	.21	.28	.29	.16	.21	.20	.19
1	1.049	.86	.27	.34	.46	.47	.26	.34	.33	.30
1¼	1.380	1.50	.47	.60	.80	.83	.45	.60	.57	.53
1½	1.610	2.04	.63	.82	1.08	1.12	.61	.82	.78	.71
2	2.067	3.36	1.04	1.34	1.78	1.85	1.01	1.34	1.28	1.18
2½	2.469	4.79	1.48	1.92	2.54	2.63	1.44	1.92	1.82	1.68
3	3.068	7.38	2.29	2.95	3.91	4.06	2.21	2.95	2.80	2.58
3½	3.548	9.90	3.07	3.96	5.25	5.44	2.97	3.96	3.76	3.47

Table 7-6. NEC Conductor Dimensions

Size AWG MCM	Types RFH-2, RH, RHH, RHW, SF-2		Types TF, T, THW, TW, RUH, RUW		Types TFN, THHN, THWN		Types FEP, FEPB, TFE, PF, PGF, PTF		Type XHHW	
	Approx. Diam. Inches	Approx. Area Sq. In.	Approx. Diam. Inches	Approx. Area Sq. In.	Approx. Diam. Inches	Approx. Area Sq. In.	Approx. Diam. Inches	Approx. Area Sq. Inches	Approx. Diam. Inches	Approx. Area Sq. In.
Col. 1	Col. 2	Col. 3	Col. 4	Col. 5	Col. 6	Col. 7	Col. 8	Col. 9	Col. 10	Col. 11
18	.146	.0167	.106	.0088	.089	.0064	.081	.0052	…	…
16	.158	.0196	.118	.0109	.100	.0079	.092	.0066	…	…
14	30 mils .171	.0230	.131	.0135	.105	.0087	.105 .105	.0087 .0087	.129	.0131
14	45 mils .204*	.0327*	.162†	.0236†	…	…	…	…	…	…
14	…	…	…	…	…	…	…	…	…	…
12	30 mils .188	.0278	.148	.0172	.122	.0117	.121 .121	.0115 .0115	.146	.0167
12	45 mils .221*	.0384*	.179†	.0251†	…	…	…	…	…	…
12	…	…	…	…	…	…	…	…	…	…
10	.242	.0460	.168	.0224	.153	.0184	.142 .142	.0159 .0159	.166	.0216
10	…	…	.199†	.0311†	…	…	…	…	…	…
8	.328	.0854	.245	.0471	.218	.0373	.206 .186	.0333 .0272	.241	.0456
8	…	…	.276†	.0598†	…	…	…	…	…	…
6	.397	.1238	.323	.0819	.257	.0519	.244 .302	.0467 .0716	.282	.0625
4	.452	.1605	.372	.1087	.328	.0845	.292 .350	.0669 .0962	.328	.0845
3	.481	.1817	.401	.1263	.356	.0995	.320 .378	.0803 .1122	.356	.0995
2	.513	.2067	.433	.1473	.388	.1182	.352 .410	.0973 .1316	.388	.1182
1	.588	.2715	.508	.2027	.450	.1590	.420	.1385	.450	.1590
0	.629	.3107	.549	.2367	.491	.1893	.462	.1676	.491	.1893
00	.675	.3578	.595	.2781	.537	.2265	.498	.1974	.537	.2265
000	.727	.4151	.647	.3288	.588	.2715	.560	.2463	.588	.2715
0000	.785	.4840	.705	.3904	.646	.3278	.618	.2999	.646	.3278

Table 7-7. NEC Values for Conduit Fill With Like Conductors

Conduit Trade Size (Inches) Type Letters	Conductor Size AWG, MCM	½	¾	1	1¼	1½	2
TW, T, RUH, RUW, XHHW (14 thru 8)	14	9	15	25	44	60	99
	12	7	12	19	35	47	78
	10	5	9	15	26	36	60
	8	2	4	7	12	17	28
RHW and RHH (without outer covering), THW	14	6	10	16	29	40	65
	12	4	8	13	24	32	53
	10	4	6	11	19	26	43
	8	1	3	5	10	13	22
TW, T, THW, RUH (6 thru 2), RUW (6 thru 2),	6	1	2	4	7	10	16
	4	1	1	3	5	7	12
	3	1	1	2	4	6	10
	2	1	1	2	4	5	9
	1		1	1	3	4	6
FEPB (6 thru 2), RHW and RHH (without outer covering)	0		1	1	2	3	5
	00		1	1	1	3	5
	000		1	1	1	2	4
	0000			1	1	1	3
THWN,	14	13	24	39	69	94	154
	12	10	18	29	51	70	114
	10	6	11	18	32	44	73
	8	3	5	9	16	22	36
THHN, FEP (14 thru 2), FEPB (14 thru 8),	6	1	4	6	11	15	26
	4	1	2	4	7	9	16
	3	1	1	3	6	8	13
	2	1	1	3	5	7	11
	1		1	1	3	5	8
XHHW (4 thru 500MCM)	0		1	1	3	4	7
	00		1	1	2	3	6
	000		1	1	1	3	5
	0000		1	1	1	2	4
XHHW	6	1	3	5	9	13	21
RHW,	14	3	6	10	18	25	41
	12	3	5	9	15	21	35
	10	2	4	7	13	18	29
	8	1	2	4	7	9	16
RHH (with outer covering)	6	1	1	2	5	6	11
	4	1	1	1	3	5	8
	3	1	1	1	3	4	7
	2		1	1	3	4	6
	1		1	1	1	3	5
	0		1	1	1	2	4
	00			1	1	1	3
	000			1	1	1	3
	0000			1	1	1	2

nor any fixed or stationary appliances, space heating, equipment, and so forth.

The square footage of the house is calculated on the basis of the outside dimensions of the structure, but does not include porches, garages, decks, crawl spaces and such. Unfinished or unused portions of the house need not be included either, but it certainly would do no harm to figure in any areas that might be finished at a later date.

Table 7-8. NEC Values for Fill With 3/8-inch Flexible Conduit

Col. A = With fitting inside conduit.
Col. B = With fitting outside conduit.

Size AWG	Types RFH-2, SF-2		Types TF, T, XHHW, AF, TW, RUH, RUW		Types TFN, THHN, THWN		Types FEP, FEPB, PF, PGF	
	A	B	A	B	A	B	A	B
18	..	3	3	7	4	8	5	8
16	..	2	2	4	3	7	4	8
14	..	..	..	4	3	7	3	7
12	..	..	..	3	..	4	..	4
10	..	..	..	..	..	2	..	3

* In addition one uninsulated grounding conductor of the same AWG size shall be permitted.

As an example, let's say that your house has 2500 square feet of living space. At 3 watts per square foot, the minimum requirement would be a total of 7500 watts. With a 115-volt supply, this would be the equivalent of 65.22 amps, or four general-purpose 20-amp single-pole branch circuits. If your figures show that your proposed general-purpose circuits fall below the 3 watts-per-square-foot level, add more circuits. This is a bare-minimum requirement, so you could even double your present number.

FEEDER CIRCUITS

The next step in this seeming marathon is to calculate the feeder circuits. If all of your branch circuits originate in one main entrance panel, you will skip these calculations. But if you have any subpanels or load centers, each feeder circuit running to them from the service equipment, or from one load center to another, has to be sized out in similar fashion to the branch circuits. And here too you must take into account certain considerations and requirements.

The starting point for figuring feeder loads is to assume that the total connected load for each feeder will be no less than the sum of the individual branch circuits, plus any other feeders served by that feeder. If there are four circuits in a subpanel with a total load of 45 amps, then the subpanel must be served with a feeder of at least 45-amp capacity. If there are two branch circuits with a total connected load of 30 amps, as well as a feeder to another subpanel carrying a load of 25 amps, the serving feeder must have a minimum ampacity of 55 amps. There are, however, exceptions that will modify this situation.

First, the continuous/noncontinuous rule applies here. If the feeder supplies a *continuous* load, the feeder rating must be 125% of that load. If there is a combination load, the feeder must have the capacity to carry all of the *noncontinuous* load plus 125% of the continuous load.

Second, you may separate the so-called lighting load that includes virtually everything except cooking appliances, dryers, air conditioning, laundry, small appliances, and electric space-heating loads. Remove the general-purpose load of each branch circuit to find the total wattage of lighting load that will be applied to the feeder in question. The first 3000 watts must be figured at 100%. Beyond that, and up to 120,000 watts (an unlikely figure for a dwelling), you may use a *demand factor* of 35%. If a load center handles a lighting load of 6000 watts, then, the load applied to the feeder will be 100% of the first 3000 watts plus 35% of the remainder, for a total of 4050 watts. Note, however, that this presupposes a unity power factor. If any conditions apply to the branch circuits lighting loads that reduce the power factor below 100% to any great degree, this must be taken into account. After these demand calculations are made, add back those loads that were not considered as part of the lighting load.

Third, where motors are involved, they should be treated the same as in the branch circuit calculations.

Fourth, the required small-appliance branch circuits must be figured at a minimum of 1500 watts each. Should they happen to be more than that, the higher figure should be used.

Fifth, the required laundry branch circuit must also be figured at 1500 watts or the computed (known) load, whichever is larger. In addition, if there is more than one laundry circuit, all of them must be figured in the same way.

Sixth, check the branch circuits to be served by the feeder to see if any of them are noncoincidental. If you find two circuit loads that are highly unlikely to be in operation *at the same time*, you can eliminate the smaller of the two. This refers to the entire circuit load, not just parts of it, and you should exercise judgement in making a decision. Should the presumed non-coincidence not work out in practice, you will have an undersized feeder and problems on your hands.

Seventh, if four or more fixed appliances are served by a single feeder (none of the appliances include a range, clothes dryer, air conditioner, or space-heating equipment), then you can use a demand factor of 75% of the total appliance nameplate ratings in computing the feeder load for those appliances. If the total comes to 10,000 watts, for instance, then you need figure only 7500 watts for that part of the total feeder load.

Eighth, where a feeder serves both space-heating and air-conditioning equipment, you can eliminate the smaller of the two loads, since they are not likely to run coincidentally.

Ninth, the feeder calculations serving cooking appliances can be figured in a similar fashion to the branch circuit calculations discussed elsewhere.

To make the figuring easier, you may wish to add the feeders to the circuit tree. You can then jot down the total connected loads for each, and using Tables 6-1 or 6-4 you can determine the conductor sizes that you will need. At the same time, determine the length of each feeder and check the voltage drop just as you did for the branch circuits. The maximum voltage drop for feeders can be no more than 3%, but note that the maximum allowable drop for feeders and branches *together* cannot exceed 5%. In other words, if you have a feeder with a 3% voltage drop feeding a subpanel that has a branch circuit with a 3% voltage drop, the total would be 6%—and you are over the limit.

BALANCING LOADS

The next step is to go over your layout or the circuit tree to see how well all of the loads are balanced. This is a rather inexact process since there is always some load diversification—some loads will be in operation while others are not, there are variable loads, and some outlets have specific uses that are uncertain and changeable. Even so, as close a balance as possible should be your goal.

The Service Entrance

To understand the purpose and reasoning behind this balancing process, we'll have to go back to the service entrance. Most homes today are served by a three-wire, single-phase power supply of either 115/230 volts or 120/240 volts, or values close to that (Fig. 7-3). There are two "line" or "hot" conductors plus a grounded neutral. The voltage between either line and the neutral is 115 volts nominal, and the voltage between the two lines is 230 volts nominal, though actual measured voltages will vary from this.

When the service-entrance conductors are wired into the main panel, one line is secured to a lug on one side of the panel and the second line is on the other, with the neutral going to its own neutral bus. The overcurrent protection devices are mounted in a row down each side of the panel, but the bus connections actually crisscross back and forth so that two top single-pole circuit slots, right and left, are tied to line A, the second slot to line B, and so on, alternating down the row. The

Fig. 7-3. Some of the more common supply systems for electrical service entrances.

object of this balancing is to arrange all of the circuits coming into the panel (ultimately joined to lines A and B) so that the total load on each line will be matched as closely as possible.

There is a reason for this balancing. In two-wire circuits, like the single-pole branch circuits in your system, each conductor carries the same current; therefore both sides of the circuit are in balance. The same is true of two-wire 230-volt circuits, like those which might feed electric space-heating baseboard units, so with no neutral wire, the current is the same in each of the two conductors.

231

In a three-wire distribution circuit, which consists of lines A and B plus the neutral, this is not true. When the currents in lines A and B are equal, there will be no current in the neutral wire. If they are unequal, the neutral will carry the difference, as shown in Fig. 7-4. This is true for any three-wire circuit, whether it is a feeder to a subpanel, a branch circuit to a piece of equipment such as a range, or a set of service-entrance conductors.

The result of a heavy unbalance in current is to produce a voltage drop on the heavily loaded line (where it can be least afforded) and an increase in voltage on the other line (where it is not wanted). This can in turn lead to overheating, operational problems, loss of efficiency, and trouble in general.

Balancing Methods

As you can see from the diagram in Fig. 7-4, a three-wire distribution circuit is actually a series-parallel circuit like those discussed in Chapter 1, and it can be broken down and the various voltage drops and other values found in the same way. You'll note, too, that if there were no resistance in the neutral conductor, there would be no unbalance. This is a possible solution in the case of severe unbalance that can't otherwise be conveniently corrected; replace the existing neutral with a much larger one.

A much better solution, however, is to minimize the problem at the outset—on paper before the installation is made. Start with the single-pole circuits and match them as closely as you can. Use those loads which you can compute from information on hand, plus the minimum loads specified in the NEC and local codes (such as 180 volt-amperes per receptacle) for those you cannot. Where you have to connect an odd number of circuits into a subpanel, and thus to a feeder, try to arrange the branch circuits so that their total load is halved between the two lines of the feeder. The two-wire, 230-volt branch circuits are no problem since they will be balanced automatically when they are connected.

Feeders should be balanced out in a similar fashion when they are connected to the voltage source at the service equipment or main panel. If you do this as you go along, the chances are excellent that while there will be some unbalance from time to time, depending upon the diversity of the loads, the condition will not be severe enough to cause any difficulties. You can make absolutely sure of this by laying out the circuits in the same manner as the one shown in Fig. 7-4, using for values those loads that you feel might cause the

Fig. 7-4. Load balance in two- and three-wire systems.

worst case of unbalance, or various combinations of possibilities. A series of calculations will then show you whether or not there is any real problem.

One simple method of finding the total effective load for single-phase two-wire circuits is to use the formula

$$I = \frac{W}{E_p \times PF}$$

Where I stands for current, W for watts, PF for power factor, and E_P is the voltage between the conductors. Note that E_P may be the source voltage, or it may be a lesser voltage if there is an IR drop present. The power factor, (see Chapter 1), which may have already been figured into the circuit load as part of an individual load on the circuit, may or may not be of sufficient magnitude to bother considering. If the circuit is single-phase three-wire, the formula is

$$I = \frac{W}{2E_G \times PF}$$

Where E_G represents the voltage between the phase or "hot" wire and the neutral.

SIZING PANELS AND LOAD CENTERS

With all of the branch and feeder circuits lined up and sized out, you can now select the proper main panel, load center, and subpanel size. Panel sizes are usually referred to in terms of the number of single pole circuits they will handle; an eight-circuit panel will hold 8 single-pole circuits, or 4 double-pole circuits, or some combination thereof. With circuit-breaker panels, you could also install 16 of the double single-pole (piggyback) breakers. In addition, panels are rated according to the ampacity of the main connecting lugs and busses; for example, as a 60-amp box, or a 100-amp box.

Panel Sizes

To arrive at the proper size for a subpanel or load center, simply add up the total load wattage of the circuits going to each panel, then divide the answer by the supply voltage. A 10,000-watt load on a panel served by a 230-volt, three-wire feeder would equal a current of 43.48 amps on each leg (line A and line B). The minimum size panel that would do the job would be the next highest standard size: 50-amp box.

A main entrance panel can be figured in the same manner. If the total connected load of a system were 24,000 watts and the supply is 240 volts, then the current draw would amount to exactly 100 amps per leg. In this case you would not choose the standard 100-amp panel because there is absolutely no safety factor; instead, the minimum would probably be a 125-amp panel. In practice it is a good idea to go up two steps, or even more, to allow for adequate future capacity, both for load centers and main panels.

You must also provide enough physical space for the number of circuits you have planned. In the subpanel example just given, for instance, your total wattage of 10,000 may include five circuits. Panels are made in multiples of two circuits, so you would need a six-circuit box. That might not be readily available, which means going to an eight-circuit box. But to gain the eight-circuit capability, you might have to use a box of 60- or maybe 100-amp capacity, with its correspondingly larger physical size to be figured into your plans. Much depends upon what is available to you at the local supply level. In any case, the prudent wireman will allow for several more circuit spaces than he has an immediate need for in order to take care of the future. The overcurrent protection devices can be added later as necessity demands and need not be supplied at the outset.

Fuses and Breakers

Sizing out the overcurrent protection devices is done in the same fashion. Remember that no overcurrent device should be of larger capacity than the conductor it protects; you cannot use a 30-amp fuse on a 20-amp line. You can, however, protect a 30-amp line with a 20-amp fuse. Assuming that you are using conductors insulated with TW or a similar insulation type, in normal service, then the #14 branch circuit will be protected with a 15-amp breaker or fuse, #12 with a 20, #10 with a 30.

Special circumstances and exceptions aside, two-pole and feeder circuits should be protected with the appropriate standard fuse size next above the load served by the circuit, but not exceeding the capacity of the conductors used. No circuit should be loaded to more than 80% of its capacity—as a rule of thumb and in some cases by regulation—so the overcurrent devices should be chosen accordingly. In cases where it is desirable to protect some particular load, a motor for instance, to very close tolerances, a special overcurrent device such as a time-delay fuse may be inserted in the line just ahead of the motor load, with the normal overcurrent device remaining at the head of the circuit as usual.

The main overcurrent device for the service-entrance conductors, which protects the entire system as a whole, is selected in the same way. Find the total connected load wattage of the system, divide the wattage by the supply voltage, and choose the next highest appropriate fuses or circuit breaker. The 24,000-watt system mentioned earlier might use the 125-amp size. In the case of the circuit breaker, it can serve as both main overcurrent protection and main disconnect. If a switch is used instead, this too would be the next appropriate standard size. Whatever numbers you come up with for a service entrance capacity, remember that for all practical purposes 100 amps is the allowable minimum for a residence.

Service-Entrance Conductors and Laterals

The minimum size for service-entrance conductors is #8 copper or #6 aluminum, and the same is true of service laterals. In any case they must be large enough to carry the load imposed upon them without heating up, and they should be sized at least one step larger than necessary for future expansion of the system. In fact, the larger sizes are the ones most commonly used. Note that in single residential services fed by a single-phase three-wire supply, some types of conductors have a special rating. Wire sizes 4, 3, 2, 1, 1/0, and 2/0 copper conductors, with three or less in a cable or raceway, with insulation of the types RH, RHH, THW, and XHHW, all of which are widely used and readily available, have ampacities of 100, 110, 125, 150, 175, and 200 amps respectively. This is irrespective of the ampacities given in the Table 6-2.

Note that though you can size the main entrance panel, service-entrance conductors, and main overcurrent protection device according to the total connected load of the whole system, this is usually unnecessary and seldom done. Diversification of normal residential loads is such that only a certain proportion will ever be activated at any given time, so the total connected load will never come on line at the same time. Figuring the total demand load of a residence can thus be accomplished in just the same way as figuring a feeder load, as explained earlier. It is a good idea, too, to calculate the total demand load placed on each leg of the supply source. For example, let's say that your final load works out to 21,000 watts, after you have applied the various demand factors. With a supply voltage of 230, the load would be a bit over 93 amps per leg, indicating the use of a 100-amp main panel. But if one leg happens to be only loaded to the extent of 75 amps, that means that the other could be overloaded to 111 amps (93 + 93 − 75). Consequently, good balance is necessary for good overcurrent protection.

Where the ampacity of the service-entrance conductors will be at least 100 amps, there is another method of calculating that can be used. First, take the largest of the following items, if present: 100% of the air-conditioning load, 65% of the central electric space-heating load, 65% of the load of three or less separately controlled electric space-heating units, or 100% of the load of four or more separately controlled electric space-heating units. To this, add 100% of the first 10 kW of everything that remains in the system, and 40% of all beyond the first 10 kW. But in considering the remaining load, make sure that the following items are included: the nameplate ratings of all motors and low power-factor loads, the nameplate ratings of all cooking appliances and other fixed appliances including four or more separately controlled electric space-heating units, 1500 watts each for all of the required 20-amp branch circuits for small appliance and laundry use, and 3 watts per square foot of living space in the house. In addition to these loads you should add any and all known loads not specifically mentioned that are a part of your system. A typical calculation, then, might look something like the following:

Electric space heating: 18 kW, 10 units	18000 watts
Range: 8.75 kW	8750
Dryer: 6 kW	6000
Water heater: 5 kW	5000
Dishwasher: 1.2 kW	1200
3 small-appliance circuits at 1500 watts each	4500
1 laundry circuit at 1500 watts	1500
2500 sq. ft. at 3 watts/sq. ft.	7500
Other known loads	6200
	57650 watts
Heating at 100%	1800
First 10 kW of remainder at 100%	10000
Remaining 30.65 kW at 40%	12260
	40260 watts

40260 watts divided by 230 volts is 175 amps. Thus, in this example a 200-amp service entrance will be required, together with commensurate overcurrent protection device and service-entrance conductors.

GROUNDING ELECTRODE

And last but far from least, we come to the grounding electrode conductor. This is the conductor that is attached to the supply side of the service-entrance equipment and runs to the grounding electrode located at the nearest convenient point. The preferred material is copper since the restrictions upon the use of aluminum for this job are such that it almost never can be used. The conductor can be either bare or insulated, so if you have a piece of cable left over from some other job this might be usable.

You'll note from Table 7-9 that the grounding electrode conductor may be of the same size as the service-entrance conductors, but it need not be. The sizes shown are minimums.

Table 7-9. NEC Grounding Electrode Conductor Sizes

Size of Largest Service-Entrance Conductor or Equivalent for Parallel Conductors		Size of Grounding Electrode Conductor	
Copper	Aluminum or Copper-Clad Aluminum	Copper	*Aluminum or Copper-Clad Aluminum
2 or smaller	0 or smaller	8	6
1 or 0	2/0 or 3/0	6	4
2/0 or 3/0	4/0 or 250 MCM	4	2
Over 3/0 thru 350 MCM	Over 250 MCM thru 500 MCM	2	0

Should the conductor be #6 or smaller, it must be enclosed in armor, conduit, or EMT for its full length from service equipment to grounding electrode. The proper sizes of raceway for this and for service-entrance conductors or service laterals are taken from the pipe fill values in Table 7-6.

ESTIMATING MATERIALS

The process of making up material takeoff sheets and doing the estimating can be a wearying one—and sometimes a head-scratcher as you try to visualize the system and foresee all of your needs. Using the plans and the specification sheets simplifies the problem, and the circuit "tree" (Fig. 7-1) will also help. You have a fixture schedule and can make up a box and device schedule and a miscellaneous schedule. Then all you have to do is count the numbers of each and list them. You have already figured the lengths of the circuits and feeders, as well as the conductor sizes, so those figures will give you the ingredients of the wire and cable listing. The same holds true for the main service-entrance equipment and the load centers and subpanels, plus the overcurrent devices. If you will require conduit or EMT, those figures can be taken from the circuit lengths and conductor requirements listed on the circuit tree.

By simply going over all of the plans and data you have worked up so far, you can make a materials list that will cover the bulk of the items you will need. Determining the hardware and all of the fittings, connectors, Wire-nuts, and miscellaneous odds and ends necessary to complete the job is a bit of a task, and it never comes out exactly right. The simplest method is to go over the layout step by step, bit by bit, jotting down everything that comes to mind as you go along, and tallying up the numbers at the finish.

Once the takeoff is complete, or as nearly so as you can reasonably make it, you can go about getting price estimates in several ways. You can consult with the local supplier or contractor from whom you plan to buy, get the unit prices from him, and complete the figures to your own satisfaction to arrive at a total job cost. Or, you can type (prevents later misunderstandings resulting from poor writing) a list of your needs for your supplier and let him do it. You will probably get back a firm price based upon supplying the entire lot, good for thirty days or some such. Or, you can make up several copies of the takeoff and put them out to bid to several suppliers. Whatever you decide, best add an extra 10% for contingencies, inflation, and good (or bad) luck. Then add another 5% to cover those items that you forgot or will want to add as the installation goes along.

Chapter 8
Special Subsystems

There are several special installations that are sometimes a part of the residential wiring system, and though unusual, it is possible that all of them might be found together on the same premises. The general wiring practices and procedures are about the same as for the normal residential general-purpose, lighting, and appliance wiring system, but in each case some further information is necessary with respect to regulations, planning, and installation.

FARMSTEADS

Farm wiring is often more complex than that for a single-family dwelling, unless the premises are unusually small. Several outbuildings, each of which may contain substantial electrical loads, may be involved, and many of those loads are likely to be relatively sizable pieces of motorized equipment unlike those found in the ordinary home. Special hazards may also be encountered.

Estimating Service Loads

To begin with the service, one or the other of two installation methods is generally used. You can locate the service equipment in the farmhouse itself, just as in any dwelling, and run feeders from there to the various outbuildings. The other possibility, especially useful where the loads are quite large, calls for the installation of a *farm service pole* at some convenient point more or less equidistant from each building, including the residence, with a service drop or lateral running to each building to be served.

In the dwelling main-entrance scheme, the total load for the dwelling itself is figured in the normal manner, just as though it stands alone (see Chapter 7). Then, the load for each farm building as a unit is computed by adding up all the load wattages and converting to amperage ($I = P/E$), then adding up all of the equipment nameplate amperage ratings, and then adding the two results together. For example, suppose there are three buildings to be served, besides the house, one having a total load of 90 amperes, another with 60 amperes, and the third with 30 amperes. First, go back through the individual loads and find the largest motor; let's say that this is a 10-hp motor with a current rating of 50 amps. This largest motor must be figured at 125% of rated current, or 63 amps, so if the motor were in the building with the 90-amp load this would then become a 103-amp load.

Using Demand Factors

The total of all the loads in the outbuildings is 193 amps, and if you wish, you can size the service-entrance equipment to this figure plus the load for the house. The usual procedure, though, is to apply demand factors. For farm loads other than dwellings, those factors are as follows: use 100% of the first 60 amperes; use 50% of the remainder, up to the next 60 amperes; and use 25% of all that is left above 120 amperes. Thus, the demand load in this case would be 60 amps plus 50% of 60 amps, plus 25% of the remaining 73 amps, making a total of 108 amps. This lower figure recognizes the fact that not everything will be running at once, and it represents the actual extra capacity needed beyond the dwelling load in the service equipment to serve the farm load.

When adding in the dwelling load, a demand factor can also be used here. The demand factors for a total farm load are as follows: 100% of the largest load; 75% of the second largest load; 65% of the third largest load; and 50% of all the remaining loads. We have already arrived at a total demand load for the farm buildings of 103 amps, and let's assume that the dwelling load is 72 amps. The total demand load, then, would be 100% of 103 amps plus 75% of 72 amps, or 157 amps.

Another way of calculating would be to leave the various loads separated; include the dwelling, list the loads in order of size, and use the total farm load demand factors. This would give us (1) 103 amps, (2) 72 amps, (3) 60 amps, and (4) 30 amps. The total demand load would then be 100% of 103 amps, plus 75% of 72 amps, plus 65% of 60 amps, plus 50% of 30 amps, a total of 211 amps. The smaller figure would be the one most often used, but remember that this allows no room for future expansion.

Sizing Conductors

These total demand load figures are the ones used to size the ampacity of the main disconnect, main overcurrent protection, the service-entrance conductors, service drop or lateral, and associated equipment, which is than laid out and installed in the normal manner (see Chapters 7, 10, and 11). In the preceding example, each of the total loads for the various outbuildings would be served by a feeder circuit running from the service equipment in the dwelling.

To find the proper ampacity of each feeder, you may use the total current draw of all the individual loads in each building. This would mean one set of conductors for 103 amps, another for 60 amps, and a third for 30 amps. In the 75°C class with encabled copper conductors, this would mean #2, #6, and #10 sizes respectively (Table 6-4). Note, however, that if the feeders are run overhead to the buildings rather than underground, the minimum size conductors that can be used is #10, in spans of up to 50 feet. If the span is over 50 feet, #8 minimum must be used. To size single open-air overhead conductors, use Table 6-3. Note too that the overall length of the feeder cable has a bearing because of its voltage drop. You can figure the voltage drop for any given conductor length and load by using the information in Chapter 7.

An alternate method of computing feeder sizes uses the same demand factors as for determining farm loads other than dwellings; that is, 100% of the first 60 amps, 50% of the next 60 amps, and 25% of the rest. For instance, let's assume that the 103-amp load previously mentioned consists of the 10-hp motor load at 125% rating, or 63 amps, another motor at 28 amps, lighting at 5 amps, and a cooler at 7 amps. The first 60 amps of the total would be calculated at 100%, while the remaining 43 amps would be calculated at 50%. The total demand load would then be 82 amps rather than 103 amps, which would require a 75°C class conductor of #4 size. Feeders for the remaining two loads of 60 and 30 amps obviously would have to be full-sized, since neither goes beyond the first 60 amps of the demand factor. Again, these would be minimum sizes leaving no room for expansion.

Where a farm service pole is used, a single service drop is run from the utility power distribution pole to the farm service pole and then into a meter and main disconnect. From this point, individual service drops or service laterals run to the various buildings, including the dwelling. In this case, the calculations are made according to the demand factors for computing total farm loads: 100% of the largest load, 75% of the next largest load, 65% of the third largest load, 50% of the remaining load. Using the same four loads as for the previous examples, the total would then be 100% of 103, plus 75% of 72, plus 65% of 60, plus 50% of 30, or 211 amps. This figure would determine the ampacity of the main disconnect and overcurrent protection, the main service drop, and the grounding system. The individual service drops or laterals from the farm service pole to the buildings are sized on the same basis as previously described.

As far as the actual installation goes, in a dwelling main service, the equipment is handled as though it were just another house. The feeders to the outbuildings if overhead are treated like service drops, and if underground like service laterals (See Chapter 10). Each outbuilding must have its own disconnect in, on, or near the structure, sized according to the ampacity of the feeder. The feeder may serve a single load, or go to a subpanel from which branch circuits are run in the usual manner within the building. The feeders must also be protected in the usual way by overcurrent devices in the distribution panel where they originate.

With a farm service pole, the situation is a bit different. Here, local codes and the requirements of the serving utility come into play: some of the details of installation vary widely from place to place. Part of the job may be done by the power company, or perhaps most of it, or possibly none of it except the main service drop. There are so many variations and possibilities that it would be impossible to cover them all. Check with the power company and the inspecting authority to find out exactly where your responsibilities lie and specifically how the installation should be put together.

HOME WORKSHOPS

Many home shops consist of nothing more than a small workbench tucked in a spare corner, with one light fixture overhead and one receptacle nearby. But with more and more leisure time available to many folks, coupled with an upsurge in arts and crafts, home projects, and do-it-yourselfing, the modern home workshop is likely to be large, well equipped with all manner of power tools and appliances, and designed to serve the hobby and craft needs of the entire family. This means that there are four prime electrical factors to be considered in wiring a shop: good grounding, good lighting, plenty of capacity, and flexibility of convenience outlets. In addition, every aspect of shop wiring should be made as safe as possible with respect to surface wiring, extension cords, attachment points, and the like.

242

Branch Circuits

Every branch circuit that feeds a shop must carry an equipment-grounding conductor. All receptacles must be of the three-prong grounding type, and all wall boxes, junction boxes device frames, covers, lighting fixture bodies, and all metallic portions of the wiring system that do not normally carry current, must be securely tied to this grounding circuit. All portable tools and appliances with metal cases or surfaces should be equipped with a three-prong grounding attachment plug, and the frames or cases of all fixed and stationary power equipment, such as a lathe, drill press, table saw, and band saw, must be grounded either by direct connection or by means of a grounding type of attachment plug. The only exceptions are those portable power tools, sometimes called *double-grounded*, that are made of a nonconducting plastic material like Cycolac. Never use metal power tools with two-prong adaptors; after all, you are carrying around a bundle of live electricity in your hands and anything can happen.

Three or four 20-amp general-purpose lighting and convenience-outlet branch circuits are not too many in an average size shop, and more may be necessary in a large one. Remember that each circuit can handle only six duplex receptacles, fewer if there are lighting loads on the same circuit. In addition there should be other circuits, both 115-volt and 230-volt, to handle specific power tool loads as necessary. To be on the safe side, don't load these circuits to more than 80% of the ampacity of the conductors.

For disconnecting equipment you can use attachment plugs for items so equipped, but direct-connected machinery should be fitted with separate disconnects located where they can be quickly and conveniently reached from the working position at the machine. Direct wiring should be done with short taps in a length of flexible conduit, run to a nearby junction box. The flex will absorb vibration from the machinery.

Grounding

A shop is fed from a subpanel or load center. Be sure that the subpanel enclosure is tied to the equipment grounding circuit, but that the equipment grounding circuit and the neutral bar in the subpanel are isolated from one another. All incoming equipment grounding conductors must be tied to the main equipment grounding conductor, whether that be an individual wire or a metallic raceway that enters the subpanel. This conductor in turn runs back to the service-entrance equipment, isolated from all other conductors, and there ties directly to the grounding electrode conductor.

Though not now required by the NEC— time will undoubtedly change that fact—and probably not required by most local codes, you can give yourself and additional measure of safety by installing GFI (ground fault interrupt) circuit breakers on all shop circuits. The possibility of ground faults is probably as high in a well-equipped shop as in a kitchen or bathroom—especially if the floor happens to be poured concrete. But the GFIs on the job, any ground fault would be quickly detected and the circuit tripped before personal danger arises.

Lighting

Adequate lighting is essential in a shop both for safety and for being able to see what you are doing in order to turn out a creditable job. If the place is dark and gloomy, you take a chance of injuring yourself. And the more intricate and concentrated your task, the more light you will need. Indeed, it would be difficult to have too much light in a shop, provided the glare is kept down.

Basically, the entire shop should be illuminated as evenly as possible with a fairly high level of background light. Additional light should be located above each general work area, such as a bench or table or machine stand. Sometimes these two light sources can be one and the same, if the shop is well laid out.

On top of this, strong local lighting should be focused on small areas where intensive work will be done, such as a model-making table, ammunition handloading bench, fly-tying bench, drafting table, weaving loom, or attached to a power jigsaw or drill press where fine cutting or milling might take place. For further information on lighting design, refer to Chapter 5.

Receptacles

The number and arrangement of receptacles is all-important in setting up a workshop that will function smoothly. There should be a place to "plug in" at every point where you might conceivably be working. This is important for convenience and peace of mind as well as the elimination of a maze of extension cords to trip over and get tangled up in. By the way, any cords you must use should have three-prong attachment plugs and carry an equipment grounding conductor.

You can start by positioning duplexes every few feet all around the perimeter of the shop, wherever the walls will not

244

be obstructed, at convenient heights. Plug strips mount on the surface and will receive an attachment plug at any location (or in some models in only a few spots); these work well when mounted on the wall behind and about 18 inches above the workbench tops. In addition, you can mount more outlets in or on the front apron of workbenches or tables.

Freestanding machinery too far from the walls to reach outlets can be served by individual floor receptacles, provided that the floor is a normally dry one. Failing this, you could use a specially designed floor extension. This is a heavy cord within a tough vinyl shield, tapered on each side so you won't trip over it, which plugs in at one end and has a receptacle on the other. You can secure it to the floor with a few dots of silicone rubber cement, which will allow later removal without ruining it. Supplying middle-of-the-floor direct-connected machinery is a bit more difficult, but can usually be accomplished by installing a "standpipe" of EMT near the machine; the service or disconnect switch and a connection box is attached to the tubing (Fig. 8-1).

Workshop wiring is often done on the surface since appearance may not mean much. And in a frame structure the tendency, if local codes allow it, is to wire with nonmetallic sheathed cable and often cheap plastic surface-mount receptacles and switches. Yet of all the places in the house where mechanical damage to the wiring system might occur, the workshop probably heads the list. A far better method, which is only a bit more expensive and time consuming and a good deal more presentable and safe, is to run all conductors in EMT and place all devices in metal enclosures. For an effective layout, this means a considerable amount of preplanning and inclusion of ways for future expansion when necessary, but the final result is worth the work.

HAZARDOUS LOCATIONS

This is a subject about which the NEC has much to say, and it includes conditions for which a good deal of specialized equipment is manufactured. Hazardous locations are broken down into two basic groups: *Class I*, where flammable gasses or vapors may be present in the air in sufficient quantity to be combustible or explosive, and *Class II*, where there may be combustible dust in the air. Both of these classes are broken down into further divisions with full treatment of different equipment and occupancies. Though most of this material is of interest primarily to commercial and industrial electricians whose work frequently involves hazardous areas, there are two particular instances that apply to residential and farmstead installations.

Fig. 8-1. Power supply and control wiring by "standpipe" to machinery located away from walls.

Combustible Dust

Combustible or ignitable dust is often found in large quantities under certain circumstances in farm operations. Though the condition may be only periodic or temporary, the danger is still very much present. Hay storage in barns, especially large ones with several lofts where hay is piled loose, is a common source of combustible dust. Another example is any building or area where feed is ground or mixed, especially in large quantities, or where grains of any sort are processed. Even with modern equipment, raising clouds of dust is difficult to prevent, and often ventilation is not sufficient to carry the particles away. This means that a spark caused by a running motor, the flip of a switch, or perhaps the removal of an attachment plug under load can result in ignition or in severe cases even explosion or implosion.

The specifications of equipment used under these conditions vary slightly with the conditions themselves. The severity or degree is dependent in turn upon the judgement of the inspecting authority. In the main, however, the first requirement for wiring any location where quantities of dust are or may be present is to keep as much wiring as possible entirely clear of the area. The second requirement is that all equipment be housed in dust-ignition-proof enclosures (Fig. 8-2) or be specifically designed and approved for dust-ignition-proof use.

All conductors must be run in rigid metallic conduit, or in type MI cable, with the appropriate fittings approved for dust-free use. In some cases, motorized equipment must be totally enclosed and equipped with ventilating ducts, while in other low-level dust areas where the equipment can be easily reached for cleaning and maintenance, standard equipment and enclosures can be used. The old and often used method of enclosing a light bulb and socket in a screw-top glass jar is out—all fixtures must be approved for this use, solidly mounted or suspended on conduit, and adequately protected from physical damage by guards or by virtue of the mounting location. Cords must be of the heavy-duty type and attachment plugs and connectors, as well as the receptacles, must be provided with effective dust seals. All equipment must be completely grounded and bonded.

Combustible Gas

The second hazardous location is encountered in many home workshops where spray painting, lacquering, or staining is done, whether by spray guns or spray cans. Either circumstance can produce large quantities of extremely volatile and flammable solvents. Unfortunately, spray painting is often done with little thought, no precaution, and totally inadequate ventilation, which results in nothing less than the preparation of a bomb that can be set off by the tiniest spark.

The first step in setting up a spray-painting booth or area is to provide plenty of forced ventilation to exhaust the vapors. The second is to exclude as much wiring as possible from that area. The third is to make sure that all of the equipment (Fig. 8-3) is designed with two principle factors in mind: the enclosures must be capable of withstanding an explosion of vapors inside itself should any vapors enter, and they must prevent any sparks from equipment within the enclosure, or

WATERTIGHT (NEMA 4) -- DUST-TIGHT (NEMA 9, CLASS II, GROUPS E, F & G HAZARDOUS LOCATIONS)

Type DH boxes feature wide flanges that provide ample contact between cover and special gasket to insure tightness under specified hazardous conditions. These boxes are designed for use in hazardous combustible atmospheres containing metal dust; carbon black, coal or coke dust; and flour, starch or grain dust.

Standard Construction:
1. Cast iron box and cover
2. Hot-dip galvanized finish
3. Mechanically attached vellum-oid gasket
4. Stainless steel cover screws
5. Flange ground before galvanizing

DUST-TIGHT (NEMA 5) – WEATHERPROOF

Type S boxes are considered the standard of junction boxes and are the most widely used type.

Standard Construction:
1. Cast iron box and cover
2. Hot-dip galvanized finish
3. Neoprene gasket
4. Stainless steel cover screws

Fig. 8-2. Dust-tight enclosures. Note that one is also watertight, the other weatherproof. Many other types of dust-tight enclosures are made for particular applications and classes of use. (Courtesy Spring City Electrical Mfg. Co.)

248

an explosion within the enclosure, from setting off an explosion outside the enclosure.

The general requirements for installation are about the same as for dust-ignition-proof wiring. Rigid conduit or type MI cable must be used, with the appropriate explosion-proof fittings and seals. Any standard devices and equipment used must be housed in explosion-proof enclosures, and all other equipment, including motorized machinery, lighting fixtures, and any other utilization equipment, must be approved as explosion-proof. Cords and attachment plugs can only be used for portable lamps or utilization equipment, and then only if of the extra-heavy duty type, properly supported, and provided with approved seals where they enter the explosion-proof enclosure to join with the serving receptacle.

Should you be faced with a hazardous location situation, your safest and best course is to discuss the installation thoroughly with your supplier and electrical inspector *before* going ahead, to make sure that you will fulfill all of the requirements.

WELDERS

One or another of the various types of welders is frequently used on farmsteads and occasionally in home workshops as well. The branch or feeder circuits that supply

DUST-TIGHT (NEMA 5) — WEATHER-
PROOF — EXPLOSION-PROOF (NEMA 7,
CLASS I, GROUP D HAZARDOUS
LOCATIONS)

Standard Construction:
1. Extra heavy cast iron box and cover
2. Hot-dip galvanized finish
3. Stainless steel cover screws
4. Mounting lugs

Type EX boxes are explosion-proof as well as dust-tight and weatherproof. All joints are the metal-against-metal type with no gasket. The wide flanges are ground to very close tolerances and with the larger number of cap screws used in assembling provide an explosion-proof enclosure in hazardous concentrations of atmospheres containing gasoline, hexane, naphtha, benzene, butane, propane, alcohol, acetone, benzol, lacquer solvent vapors or natural gases.

Fig. 8-3. Explosion-proof enclosure. Other types are made for different applications and classes of use. (Courtesy Spring City Electrical Mfg. Co.)

Table 8-1. NEC Duty Cycle Factors for Welders

DUTY CYCLE%	COL. A	COL. B	COL. C
100	1.00	1.00	--
90	0.95	0.96	--
80	0.89	0.91	--
70	0.84	0.86	--
60	0.78	0.81	--
50	0.71	0.75	0.71
40	0.63	0.69	0.63
30	0.55	0.62	0.55
25	0.55	0.62	0.50
20	0.45	0.55	0.45
15	0.45	0.55	0.39
10	0.45	0.55	0.32
7.5	0.45	0.55	0.27
5.0	0.45	0.55	0.22
less	0.45	0.55	0.22

them are installed in the usual manner, but the size of the conductors and overcurrent protection devices is determined in a somewhat different manner than other equipment.

Each welder has a nameplate attached that specifies, among other things, the rated primary amperage and the duty cycle or time rating of that particular machine. If the welder is of the AC transformer or DC rectifier arc type, first determine the duty cycle percentage, and then find the corresponding multiplying factor in column A of Table 8-1. Multiply this factor times the current rating on the welder nameplate, and then choose the appropriate conductor size from Table 6-4, using the next highest ampacity above your answer. Usually a 60°C class of conductor is adequate, and the wiring may be encabled or run in conduit or EMT, depending upon local regulations.

The overcurrent protection device may be a breaker or fuses, but cannot be rated at more than 200% of the nameplate rated current. If a disconnecting means is not provided on the unit itself, one must be installed. The overcurrent protection device might possibly serve for the disconnect, but if this is not nearby, a switch or circuit breaker should be located where it is easily visible and accessible from the work area.

If the welder is of the motor-generator type, much the same situation applies, except that the duty cycle multiplying factors are taken from column B of Table 8-1. The disconnecting means can be a circuit breaker, but if a switch is used it must be of the motor-circuit type.

Resistance welders are figured a bit differently. If the unit is not automatic, being manually operated at various duty cycles and actual primary currents (not the nameplate rated current), then the supply conductors must be sized at not less than 50% of the nameplate current rating. But if the welder is always, or nearly always, used at one particular actual primary current and duty cycle setting, then choose the proper duty cycle multiplier from column C in Table 8-1 and multiply that times the nameplate rated current. Choose the next heavier size conductors from Table 6-4. The overcurrent protection device should have a rating of not more than 300% of the rated primary current, and the disconnecting means can be a circuit breaker or a switch, rated at not less than the ampacity of the supply conductors.

In all cases, to find the nameplate current rating of the unit, multiply the rated kVA by 1000, and then divide by the nameplate voltage rating. The actual primary current varies and is determined by whatever heat tap or control setting is being used at the moment. The duty cycle is established by the design of the machine, which involves the number of electrical cycles during which the unit actually welds followed by the number of cycles it is idle. A particular unit might operate for 10 cycles (out of 60 per second), for instance, and then idle for the next 10. This would be a duty cycle of 50%.

MOBILE HOMES

There has been, and probably will continue to be, considerable confusion and argument about mobile home wiring. Part of this has been caused by attempting to define exactly what a mobile home is. According to the NEC, it is a factory-assembled unit designed to be readily movable upon its own running gear, and for use as a dwelling with no foundations. Not all communities agree with this, however. Then too, mobile homes may be set on cement block piers with the running gear removed, or otherwise made semi-permanent, or perhaps set upon a foundation with the running gear left intact. Or, the unit may be permanently installed so that it obviously cannot be moved without a major upheaval, and yet still be a mobile home under local interpretation.

So the line between a mobile home and a permanent dwelling changes with the local authority, and with it the electrical regulations also change. One thing seems reasonably certain: the so-called prefabs and modulars are generally considered as permanent dwellings and are electrically treated as such.

Another problem arises in that mobile homes are completely wired before leaving the factory. This means that when they arrive at their destination, local inspectors can find little to effectively inspect. Some units are accompanied by UL listing seals or similar affadavits that usually satisfy the authorities, but others do not. It is best to check with the manufacturer of your particular unit to avoid becoming involved in a squabble.

General Rules

There are a few general observations that apply with widespread equality to mobile homes. For instance, the distribution panel in a mobile home is just that; it cannot be considered as service-entrance equipment. The service-entrance equipment must be located outside and adjacent to the home, and it usually consists of a main overcurrent protection device, meter, main disconnect, and a feeder to the distribution panel inside the home. This in turn means that equipment grounding conductors must be isolated from the supply neutral in all cases, but the distribution panel enclosure must be a part of the equipment grounding circuit that runs isolated to the service-entrance equipment and eventually to the grounding electrode conductor.

Grounding is also most important, including not only all of the enclosures, device frames, appliance frames, and other non-current carrying metallic parts of the wiring system in the mobile home, but also the frame or chassis and the skin or shell (if metallic) or the mobile home itself. This is accomplished by tying everything to the equipment grounding bus in the distribution panel, which in turn is grounded by connection to the equipment grounding conductor in the serving feeder cable, or in the case of connection by flexible cord, by the green grounding conductor in the cord assembly. Nothing, under any circumstances, can be attached to the neutral conductor of the feeder.

In cases where the mobile home is supplied by means of a flexible cord, that cord must be a four conductor type especially designed and marked for mobile home use. The point of attachment is to an outside weatherproof socket mounted in the skin of the mobile home, from which point another cord or cable runs back to a permanent connection within the distribution panel. The length of the cord cannot be less than 20 feet nor more than 36 1/2 feet, and extension-type cords or jury-rigged hookups are not allowable. In some cases a mobile home is fitted out with two such cords, but they must run to two separate distribution panels that cannot be interconnected in any way.

Making Additions

Additions to mobile homes such as porches, enclosed entries, storage lean-tos, carports, and the like, should never be "wired" with extension cords. Instead, provide circuit extensions or new circuits direct from the distribution panel as needed. The same hold true for unattached outbuildings, which usually are best served with an underground circuit. Existing circuits including outdoor outlets should be protected by GFI circuit breakers. Outdoor fixtures must be weatherproof. And if you add new outdoor receptacles, they must be both weatherproof and GFI protected, either at the individual receptacle or at the head of the serving circuit.

Older mobile homes, especially smaller ones, are often woefully inadequate in the electrical department. Not only were many systems skimpy to begin with, but they may have been attacked by a bevy of back-yard mechanics down through the years. To make matters worse, rewiring, like remodeling, in a mobile home can be a difficult process because of close quarters, cramped spaces, and the construction methods used. In general, however, additions and rewiring in a mobile home follow the same rules as for any dwelling, and the principles and methods of doing the work also remain the same. If you are faced with this kind of a task, first investigate to see how you might be able to accomplish the job, and then check with the local inspecting authority to determine the specifics and any possible restrictions you have to contend with. If you are armed with a set of drawings and specifications that show just what you have in mind, very likely some reasonable solution can be found between you and the inspector and the material supplier.

When major additions are made to a mobile home, such that the extra living space is substantial and perhaps exceeds that of the mobile home itself, or when the mobile home becomes essentially a core for a larger building, then the entire structure will undoubtedly be considered a permanent dwelling under any local ordinances. All calculations and installations would then be done as in any single-family dwelling, under the NEC and other applicable local codes. The most practical method of wiring is to install a complete set of service-entrance equipment at some convenient point, wiring from that to the new sections of the building, and leaving the original mobile home distribution panel intact and served by a feeder from the main panel. The original mobile home wiring can be left intact as well, provided that it is adequate and in good shape.

MOTORS

Most of the motors found in the average residence are small. They are often an integral part of an appliance such as a refrigerator or a clothes washer, or used for short durations in power tools. And by the large, they are simply hooked up or plugged into a handy branch circuit, then forgotten. Most of the time we get away with this procedure because the loads are small, diversified, and usually in operation for only short periods of time. Also, domestic appliances are fairly well protected internally, and the NEC and local codes have regulated their installation so as to minimize any hazards and problems.

Motors and motorized equipment are sometimes problematical, however largely because they are heat-producing machines often subject to varying workloads. Any motor can, for instance, commence to bog down under its workload, generate more and more heat, and burst into flame long before a branch circuit overcurrent protection device trips out. Often special types of protection are needed along with control methods and controllers, specially designed circuitry for feeders, branches, starting, and control.

Even a partial discussion of the various kinds of motors and how they are designed, constructed, and operated would require a large volume. The intricacies of motor control, which includes circuit requirements, overload protection, electrical and mechanical pilot devices and circuitry, contactors and controllers, motor starting equipment and circuitry, and speed and positioning controls, would fill another.

Most common household motors can be better installed and protected than they usually are. Commercial and industrial equipment, which frequently finds its way into the residence or farmstead, almost always requires special and specific treatment, and invariably deserves as good safeguarding and protection as possible. After all, motor replacement is expensive, downtime exasperating at the least, loss of property from fire disastrous, and life irreplaceable. If your wiring plans include the installation of motors not in the ordinary appliance and power tool category, or if you are simply interested in the subject, familiarize yourself with Article 430 *Motors, Motor Circuits and Controllers* in the NEC. Refer also to the books in your local library on the same subjects.

POOLS AND FOUNTAINS

More and more homes these days are being equipped with various sorts of swimming pools, all the way from little plastic wading tubs for children to large and luxurious Olympic-size numbers. Fountains both tiny and huge, indoor and out, are also enjoying a resurgence, especially since complete fountain "kits" for all purposes and pocketbooks have come into the general consumer market with a great splash. And with both, a full complement of requirements and regulations for proper installation has also appeared. Incidentally, therapeutic pools and all pool and fountain associated equipment, whether fixed, portable, permanent or storable, are included.

General Precautions

To begin with, all of the usual considerations in normal residential wiring and wiring practices apply here. Also, any and all equipment used must be approved for the purpose. Most of this apparatus, at least that turned out by reputable manufacturers, bears the UL seal, and this is almost universally recognized and accepted by inspecting authorities.

As you might expect, a prime consideration with pools of all sorts is to insure, insofar as is humanly possible, that no user can accidentally be shocked or electrocuted. Obviously, in this situation quantities of people, water, and electricity do come into proximity at the same location. And just as obviously, the three don't mix well at all. This means that no receptacles can be located within 10 feet of the inside wall of the pool. In addition, any receptacles located within 15 feet of the inside pool wall must be equipped with a GFI breaker, either at the individual outlet or protecting the serving branch circuit. All outside receptacles (which must be weatherproof) installed during the past few years should by regulation have been protected this way. In the event that there are existing receptacles installed prior to that regulation and without ground-fault protection, they must be eliminated or converted.

Lighting Fixtures

There are several requirements for lighting fixtures near pools. Any fixture that is not less than 5 feet but not more than 16 feet on a horizontal line from the inside wall of the pool must be protected by a GFI breaker wired into the branch circuit at the distribution panel. Any existing fixture less than 5 feet from the inside pool wall must be at least 5 feet above the ground or deck, rigidly attached and supported, and also GFI protected in the branch circuit. If new fixtures are installed within 5 feet of the inside pool wall, or out over the surface of the water, they must be a minimum of 12 feet above the ground, deck, or water surface, and also GFI protected in the branch circuit. Any fixture, new or old, or any non-current-carrying metallic part thereof, if less than 16 feet from the surface of the water, must be GFI protected in the branch circuit.

Flexible cords may be used, with attachment plugs, to connect lighting fixtures or other equipment, fixed or stationary, for ease of service or removal. But note that the cords can be only 3 feet long and must be equipped with a #12 equipment-grounding conductor rather than the smaller sizes commonly used. Note too that this applies to equipment which draws 20 amps or less; heavier loads must be direct-wired.

One other general point: proper clearances must be kept between pools of any kind and overhead electrical conductors of any kind. A pool cannot be placed under them, nor can conductors be run over an existing pool, regardless of height. In this instance, a pool includes itself plus an area 10 feet from its perimeter, including any diving boards, towers, or platforms.

Permanent pools, whether for swimming or decoration, have a few additional requirements. Underwater lighting fixtures used in pools must always be approved for that use. There are two distinct types of installations. The *wet niche* fixture is a self-enclosed unit designed to be mounted in an open forming shell set into the pool structure, and it is surrounded by water. The *dry niche* type is also mounted in a forming shell, but the shell is watertight and provided with a sealed lens, so that the fixture itself remains dry. Either type must be located at least 18 inches below the normal water surface, and if facing upwards, must be protected with a guard so that a swimmer cannot touch the unit itself. If the fixtures operate at over 15 volts, the supplying branch circuit must be equipped with a GFI circuit breaker.

Forming shells for holding fixtures, around which the concrete structure of the pool is poured, must be approved and made of noncorrosive metal. Junction boxes connected to these shells can be formed of brass of some other corrosion-resistant metal or a suitable type of plastic. They have to be located at least 8 inches above grade level, pool water surface or deck level, whichever is highest, and at least 4 feet from the inside wall of the pool unless separated from the pool by a solid, permanent barrier. Obviously these above-ground junction boxes must be located where they won't be tripped over or walked into. If the supply voltage is 15 or less, then flush deck-mounted boxes can be used, but they must be filled with potting compound. All junction boxes must be equipped with a number of grounding screws. And where transformers or individual GFI devices are used, their installation follows the same rules.

Electrical Wiring

All wiring in the immediate pool area is done with brass or other noncorrosive rigid conduit, or approved rigid non-metallic conduit, threaded into the junction boxes and forming shells. If the conduit is nonmetallic, a #8 copper wire must be run with the other conductors to serve as an equipment grounding conductor. Every bit of metal must be securely and effectively grounded and bonded. Use a #8 copper wire to bond metal diving structures, ladders, reinforcing steel used in the construction of the pool, forming shells, and conduit or metallic piping or fixed metallic parts within 5 feet of the inner walls of the pool, all equipment used in the pool water circulating system, and so forth. Junction boxes and enclosures, underwater fixtures, all ot the electrical equipment used in recirculating and cleaning the pool water, and any other electrical equipment within 5 feet of the inner pool wall must be tied into the equipment grounding circuit, which in turn runs back to the serving distribution or main entrance panel.

It is worth noting that in most areas the pool electrical installations, which often are rather complex, are expected to be installed with above average care and attention to details. Usually local codes apply to pool construction and wiring, and these installations are quite likely to be inspected to a fare-thee-well, since the safeguarding of human life is paramount at poolside. Well-formulated and heavily detailed plans and wiring diagrams made up prior to the construction of the pool often save a lot of time and headaches.

Fountains

Electrical installations in or around fountains are usually not nearly so complex or problematical as those for pools. All of the supplying branch circuits involved must be fitted with GFI circuit breakers where submersible equipment, including lighting fixtures, is used. The only exception is if the supply voltage is 15 or less, produced by the properly approved transformer. Submersible lights must be below water level and guarded if they point upward, and they cannot be embedded in the fountain structure. All submersible equipment must be readily removable for service, but must also be attached securely. All metal parts, of course, have to be corrosion-resistant.

Unlike the situation with pools, flexible cords can be used in fountains to connect the equipment, provided that the cord is approved for the purpose and not more than 10 feet long. If the cords travel beyond the perimeter of the fountain, though, they must be run in a protective raceway. Junction boxes and other enclosures are permitted below water level provided that they are approved, equipped with the right seals and fittings, and properly anchored. The same holds true when they are in wet locations where spray might hit them. Underwater junction boxes must be filled with potting compound after connections are made and tested. Depending upon what type of equipment is used, a low-water cutoff switch may be required to prevent overheating and burnout if the water level becomes too low to serve as a pump coolant.

All of the electrical equipment within 5 feet of the inner wall of the fountain, all of the water recirculating equipment, and any subpanels supplying circuits to the fountain must be securely and effectively tied to the equipment grounding circuit. As with all subpanels, the equipment grounding circuit must be kept isolated from the neutral conductor and bus until it joins the grounding electrode conductor at the service equipment. All of the piping systems that serve the fountain must be bonded and tied back to the equipment grounding circuit. Of course, all cord-connected equipment must also be grounded via a separate grounding conductor in the cord assembly, with attachment by three-prong grounding plug on a two-wire circuit and a four-prong grounding plug on a three-wire circuit. And as with pools, wiring is done through corrosion-resistant rigid metal conduit or approved rigid nonmetallic conduit, with the proper fittings and seals.

Part 3:
Installation

Chapter 9
Tools and Equipment

Once you have finished all of the drawings and sketches to your satisfaction and have completed the material take-offs and checked and double-checked all of your work, the next step is to proceed with the actual installation of the wiring system. At this point you are faced with four principal alternatives, or perhaps some combination of them. For instance, you can elect to make the entire installation yourself, in which case you will need a full complement of tools and equipment to work with.

You might decide to do only the *roughing in*—the preliminary work of running wires, setting boxes and panels, and the like. Then you can hire a professional electrician to do the finish work and make all the final connections and tests. In this case, you will still need almost as full a tool box as for the complete installation.

You could also hire a professional electrician to take care of all the roughing in and other preliminary details, and then install devices, fixtures, appliances and finish trim, and make the final connections and tests yourself. This would require only a modest assortment of tools and testing gear.

The final possibility is to turn all of the paperwork over to an electrical contractor and turn him loose on the job. For this, the only tool you will need is a ball-point pen to make out a fat check to the contractor when the job is complete. If you make this decision, the remainder of this chapter will probably hold little interest for you.

Those who go ahead with all or part of their electrical systems will need a good assortment of hand tools and a certain amount of test equipment. There are a few power tools that will ease the work load a great deal, and also a number of specialized tools used by professional electricians that are helpful from time to time. The same holds true, of course, for extensive electrical remodeling and updating projects. In fact, working on existing structures usually requires an even greater variety of equipment.

HANDTOOLS

Trying to work without the right gear is always frustrating and time consuming, so you might just as well spend a few hours at the hardware store and outfit yourself properly before you start the job. When you make your purchases, don't scrimp. Cheap, poor quality tools do not hold up, they waste time, and the finished job is often not up to par. Remember that you will be saving a considerable amount of cash by doing your own work, so spend a few dollars extra from those savings to buy quality equipment. After all, good tools are a good investment and they will be with you for a long time, useful for many another project. Look for quality brand names, good construction and workmanship, solid warranties, and heavy-duty capacity—you will have few disappointments.

Screwdrivers

A variety of screwdrivers, even a dozen of them, is essential (Fig. 9-1). You probably could scrape by with only three or four, but you will wish you had more. Electricians generally use plastic-handled screwdrivers because they are tough and will withstand a lot of abuse (which they usually get), and the grip is electrically insulated from the metal blade, thus eliminating the possibility of shock. Though not as rugged, wooden-handled screwdrivers are effective from that standpoint too, provided that they do not have a metal shank that extends through or along the handle as some carpenter's screwdrivers do. In any case, the blades and shanks should be of quality forged carbon steel, not a soft, stamped-out material. The tips need not be hollow ground, but such screwdrivers are nice to work with.

Standard screwdriver sizes are designated both by the length of the shank, from the tip to the point where the handle starts, and the width of the tip. You will have need of a 1/8-inch × 2-inch (pocket-sized), a 3/16-inch × 3-inch (or 4-inch), a 1/4-inch × 4-inch, and a larger handled 1/4-inch × 6-inch. In addition, there are at least two other sizes of

Fig. 9-1 A good selection of screwdrivers, preferably with insulated handles, is essential to the wireman's tool kit. (Courtesy Utica Tool Co., Inc.)

standard screwdrivers that are often useful: the 3/8-inch × 8-inch (or 9-inch) for heavy work and a 3/16-inch × 8-inch (or 9-inch) for getting into tight spots. A *screwholding* screwdriver—one with a split tip and pressure lock or with spring steel screw-head clips—is not often used, but when it comes to starting a small screw in a spot too small for your fingers, this item is worth its weight in gold.

The chances are excellent that you will also need at least one *Phillips* screwdriver for those screws that have the cone-shaped, crossed slot instead of the usual straight slot. Phillips screwdrivers are designated by number. The smaller the number the smaller the tip, and usually the shorter the shank as well. For most purposes, a #2 with a 4-inch shank will be sufficient, but a #1 is helpful for smaller screws.

There are numerous special screwdrivers, but you probably would have little use for any except perhaps standard-bit "stubbies," or maybe an angle screwdriver. Most electricians also carry one more type, of a sort you will not find in any store. This screwdriver is usually bent, battered, burned, and has a tip that looks more like a tire iron. This sad looking device is variously used for prying things apart, jamming things together, levering heavy objects, applying glue, chipping concrete, poking holes in plaster walls, testing wires to see if they are "live" by short-circuiting them, bashing knock-out slugs out of electrical boxes, opening cans, and the like. They are handy items to have around.

Hammers

Another essential item is a hammer—a good one with a smoothly forged head that will not chip. Most workmen seem to favor the rubber-handled variety with a steel shank, simply because they are practically indestructible. A standard carpenter's hammer with moderately curved claws will do the job nicely. A 16-ounce head is about the right weight, giving a good overall size and balance. A smaller hammer will not do the job well, while a larger one is clumsy, awkward, and will wear you to a frazzle.

Once in a while a short-handled two-pounder comes in handy, but you probably can get by nicely without one unless you will be driving a number of screw anchors in concrete or cement blocks. You will need an eight-pound sledge for driving ground rods, but there seems little point in buying one unless you can foresee some future use for it. Rent or borrow one instead.

Saws

A standard *carpenter's handsaw* is a useful tool and has a blade about 2 feet long with 8 points (teeth) to the inch. You will also need a sturdy *hacksaw* and few blades to fit it. Use fine-toothed metal-cutting blades for steel, but change to a somewhat coarser type for sawing off heavy copper or aluminum cable.

The *keyhole saw*, with its long, pointed blade, is a must for cutting in wall boxes and such. The swivel-handle type is by far the most useful and versatile. With this saw, all that is required to start a cut in the middle of a blank surface is a half-inch hole. The most useful blade is a fairly coarse-toothed type designed for use in wood and similar soft materials. Plaster or gypsum wall board is so abrasive that it will dull any blade in short order, but even an old blade that is as dull as a hoe seems to work nearly as well in these two substances. The trick, then, is to use only one blade for plaster; keep another sharp blade on hand for wood work.

Pliers

Next to screwdrivers, the various sorts of pliers (Fig. 9-2) are the most widely used tools in an electrician's kit. You will not be able to get along without a pair of *wide-jaw diagonal cutting pliers*, usually referred to as *diagonals* or "dikes." The wide-jaw type provides the leverage necessary to cut through heavy wire or three-conductor nonmetallic cable with one chomp. Some electricians also like to use the somewhat smaller and lighter *narrow-jaw diagonals*, which are easier to

263

Fig. 9-2. You will have a steady use for at least two or three of the many types of pliers available. (Courtesy Utica Tool Co., Inc.)

use in confined spaces, but do not have the cutting power of the wide-jaw type.

Though not necessary to have, a pair of *needle-nose pliers*, either with or without a cutting blade, in about a six-inch length often proves useful. And the *right-angled needle-nose pliers* of about the same size can get you out of tight spots. You should also have a pair of standard *slip-joint pliers* on hand, for they seem to have a thousand uses.

A pair of *mechanics pliers*, sometimes called *water-pump pliers* or *pipe pliers*, is about the best tool to use for snugging up locknuts. There are several incremental adjustments on these pliers, which are available in several sizes, so they can be used for nearly any tightening job that does not need a high amount of torque. Two pairs working in opposition to one another work nicely for tightening up raintight fittings on electrical metallic tubing. The largest sizes, 16-inches or so, will even handle 2-inch fittings without danger of crushing either fitting or tubing.

Wrenches

Certainly a selection of various types of wrenches would be nice to have around, but if you do not already own some, there is probably no need to purchase them. Mechanics *open-end* and *box-end wrenches* are quite useful in some circumstances, but a small *Crescent* (adjustable) *wrench* of about six inches in length will most likely see you through your installation. There are rare occasions when only a mechanic's *socket wrench* of some particular size will do the necessary job, but the best course is simply to wait until the occasion arises; it may not.

If you will be installing any amount of rigid conduit during the course of wiring your house, you will need a pair of pipe wrenches. The *Stillson pipe wrench* is the most common type, with its continuously adjustable toothed jaws, and it works well. The *chain pipe wrench*, however, is lighter, will grip any sort of odd shape, causes less damage to the surface it grips, and has a much wider range of adjustment. The cost is nearly the same as a Stillson, and its utility value is considerably higher.

Miscellaneous

There are numerous other small handtools that you should add to your kit, some essential and others merely helpful (Fig. 9-3). You will need a measuring device of some sort, and the sliding *steel tape measure* (M) with a lock and power return, and a belt clip seems the most useful. A *punch* or *drift* (I) is a handy item, and so is a *cold chisel* (H). You will need one or more sizes of *wood chisels* (O) if your plans include mortised electric locksets or door switches. A small *spirit level* (L) and a *carpenter's square* (P) will help you keep things straight. And the first time you bore a hole through a joist over your head you will notice a definite need for *safety goggles*.

Pipe Cutters and Threaders

EMT can be cut with a *pipe cutter* (Fig. 9-3R), but this leaves a sharp inside ridge of rolled metal that is difficult to ream out. Instead, use a hacksaw, then a tapered *hand reamer* to get rid of the burrs, and a flat metal-cutting file to knock off the ragged outside edge of the cut.

A *pipe vise* is handy but not necessary for working EMT, but with rigid conduit the reverse is true, particularly in the larger sizes. Conduit must be threaded, and this means that it must be firmly secured in a pipe vise. To do the threading, you will need a *die stock* and the appropriate dies for whatever size of pipe you are using. Once the pipe size exceeds one inch, however, a *power threader* is in order. The simplest way to handle the whole situation is to make up a list of the pieces you need and let your supplier take it from there.

As to the cutting of conduit, a hacksaw will do the job, but a pipe cutter makes the task much easier. The inside must then be reamed, just as with EMT, to eliminate any burrs and sharp edges. The outside of the cut will be taken care of during

Fig. 9-3. There are a number of miscellaneous hand tools that may prove useful to you. Shown here are (A) hacksaw. (B) mason's trowel. (C) bitstock. (D) long auger bits. (E) short auger bits. (F) expansive bit. (G) prick punch. (H) cold chisel. (I) drift punch. (J) putty knife. (K) metal file. (L spirit level. (M) measuring tape. (N) Yankee screwdriver. (O) wood chisels. (P) try square. (Q) scratch awl. (R) tubing cutter.

the threading process. If your supplier is not equipped for cutting and threading conduit, you can probably enlist the aid of a local plumber. The equipment used, the pipe sizes, and the threads are the same for both trades. Remember, though, that pipe used for plumbing purposes is not acceptable for electrical installations under the NEC.

Plaster Tools

Depending upon the specifics of your particular job, you may want a small *putty knife* (Fig. 9-3J) for patching plaster or wallboard, or for applying roofing tar to a through-the-roof service mast flashing sleeve, or "roof jack." A small *mason's trowel* (Fig. 9-3B) can be used to patch concrete or cement blocks in the event of an oversize conduit or cable hole. If you will be working with armored cable, a pair of compound-action *tin snips* with narrow jaws will work more easily than diagonals for cutting the tough armor sheath.

266

Bits and Braces

If electricity is not yet available at the job site, or if you feel that you need a lot of exercise, you can use a carpenter's *bitbrace* or *bitstock* (Fig. 9-3C) to bore the holes in the framing members that will carry the cables. A bitstock is a time-consuming, knuckle-busting, sweat-producing machine that might have been designed during the days of the Spanish Inquisition, especially when it comes to boring several hundred holes through tough spruce or fir, but it will do the job. In the early days of this century before the advent of the electric drill, all buildings were wired with hand-drilled holes, but that doesn't make the process any easier.

The proper bits (Fig. 9-3D, E, F) to use with a bitstock are called *auger bits* and they are available in sizes of 1/4 to 1 inch in diameter, in steps of 1/16 inch. You probably will need only three sizes: 1/4 inch, 1/2 inch, and 3/4 inch. The small size is good for running a single low-voltage wire; the other two will cover most of the remaining needs. However, there would be no harm in measuring the diameter of the cables you actually will be using and then sizing the bits to suit; there are many variations in the nominal sizes of cable on today's market.

There is one job that a bitstock does quite effectively, and that is boring large-diameter holes from an inch up to three inches. Most jobs require only a few such large holes, and drilling them manually is generally the most practical method. For this you will need a two-piece expansive bit (Fig. 9-3F), which has an adjustable blade at its tip. A scale is marked right on the blade so setting the bit for the proper size hole is merely a matter of loosening a setscrew, lining up the right indicating marks, and tightening the screw again.

When selecting a bitstock, make sure that the grips are fitted with ball bearings and that the chuck will ratchet in either direction, for drilling partial revolutions when quarters are too cramped to make a full 360-degree sweep with the stock. This situation crops up much more often than you might think, as when you must drill through a floor right next to a wall.

Hand Drills

For manually drilling holes smaller than 1/4 inch in diameter, you will need a *hand drill*. The larger versions of these are called *breast drills*. (Some models come complete with an assortment of small bits.) These tools, which operate much the same as an eggbeater, are amazingly effective and can be used to drill almost any material.

The *Yankee screwdriver* (Fig. 9-3N) or automatic-return screwdriver, is another tool which has a variety of uses. They usually come equipped with a set of screwdriver and drill bits that are slipped into a collect shuck. These are handy for drilling the occasional screw starter hole in wood or similar relatively soft materials, and they can drive and remove wood and metal screws with much less torture than a regular screwdriver.

You may have to drill a couple of small holes in a masonry surface in order to mount a box or pipe clips. If power is not available or you do not have the necessary power equipment, you will have to use a *star drill*. This tool is almost identical with the one that miners used back in the early hard-rock mining days for drilling powder holes, and it works quite well. A star drill is simply a heavy steel rod with one end formed into a four-bladed, shallow-pointed bit that looks something like a squat Phillips screwdriver tip. With one hand you hold the drill upright against the surface to be drilled, then you rap the striking end with a hand sledge held in the other hand. After each blow with the sledge, turn the bit 50 or 60 degrees, and you will have a hole in a surprisingly short time.

Drilling metals or other hard materials usually requires a slight indentation to hold the tip of the bit on place. This is true regardless of the type of drill or bit being used. Without this indentation, the bit will slip and wander away, resulting in an off-center or crooked hole. By centering the point of a *prick punch* or *center punch* where you propose to drill, giving it a sharp rap with a hammer, you can avoid the problem. Center punching usually is not necessary in wood, though it is sometimes helpful. You can make a starter hole in soft wood by driving a scratch awl, or even an icepick, into the wood a short way; then simply drive the screw in. Hard or splintery woods usually need a drilled pilot hole in order to avoid splitting.

POWER TOOLS

As you may have gathered from the foregoing discussion of hand drills and their accessories, an electric drill is probably the most useful power tool of all for an electrician. They save so much time and make the job so much easier that it is almost unthinkable to go without if there is any choice.

Electric Drills

The standard 1/4-inch size electric drill (1/4-inch refers to the maximum size bit shaft that can be clamped in the shuck) is not the best one to use. In fact, it is doubtful if you could complete an entire residential wiring operation, especially a large one, without burning out at least one drill. They simply do not have the capacity or the power to handle big jobs on a continuous-use basis. The bearings in the cheap home-shop models are usually of the bronze bushing variety that will not take constant strain. Few 1/4-inch drills have enough power to drive large wood bits, and they bog down easily. In addition, most of them will overheat rapidly, even to the point of becoming too hot to hold, after boring only a few 3/4-inch holes. But when used for their intended purpose, drilling holes of 1/4-inch size or less, principally with twist drills, they do an admirable job.

A 3/8-inch capacity drill is the smallest you should consider buying for use on a complete residential wiring job. It should be of heavy-duty construction and rating, 1/2-horsepower at least, with ball bearings throughout. A reversing feature is sometimes most helpful, as when a bit gets caught deep within a hole, and a variable-speed capability makes drilling a gread deal easier, both on the drill and the driller. A good 3/8-inch drill will easily handle wood-boring bits up to an inch in diameter, and with care can be used with bits as big as 1 1/4-inches. At the same time, you can use it with twist drills as small as #50, if you have steady hands, or as large as 3/4-inch with 1/2-inch shanks. They have the power to drive large *hole saws, rotary files*, and *rasps*. At the same time, they are small and light enough to be easily handled and maneuvered in tight spots.

There are several compact 1/2-inch capacity drills on the market that you might like to consider. Most of them, however, have no more and sometimes less power than a good 3/8-inch drill. About all these can offer, for the most part, is a chuck of 1/2-inch capacity. But full-sized models are another story. These drills, the better ones, are tough and solid, have more power than most people would ever need, and are generally longlived. It is likely there are no tough jobs around the house that one of these rigs could not handle with great ease, unless you are building your home from old bridge girders. However, they are heavy, tiring to use, and too large to get into odd corners and crannies. And when the drill bit catches—as it invariably does at some point—the 1/2-inch drill motor does not stop. If keeps right on cranking, with enough torque to literally snap the wrist of the unwary operator, or twirl him around in a complete circle before he can let go of the trigger switch.

When selecting a 1/2-inch drill (all of which are fairly expensive) you might just as well get the best you can afford.

Look for ball and needle bearings, a rating of at least 1/2 horsepower but preferably 3/4 horsepower, stout construction with a removable top handle for extra control, and the reversing feature. This latter is almost essential, for otherwise you will find yourself trying to crank out a trapped bit with a pair of pliers. If you are faced with a huge number of holes to be bored, or thick beams, logs, stone or concrete to be drilled, this is the drill for you.

Drills and Bits

Metals and very hard materials should be drilled with high-speed drills made from high-quality heat-treated steel. AISI type M-2 steel is probably about the best, and a chrome-vanadium steel is far preferable to high-carbon steel. Buy good ones from a reputable dealer because the cheap varieties will dull unbelievably fast and break all too easily, usually down inside a hole you have almost finished drilling.

There are three different sizing systems of twist drills, apart from the metrics. First there is the *fractional* system, which constitutes the most familiar and widely used type. These start at one-sixteenth inch and go up the scale in sixty-fourths to a maximum size of one-half inch, then further up in varying shank and tip diameters to over an inch.

The other two systems are primarily used by machinists and tool-and-die makers. One is the *number* system, which starts at #80 (0.0135 inch) and goes up to #1 (0.228 inch). The second is the *letter* system, with the drills designated A through Z.

Fractional drills will probably satisfy most of your requirements, but if for some reason you need tapped holes in metal, you will also need one or other of the number or letter drills. The proper combination can be found on a tap drill chart. For instance, if you want a 75% thread-height tap for a #6-32 NC (National Coarse thread) machine screw, you will have to drill a hole with a #36 drill, then tap with a #6-32 tap. A 5/16-18 NC bolt would require a hole drilled with an F drill. A 7/16-20 NF (National Fine thread) bolt needs a hole drilled with a 25/64-inch drill.

Fortunately, boring holes in wood requires much less fussing around. The most commonly used bits for this purpose are known variously as *speed bits*, *spade-drill bits*, or simply *woodboring bits*. A speed-bit consists of a flat, thin blade of the same width as the diameter of the hole it bores, with two slightly angled, opposing, cutting edges and a centering/starting tip. They are designed to be used in power drills and cut quickly and cleanly as long as you remember to keep them lined up with the hole. They are quite inexpensive, available in diameters from 3/8 inch to 1 1/2 inch, in 1/16-inch steps, and can readily be resharpened several times with a small flat metal file before they become useless.

You can also buy power wood bits, which look much like the auger bits used with bitbraces. These do cut a somewhat cleaner hole than a speed-bit and will bore straight without chattering or gouging. On the other hand, complete cleanliness of the holes has little significance in roughing in electrical wiring, and the cost of this type of bit is somewhat higher. Also, the variety of sizes is smaller.

Speed-bits are not very long; by the time they are chucked in the drill, a hole depth of only a little over four inches can be bored. Obviously this is somewhat inconvenient if you have to drill through a six-inch beam, especially if you can't get to the opposite side of the beam to drill back through. The solution to this problem is called a *bit extension*—a slender steel rod in which the bit is secured with an Allen setscrew. With this device chucked in the drill, the effective working length is extended to about 16 inches, enough for almost any task. If necessary, two extensions can be coupled together, but the drill should be run at low speeds to prevent whip.

Drilling masonry materials, such as brick, concrete, cement or cinder block, quarry tile and the like, or any type of stone-work, requires the use of special bits called *masonry drills*. These bits, which are made in the smaller sizes, consist of a tough, short steel shank with deep spiral grooves to carry away the drilling dust, and a two-bladed, shallow-angled carbide tip. They are made for use in a power drill running at low speed. The tips will chip upon sudden hard impact, but otherwise they do a fine, if somewhat slow job.

Ceramic tile or any similar material with a glazed vitreous surface must be drilled with yet another type of bit. Instead of a carbide tip, this bit has a diamond tip, and the drilling process is a tricky one. Because of this and the fact that such drills are obviously expensive, drilling tile is best avoided unless absolutely necessary.

Not infrequently it is necessary to drill a large hole through masonry or stone walls, a miserable chore at best. But sometimes this is the only way to get from here to there with a conduit run, service raceway or cable, or an underground line. If the hole needed is not too large—in the neighborhood of one to four inches, for instance—an electric *impact drill* will usually do the job. This is an expensive rig, so the best course is to rent. If the hole must be quite large, or the material is particularly tough or thick, you may have to go to a full-fledged *pneumatic jack-hammer*. In this case, the easiest way out is to hire a local contractor who has the equipment and the knowhow.

In everything but masonry you can drill large size holes (over an inch) with *hole saws*. These come in a wide range of sizes and are designed to be used interchangeably on a single mandrel. They cut a neat clean hole, but unfortunately can only be used in materials up to 1 or 1 1/2 inches in thickness, depending upon the manufacturer. Of course, that depth can be doubled if you can bore in from the reverse side of the workpiece and meet the original bore with some degree of precision.

Saws

In most residential electrical applications, there are only a couple of power saws that are of much use, and you undoubtedly could get along very well by staying entirely with handsaws.

An electric *saber saw* or *jigsaw* is helpful in cutting out holes for wallboxes and the like, especially when you have to cut a great many of them into wood paneling or planking. They also make short work of large cutouts such as those needed for some types of recessed lighting fixtures or flush-mount panel boxes. The one major drawback is that a saber saw cannot be used in close quarters where there is insufficient room to move the saw around, so there are many instances where a handsaw has to be used anyway.

The *reciprocating saw* is a similar but somewhat more versatile (and expensive) saw. It is bigger and heavier than a saber saw, as well as being more powerful, and uses larger blades. But because of its design, with the blade protruding from the front like a logger's chain saw, it is quite maneuverable and easily controlled. In addition to the same work that a saber saw will do, you can use a "recip" saw to cut heavy cable, electrical metallic tubing, and rigid conduit with remarkable ease, provided that the workpiece is securely clamped down.

SPECIALTY TOOLS

There are a few bits and pieces of equipment of a specialized nature often used by professional electricians that will help make your installation chores easier. Though seldom found in any but the most well-stocked hardware store, you should be able to purchase them through a good electrical supply firm; if not directly out of stock, then at least through their catalogs.

Knives

The *electrician's knife* is an overgrown jackknife with only, as a rule, one blade. The grip is large and sturdy, of either wood or notched plastic, and the frame is heavily constructed to withstand a lot of hard use. The folding blade is large and heavy with a broad, hooked end that curves into a sharp point; a sturdy metal loop is attached to one end of the frame so the knife can be hung on a belt hook or tool pouch clip. This is the tool you will need for slitting the tough plastic outer covering of nonmetallic cable, or the jackets of the larger cables such as those of the SE type—that is the purpose of the hooked tip. In addition, you will need this knife to strip away the insulation from each conductor before making a connection.

In practice, you can actually use any sort of knife that will do the job; the electrician's knife just happens to be the one that is best suited. A linoleum or carpet-layer's knife is similar in shape but less convenient because it does not fold up for quick storage in a pocket, nor does it have the loop hanger. In addition, the blade is a bit thin and easily nicked or bent, and it tends to waver off course under heavy pressure when cutting tough insulation coverings.

A standard *utility knife* of the sort you can find in most hardware stores will also work if used with care. This knife has the advantage of having replaceable blades, but the blades are razor sharp and must be used with care in order to avoid damaging the conductors. They are also easily broken or nicked and are really too flexible to control easily.

Strippers

Some electricians simply use their diagonal cutters to strip the insulation from the smaller conductors by first scoring a complete circle around the wire, about an inch from the end, then pushing the material ahead until it breaks away and flies off. This, however, is not a good practice since the wire itself can be scored or nicked, as a result it will break off if bent a couple of times or twisted too far in the process of making connections. You are much better off using a wire stripper (Fig. 9-4) designed for the purpose. It will give you a trouble-free strip every time, not to mention the fact that the process is much easier.

Strippers are readily available in a number of different models. This will usually accept a range of wire sizes from #10 copper to #16 copper, or from #12 to #18. Some models also have a *ripper tooth* for slitting the outside casings of cables. Less expensive types, which look somewhat like a pair of

Fig. 9-4. A mechanical wire stripper does a faster, neater and less problematical job of stripping than any other method. (Courtesy Greenlee Tool Co.)

pliers, score the insulation cleanly at a squeeze of the handles, but do not touch the conductor itself; the cut end can then be pushed off. Fancier models automatically grip the wire, slice the insulation, and when the handles are released, pop the scrap of insulation away like a rifle ejecting a spent shell casing. Whatever type you choose, you will find a stripper to be a time and energy saver. And in a complete residential installation, there is a horrendous number of conductor ends to strip.

Cable Cutter

Armored cable can be mean to work with. As previously noted, the heavy spiraled metal sheath can be cut with tin snips; diagonals will also do the job, although with somewhat more difficulty. Either method, however, leaves sharp or bent edges that you must then fiddle around with and clean up.

There is a special plier and cutter combination that makes the job easier and neater. Sharp, slender jaws cut and trim the material cleanly. Then you simply insert the cable into a specially formed circular inset between the handles, and with a couple of easy squeezes, you re-form the bent sheath.

Taps

Sometimes you will find wall boxes or other items with threaded holes that, for whatever the reason, are fouled up and will not accept a screw. Or you may inadvertently start a screw crooked or wind it down too tight, stripping out the threads. This requires either replacement or rethreading; the latter course is usually the most practical. The answer to the problem is a set of taps in the more common sizes, together with a tap holder.

The electrician's *tap tree* is much handier and a good deal less expensive, though. This device resembles a screwdriver, but the shank is a series of continuous taps starting with a small size and graduating to the larger, usually in the four most commonly used sizes in electrical work. All you have to do is insert the shank into the stripped hole until you arrive at the matching size, then turn the tool clockwise with care until the threads are restored. If you are unsure of identifying all the various thread sizes, outfit yourself with a thread gauge, then you will always be positive.

Benders

You do not need to use fittings for every bend or change in direction you must make with EMT or rigid conduit; the pipe can be bent. Special tools are made for this purpose, both manual and hydraulic (Fig. 9-5). The manual sort is used for small pipe size and type. For instance, if you are using 1/2-inch EMT you will need a 1/2-inch EMT bender (or *hickey* as it is sometimes called). Rigid conduit in 3/4-inch trade size requires a 3/4-inch rigid pipe bender, and so forth.

Some benders are designed to handle two sizes; 1/2 inch and 3/4 inch, for instance. They are designed to cradle the pipe securely in a half-moon shaped flange so that when pressure is applied on the handle, the pipe actually stretches and bends without kinking. They are easy to use after a little practice and will bend to any angle up to 90 degrees.

Fig. 9-5. Special equipment, either manual or powered, is used to bend EMT and rigid conduit. (Courtesy Greenlee Tool Co.)

Larger pipe sizes have to be bent hydraulically. This is a common enough procedure in commercial and industrial jobs, but seldom necessary in residential work. Here, even when large pipe is used, the piping can be planned to use fittings and preformed sweep ells. If for some reason you should need a special bend in large diameter pipe, your only logical recourse would be to have someone else do the bending for you. Hydraulic benders are most expensive, and only the larger suppliers and contractors have them.

Snakes

Another item familiar but hardly dear to all electricians is the snake, which can be every bit as mean as the name implies. Also called a *fish* or *fishwire*, this is nonetheless an essential piece of equipment wherever conductors must be run in pipe or when rewiring existing structures. Most snakes, which range in size from 25 to as much as 30 feet in length, are a flat ribbon of springy, tough steel from 1/8 to 1/2 inch wide. Some of the short ones self-store in a special closed reel, but most are kept in a tied coil and unrolled directly into the pipe or cavity to be fished. If let loose all at once, a new snake that still has its full springiness will uncoil in all directions and fill the room before the user has a chance to duck.

Snakes are used in two principle ways. When conductors are run in a pipe, something must be used to pull them through; the wire itself usually does not have enough rigidity to be pushed through except in runs only a few feet long. So the snake is pushed through the pipe from one end of the pipeline. The conductors are looped and taped to a hook on the end of the snake, and then drawn back through the pipe.

The second use most often occurs in rewiring but also sometimes in new construction, especially if building is nearly complete. This involves the fishing of wires through the spaces between floor joists, closed rafters or wall studs, through shallow crawl spaces, or any other inaccessible area to get from one point to another.

Sometimes a single snake can be pushed, jiggled, turned, or otherwise wiggled around and make it pop out of a hole a good many feet away. Then the cable can be tied on and pulled through. On other more frustrating occasions, two men using two snakes, each with a large hook bent into the end, must start at opposite ends of the proposed path of the cable and try to connect with each other somewhere in the middle. One snake then pulls the other out of the cavity. The wire is tied to the first snake and pulled back through. The process requires a great deal of time and patience, not to mention a little luck, but sometimes it is the only way to run concealed wiring.

Stud Drivers

In the event that you will be installing EMT or conduit, or mounting boxes on concrete or other masonry surfaces, you might want to use an *impact driver*. This is a small hand-held tool that sets studs or special short nails. The driver is loaded with a special stud and then struck repreatedly with a two-pound sledge. Boxes or pipe clamps are then mounted on the studs, and secured with a nut. Nails, of course, are first run through holes in the boxes or clamps and then driven home, a somewhat more awkward procedure that also means the boxes or clamps are not readily removeable.

If your plans call for a great deal of this sort of work, however, use a *Ramset gun* or something similar. This is an expensive item, so you might want to rent one if possible. It must also be used with great caution, for it can be a lethal weapon—it actually is a gun. The user first loads a special stud or nail into the end of the barrel, then a .22 caliber blank cartridge into the chamber. The cartridges are color coded for various charge strengths, the choice of which is determined by the relative density of the material into which the stud or nail will be driven. Brick requires one charge, poured concrete another, and so on. Then the user positions the tip of the stud against the surface, lowers a shield, and pulls the trigger. The force of the charge drives the stud home and if the charge has been correctly figured, to exactly the right depth.

Drills

Roughing in the wiring in a large frame house usually means drilling literally hundreds of holes through which the cables are run. Many of them are bored in floor joists or other members that are placed on 16-inch centers, just a bit too close together to allow a drill and bit to fit between them. The result is that all the holes are drilled at an angle to the course of the wire, which leaves sharp edges and makes pulling the cables difficult. This is not only aggravating, but can also cause damage to the wires themselves. To get around this problem, many electricians use a special power drill with a *right-angle* head and chuck. The drill easily fits into the space between studs or joists, and all the holes can then be bored straight and true, parallel to the course of the wires.

The wireman who elects to use a bitstock and auger bits for his hole drilling may be faced at some point with the problem of too short a bit. The solution is to use *electrician's* auger bits instead of the standard ones. These are usually about 18 inches long and shaped somewhat differently than standard bits. They do a fine job, but there are two drawbacks:

they bend easily and so must be used with care, and they are available only in a limited range of sizes.

Stud Locater

There are times, usually in rewiring work but occasionally in new work as well, when it is necessary to locate a stud or some other structural member in a wall or floor. Or more properly, to locate the space between two such structural members so that a hole can be drilled and wire drawn into the cavity, or a box or device mounted without hitting such a member.

A stud locater is nothing more than a magnetized needle mounted upon a swivel pin and enclosed in a protective case. This will often do a better job than the time-honored (and chancy) procedure of thumping on the wall or floor and listening for a hollow sound. As you move the locater slowly along the finished surface, the needle will abruptly swing as it passes a nail, thus spotting the stud or rafter to which the covering is attached. This device will not work in all cases; brick veneer is too thick to read through, and a plaster wall backed with expanded steel mesh would throw it completely out of whack. But for the dollar or so involved, it is a handy gadget to have around.

Punches

Most electrical boxes have sufficient numbers and sizes of knockout openings built into them by the manufacturer. But there are occasions when one more is needed, or a small hole needs to be enlarged. The tools for this sort of operation are called, fittingly enough, *knockout punches* (Fig. 9-6) and they are available in round sizes to match pipe fittings or cable connectors from 1/2-inch to 4-inch trade size. They are also made in geometric shapes, convenience outlet shape, and numerous others. They consist of three parts: a hollow die, a cutter head that slips down into the die, and a center bolt. To cut a clean, true hole, first drill a pilot hole for the center bolt. Then slip the bolt through the cutter and on through the pilot hole, threading the die onto the end of the bolt. A few turns with a wrench on the bolt head, and the cut is made.

Large size punches, those above 1 1/4 inch or so, are usually operated with a hydraulic ram arrangement. A special piston on a slave cylinder replaces the center bolt, and the cutter and die are drawn together by pumping a larger master cylinder. This tremendous mechanical advantage makes short work of even the toughest punching jobs. Cost of the equipment is rather high, though, so unless you are faced with a large job

Fig. 9-6. Knockout punches are used in standard trade sizes to create or enlarge holes for conduit and fittings in electrical enclosures. Both manual and hydraulic types are available. (Courtesy Greenlee Tool Co.)

of exceptional complexity, or think that you will have continued use for a hydraulic knockout set in the future, or simply like to collect tools, this would be an unnecessary investment.

Stakers

There is another type of tool costing only a few dollars that often is an asset to even the part-time electrician's tool box and is equally useful in either repairs or new work. This is a special plier-like device called a *staker, stake-on* tool, or *terminal pliers* (Fig. 9-7). There are several brands available, each with slight variations, but their purpose is to crimp solderless wire terminal ends onto conductors. The terminals themselves come in many varieties, including splice connectors, both insulated and non-insulated. Usually one type of staker is used for the insulated type of terminal ends and another for the noninsulated.

While not essential for a straightforward home wiring project, the need for a staker and a selection of terminal ends does arise sometimes. The home handyman will find considerable use for a kit in appliance repair, construction of special panels or electromechanical devices, and automotive electrical work.

279

Fig. 9-7. Special pliers, sometimes called "stakers", are used to apply solderless connectors to conductor ends. Usually one type is needed for insulated connectors and another for noninsulated types.

Keys

Some pressure connectors in panel boxes are fitted with Allen setscrews, either along with or instead of the usual setscrews with a screwdriver slot. Instead of a slot there is a hexagonal depression in the body of the screw. To work with these you will need special wrenches called *Allen wrenches* or *hex keys*. You can buy individual sizes—they come in both decimal and numerical fractions of an inch—or complete sets. The price of even a large set is so low that having a complete array makes more sense from the standpoint of convenience and flexibility.

Soldering Equipment

Though most electrical connections are no longer soldered and then taped as they were years ago, you still may find a need for a soldering iron. This is especially true in some aspects of repair work. Electric soldering irons are available in a great number of varieties, and one works about as well as the next. Usually a small pencil type iron will do an adequate job and is less cumbersome than the large 100-watt and up types. One of the handiest is the soldering gun, which is held like a pistol and features a trigger switch, almost instant heating of the element, and the additional nicety of a small light bulb aimed directly at the work point. Where power is not readily available, you can use a propane utility torch equipped with a soldering iron head on the burner.

Pouch

One of the handiest gadgets that you can supply yourself with, though certainly not a necessity, is an electrician's tool pouch (Fig. 9-8). There are several variations on this theme, but most pouches are constructed of heavy leather, with a sizable single inner compartment and four or five outer pockets. Some also have a leather loop in which a hammer can be hung, a small spring clip for a knife, and a T-chain to hang a roll of tape on. They have either a belt loop or belt slots, and most workmen prefer to use them with a heavy slip-buckle web belt for easy removal without using the belt on their pants.

By using a tool pouch, you will always have with you the several tools that you need constantly: two or three screwdrivers, hammer, waterpump pliers, diagonals, knife, or whatever other combination the job of the moment dictates. This saves a lot of wandering back and forth looking for whatever you last laid down, not mention wear and tear on hip pockets.

Fig. 9-8. A loaded tool pouch on your hip will save time and frustration, and make the job a bit easier.

Medical

Oh, yes, there are two more items that should be a part of every sparky's kit—a pair of tweezers and a box of Band-aids. You will be amazed at the number of wood splinters and metal slivers that will find their way into your fingers. The Band-aids are also helpful with the usual scrapes, cuts, and blisters common to every electrician.

TESTING EQUIPMENT

Once the installation or certain segments of it are complete, you will need testing equipment to check out your work before energizing the system. Testing is also essential in troubleshooting, of course, and some of the same equipment can be used for circuit tracing, to see what goes where for purposes of remodeling, adding on, or general up-grading. Most of this equipment is inexpensive to buy, some of it you can put together yourself either from scratch or in kit form, and all of it is easy enough to use.

Voltage Testers

One of the simplest items is the pigtail amp (Fig. 9-9G). All you need is a so-called pigtail socket, which is a light socket with two short insulated leads, or *pigtails*, molded into it; and a light bulb that has a base the same size as the socket, a voltage rating the same as the circuit or device to be tested, and of any wattage. Usually a 7 1/2-watt bulb is used because it is small, throws little heat, and is not quite so prone to breakage. By touching one lead to a neutral and the other to a hot wire, you can quickly see whether or not the circuit or device is dead or energized.

A variation on this tester is the plug lamp (Fig. 9-9F), which consists of a light socket with plug prongs protruding from the back in place of leads. Obviously this can only be used for checking standard convenience receptacles, but it affords an easy way to check a long series of newly installed outlets at a rapid pace. Of course, nearly anything can be used for this purpose, such as an electric drill, a small table lamp, any item of the proper voltage that will plug in and light up or make a noise.

Most electrical supply houses carry neon test lights (Fig. 9-9D). These are small, molded-plastic items with an enclosed neon bulb, two leads that end in probe tips with small plastic grips on them. These are more advantageous than either of the above test lights because they are rugged, small, easily carried and used without fear of breakage, and have the capability of testing a wide range of voltages instead of the test lamp's single voltage. Most of them are rated for 90 through 500 volts, so any high-voltage system encountered in residential use can be readily checked. Their use is identical with the test lamps—touch one lead to a neutral and the other to a hot, or plug the leads into the lots of a convenience outlet, and see if the bulb lights. One drawback is that the light is so feeble, only a glow, that it cannot be seen from a distance.

For low voltages, up to 24 volts or so, you will need a different sort of tester. Low-voltage testers (Fig. 9-9B) are equipped with a small incandescent bulb, and in some of them the bulb can be replaced. This can be a handy feature when the probes are inadvertantly placed across a line carrying a voltage higher than the maximum rating of the bulb. Voltages much lower than the maximum rating do no harm—the bulb simply burns less brightly as the voltage goes down. Low-voltage testers may be more readily available at your automotive parts supply house than your electrical dealer. Check the voltage rating before you buy, though, because some of them are good for only 12 volts, even though a few automotive electrical systems operate on 24 volts.

Fig. 9-9. Various types of simple testers. (A) pocket penlight continuity tester with plug-in lead. (B) low-voltage tester. (C) neon receptacle tester. (D) neon voltage tester. (E) continuity tester with battery, lamp, and probe. (F) plug-in socket and bulb. (G) pigtail socket with bulb.

Continuity Testers

Checking a circuit that is not energized is done with a continuity tester (Fig. 9-9A and E). This device will check a circuit to determine whether it is *open* (the electrical flow is incomplete) or whether it is *closed* or *short-circuited* (the electrical path is continuous).

There are a number of styles of continuity testers, but they all work the same way. They are battery powered, usually with one or two penlight cells (A size), and contain a suitable bulb. At one end there is a long metal probe, at the other a test lead with another probe or an alligator clip. The two probes are touched to opposite sides of the circuit to be tested, and if the circuit is complete and the path which the electricity should follow is intact, the bulb will light up. This is an invaluable aid, especially in troubleshooting, but it must be used with some caution. If the probes are placed on an energized circuit, you will have a lively time, at least for the instant it takes to burn out the low-voltage bulb.

There is another device, somewhat elderly but still very much in use, that you can use for continuity checking—a bell or buzzer. Widespread use of this tester, usually home-brewed in origin, is what gave rise to the old expression, "ringing out the circuit." When you ring out a circuit, you simply check it with a bell to see if it is either open or closed, depending upon the requirements. You can make a circuit ringer by mounting a small low-voltage buzzer or doorbell in any convenient box or case, together with a suitable battery. Attach one lead from one side of the battery to one terminal of the bell, then connect a test lead or probe to the two remaining battery and bell terminals. With a probe on each side of the circuit to be tested, the bell will ring if the circuit is closed and will not if it is open. Simple.

Outlet Testers

Most of the 115-volt convenience outlets installed today are of the grounding or three-prong type, rather than two-prong. These are connected with three separate conductors. If any of the three get mixed up somewhere along the circuit, though the outlet may still function, a potential hazard exists from the improper connections. To check this out, use a plug-in indicating circuit analyzer (Fig. 9-9C). This little gadget looks like an oversized three-prong grounding plug. You simply plug it into each and every receptacle on your circuits, and it will indicate by means of a series of tiny neon lights that everything is all right, or that you have an open circuit, or that you have reversed some conductors.

Voltmeters and Ammeters

Elementary testers such as the preceding are convenient and easy to use, but there are many occasions when a great deal more information is needed beside a simple on/off or go/no-go answer. Then you must turn to measuring instruments of one sort or another.

One perfectly reasonable method is to use meters or gauges similar to those installed in your automobile to do the job. Meters of all sorts are readily available through electronics supply outlets or catalogs, and some electrical suppliers carry a limited selection. All you need do is choose the ones applicable to your uses. You can mount each meter in a small individual case and attach the necessary test leads, or if you wish you can make up a complete test kit with carrying case.

You will need a *voltmeter* for reading voltages, in a range of 0−250 or 0−300 volts for residential work. Current flow is measured with an *ammeter*, and you will have to size this meter for the highest current draw you feel you will be measuring. Though the total demand load of a residence might reach as high as 180 amps or even a bit more, not many household appliances or circuits pull more than 45 or 50 amps. One exception to this would be a particularly large range, which might draw 70 amps with everything running wide open.

Multimeters

An easier and more convenient method of making voltage and amperage measurements, with the added benefit of making resistance measurements as well, is to equip yourself with a so-called *multimeter* or *multitester* (Fig. 9-10). These instruments are made in a great variety of sizes and capacity ranges, and are priced anywhere from less than twenty dollars to well over a hundred. Also known as a VOM, for *volt-ohm-milliameter*, this instrument consists of a small case that holds the electronic inner workings and a battery. A multiple-scale meter is mounted on the face, together with switches for changing the measuring modes and scales, and terminal posts for attaching test leads or probes. Once you become familiar with a multitester and learn all of the many test procedures involved, you will find that you can get an answer for nearly any problem by reading the meter.

Clamp-Around Meters

Another most useful type of instrument, which is widely used by electricians everywhere, is the snap-around meter (Fig. 9-11). One of the best known is the Amprobe, but there

Fig. 9-10. Small, readily portable volt-ohm-milliameter. (Courtesy Simpson Electric Co.)

are other brands as well. This is a small hand-held meter which sprouts a pair of jaws from the top, one of which is movable. To measure current flow on the ammeter scale, you simply slip the jaws around a hot conductor, while it is in operation and without any disconnecting or jury-rigging, and read the result from the scale. Test probes are provided for reading voltage on other scales, and depending upon the model of the instrument and the accessories used with it other measurements and tests can be made as well.

Megohmmeters

There is one more instrument which, while you will not be inclined to purchase one yourself, should be mentioned. This is a special type of ohmmeter used to measure the resistance of the system ground of any electrical installation, residential or otherwise. This instrument is called a megohmmeter because it can measure millions of ohms, but it is not commonly found even in the bigger electrical contracting shops, largely because of their high cost and limited use. You may have to ask you local power company to make the tests for you.

The Megger is the brandname of one old standby, and there are other brands as well; the Vibraground, for instance. These instruments use vibrators to pump high-voltage pulsating alternating current into the ground system, and then measure the resulting difficulty (resistance) that the charge has in dissipating into the earth. A standard DC ohmmeter can't be used because of the possibility of encountering DC electrical components from chemical reactions or electrolysis in the earth which would throw the reading out of kilter.

HARDWARE AND SUPPLIES

Every electrical installation requires a certain amount of miscellaneous hardware and supplies not provided with the

Fig. 9-11. Snap-around volt-ohm-ammeter. (Courtesy Simpson Electric Co.)

Fig. 9-12. Connection hardware. Across the top (A) is a row of assorted wirenuts. (B) a pair of screw lug connectors. (C) split-bolt connector. (D) two grounding clips. (E) an assortment of insulated and noninsulated solderless connectors.

component parts of the system itself. All of this is easy enough to come by at hardware and electrical supply houses. The cost is so low that it usually pays to buy most of the items in bulk rather than by the piece, from a standpoint of convenience if nothing else.

Nails and Screws

Some workmen use nails to attach boxes and pipe straps or clips to wood framing. If you decide to follow this course, two sizes will probably do the job: 12-penny common or box nails for the long size, and 1-inch roofing nails for the short. In most instances, however, screws do a more satisfactory job—3/4-inch and 1-inch lengths in either #6 or #8 sizes are satisfactory for most applications, using either round-head wood or pan-head sheet-metal screws. To mount wall boxes by means of their mounting ears, use 3/4-inch #4 flat-head wood screws.

Wire Connectors

You will need a good supply of *wire nuts* or *crimp connectors* (Fig. 9-12A), which you can buy by the box of 100. Different manufacturers use somewhat different sizing codes, but all are rated according to the sizes of the conductors that they will accept, and sometimes the quantity as well. You will undoubtedly have use for two main sizes, designated for #10 and for #12 conductors. There is also a tiny size used for #6 and thinner conductors found in low-voltage wiring. Your supplier can acquaint you with the specifics and help you to choose the proper types.

You may also wish to investigate the great array of solderless terminals (Fig. 9-12E) your dealer will have in stock, to determine whether or not they will be of use to you in your particular installation. Sometimes having an assortment of terminals on hand can save some trips to the store.

Making splice connections in large conductors, #8 and up, is accomplished by the use of split-bolt connectors (Fig. 9-12C). You will need few if any of these, and should buy only to exactly fill your requirements.

Electrical Tape

You will find a good many needs for electrical tape. Use it to secure a cable to a fish wire, to tape up a soldered connection, to wrap a screwdriver shank or plier handles for insulation, or whatever else seems appropriate. Electrical tape is also used as the final wrap when a splice connection is made with a split-bolt connector. The initial wrap should be made with *rubber tape* or *splice tape*—a specially compounded thick and sticky material made for the purpose.

Friction tape, by the way, is not widely used any more since plastic electrical tape has better all-around characteristics. There are two instances, though, where plastic tape does not work very well: at high and low temperatures. Heat melts the stickum adhesive. And at below freezing temperatures the tape becomes brittle, will not wrap properly and may break. Should you encounter either of these two conditions, you can buy special tapes that will do the job.

You might possibly need *marking tape*. This is a thin plastic tape that comes in assorted colors and is meant for identification only. When several individual conductors are run in conduit, it often is desirable and sometimes mandatory to *color code* them so that there will be no confusion as to phase or purpose and no chance of a mixup in connections.

Grounding Conductors

Wall boxes, fixture boxes, junction boxes, switch boxes and the like must all be bonded together with an equipment grounding conductor. This conductor must be securely attached to each enclosure, and this in turn requires a supply of special green-lacquered *grounding clips* (Fig. 9-12D), assuming that you are using a nonmetallic type of cable in your branch circuits. These clips are forced over the edge of the enclosure to clamp the grounding conductor down, making a good solid contact. You will need at least one and probably two (for ease of installation) for every box.

In lieu of the clips, you can use special broad-headed green-colored machine screws called, appropriately enough, *grounding screws*. Most enclosures have at least one tapped hole for a grounding screw. If there are none, you will have to drill and tap the necessary holes yourself.

Box Hardware

Wall boxes are usually mounted to a stud in new work, but sometimes there is good reason to install a wall box between studs where there is no solid mounting surface. This often occurs in remodeling, too, where boxes must be cut into an existing plaster or wallboard surface. Should you be faced with this situation, you will need a suitable number of *plaster clips*, two for each box. A plaster clip is a flat piece of thin and fairly soft metal stamped out in the shape of an upside-down F, with the vertical member extended at the top. After the box with its plaster ears is inserted into the hole in the wall, a plaster clip is slipped in on each side and crimped around the box edges in front, holding it in place from behind.

An alternative, though more expensive method, for mounting boxes is to use *adjustable bar hangers*. These must be installed on the framing members before the finished walls are put up, but they will support the box anywhere between two studs. Similarly, certain types of lighting fixtures can be mounted between joists or rafters with the use of adjustable fixture bars, or bar hangers.

Miscellaneous Supplies

You may have some special circumstances that will dictate the use of other supplies. For instance, you will need *roofing tar* or *mastic* to seal a through-the-roof service mast, and *duct seal* or *caulking* to seal up around conduits or cables that pass through foundation walls.

If you will be connecting copper to aluminum cables, or connecting aluminum wire onto solderless connectors onto terminals not specifically designed for use with aluminum, you will need a supply of a special *noncorrosive paste* that will inhibit the inevitable detrimental reaction between the two dissimilar metals.

Soldered connections will obviously require the use of *solder*, but never the *acid-core* type, please. Use instead the *rosin-cored* type favored by electronics workers. If you prefer plain solder, make sure that the flux or paste you use is absolutely noncorrosive.

Mounting material on masonry surfaces means that you will need *lead shields* or some other sort of *anchors* made for that purpose, in suitable size and quantity. Surface mounting on hollow walls means *Molly bolts* or *toggle bolts*, and so on. If you run into any peculiar problems, your supplier can help you out.

Chapter 10

Roughing In

With your completed layout clutched in one hand and a set of specifications in the other, and a well-stocked tool box and a heap of materials parked beside you, there is nothing left to do but get on with the sweaty part of the job—the installation.

TEMPORARY POWER SERVICE

Usually, though not necessarily, the first step is to get power to the construction site as quickly as possible. This means setting up some sort of temporary system to tide you over until the permanent installation can be put into operation, at least partially. In the past, temporary services were often put together in a haphazard fashion, with the greater percentage of them being astonishingly unsafe for use. This is still true today in many areas where inspection is either lax or lacking, but happily conditions are improving. Temporary wiring of all types is now regulated by the NEC, so certain rules must be followed. They should be, in any event, because it makes little sense to jeopardize either your own safety or that of your dream home.

There are any number of ways to build up a temporary service, and much depends upon its specific location with respect to the construction area. The basic ingredients consist of a meter box and meter, a main disconnecting means, overcurrent and usually ground-fault protection, a small number of feeder or branch circuits, and a few receptacles.

The principal points to take note of are:

1—The service must be installed under the same regulations that govern a permanent service installation.

2—Feeders and branch circuits must be protected electrically and physically, whether they are cables or extension cords, in the same fashion as those in a permanent installation.

3—A disconnecting means must be provided.

4—All receptacles must be of the grounding type and unless they are part of the permanent structure they must be backed by ground fault circuit interrupters.

5—In addition, any lamps used during the construction process must be either seven feet above the normal working surface or have a suitable guard or cage around them for protection from contact or breakage.

Your first move should probably be to talk with your power company, local supplier, and electrical inspector or building department personnel. From them you can find out what the usual procedures are in your area for setting up a temporary service; then follow their recommendations. With this preliminary knowledge in mind, you can determine what component parts from the permanent installation might be usable in the temporary hookup. Or you might be able to borrow a complete made-up temporary service all ready to go; some suppliers and power companies keep them on hand. Renting one from a local contractor is another possibility.

Power requirements for a temporary service are usually rather small—just enough to feed a portable saw or two, perhaps a table saw or radial saw, drills, and a few lights. Figure 10-1 shows one possibility for a light-duty temporary service. The minimum size is #8 service conductors in an SE cable, or some other size and type of the power company's choosing. These feed a 60-amp meter and then drop into a small 50-amp two-circuit subpanel. The cable could be used later for a short range hookup, while the subpanel might be mounted in the garage or workshop for service there. One 20-amp GFI circuit breaker is mounted in the subpanel, with the other circuit slot left covered. The breaker can be used later to protect a branch circuit that includes outside receptacles. A utility box with a duplex receptacle is close-nippled to the bottom of the subpanel, with the duplex wired to the breaker to form a single branch circuit. A suitable grounding electrode conductor is provided, with a ground rod driven below the service. The heavy-duty extension cords can then be run from the duplex to the job site.

SERVICE-ENTRANCE CABLE

METER BOX

MOUNTING BOARD

PLYWOOD BOX AND DOOR

GFI C.B. BOX

DUPLEX

GROUND

HOLES FOR EXTENSION CORDS

Fig. 10-1. One possibility for a temporary service arrangement.

The breaker provides both ground fault and overcurrent protection, and acts as a disconnect as well. There is a further disconnect at the attachment plug on the extension cord, of course. To prevent pilferage the cords can be taken up and stored away when work is not going on. The cords, by the way, should be at least #16 and preferably heavier if they are very long. The voltage drop on long extensions can be considerable and can cause a drill or saw to burn out all too easily under a heavy load.

If the subpanel and receptacle housing are not of the weatherproof type, they must be installed in a weatherproof cabinet built for the occasion. The service shown is surrounded by a tight plywood box with a hinged door for access. Slots in the bottom allow the extension cord to pass through with the door remaining closed. A hasp and padlock can be added if necessary. Many of these details will depend upon the decisions of the local governing authority, but in any case the equipment should not be unprotected from the elements.

There are plenty of variations on this theme, of course. For instance, you might opt to wire a #12 Romex cable directly into the subpanel, roll the cable out to the job site, and install a handy box and duplex at some convenient point on the site. In this case, since the cable is a branch circuit conductor, it *cannot* lie about on the floor; it should be secured up out of the way. If you use a fuse panel at the temporary entrance, then you will need a disconnecting means, as well as GFIs to protect any receptacles. You could also use a disconnect switch with a feeder line to a subpanel mounted in the partly constructed house, or any one of a hundred other combinations, provided that you follow all the necessary requirements.

In a situation where there is a considerable distance from the power distribution pole to the house and a long service lateral is to be run, this can be done even before construction starts (Fig. 10-2). The permanent meter box, utility disconnect, and overcurrent device (if there is one) is mounted at the nearest power pole. The permanent lateral is then run to a point just short of the construction site. Here it can be joined to a temporary service equipment installation, complete with ground, mounted on a post or braced panel. When the permanent service equipment is ready, the lateral can be disconnected, shifted over into the house, and made into the main disconnect with little loss of time.

ROUGHING-IN THE SERVICE

The first phase of the permanent installation is called *roughing in*. The manner and sequence in which this is accomplished depends upon a number of things. The specific type of construction and the design of the structure has a bearing, and so does the method and sequence of the construction process itself since the activities of the various

295

Fig. 10-2. Permanent service lateral started and attached to temporary service, to be completed after removal of temporary service.

PERMANENT METER AND RACEWAY

PROTECTIVE BOX OVER EXTRA CABLE

TRENCH FILLED

PERMANENT CABLE

TEMPORARY SERVICE ON POST

CLOSE TO WORK AREA

TRENCH OPEN

HOUSE CONSTRUCTION

FOUNDATION

trades must usually be closely coordinated. The weather may play a part, as may the personal convictions or fancies of the wireman. Actually, it makes little difference, so long as the job is done right. But since standard frame construction is probably the most common, we will use that in our principle example, with side notes on other types as necessary. And though you may want to jump around from point to point during the installation, we will begin at the service entrance and work our way through to the branch circuit ends in somewhat of a logical progression.

Preliminaries

The roughing in should start when the building is relatively tight to the weather, both for the sake of comfortable working conditions and the protection of the materials being installed. In a frame or masonry-veneered house, this point usually occurs when the shell of the building is fairly complete; the subsheathing and interior stud walls are installed but still open. No finish materials have been put in, and preferably there should be no insulation, either. Trying to work around insulation is difficult and time-consuming, and invariably leads to damage of the vapor barrier and loosening of the material, so repairs must be made later. You will find more information on the importance of a perfect insulating job to all electric heat installations in Chapter 4.

Setting the service is the first step and should start with a conference with the engineers at the local power company to make sure of all of their requirements, and also to obtain the meter box if they will be supplying one. Mount the meter box at about five feet (or eye level) from the finished grade, and make sure that it is secure. Its location, which theoretically you have already worked out, should be such that all necessary clearances for the service drop are within limitations, and the meter is out of harm's way. Though usually mounted on the side of the building, in the case of many underground services and even some overhead services, the meter is mounted on the nearest power distribution pole.

Service-Entrance Conductors

Now you can ready the service-entrance conductors. If a service cable is to be used, cut off a suitable length to reach from the top terminals of the meter box up to the point of attachment of the service drop on the house, plus ample footage for a big drip loop, plus enough slack to make connections at both ends. Slip the cable through the weatherproof squeeze-connector on the top hub of the meter

SERVICE-ENTRANCE
CABLE

HUB AND
CONNECTOR

BLACK   RED

BARE OR WHITE

Fig. 10-3. Meter box connections.

box, leaving plenty of length to make the connections, then slit the cable jacket. Carefully peel off about an inch of the insulative covering on the conductors—no nicks or gouges in the metal, please—and secure them in the connecting lugs (Fig. 10-3). Make sure that the lugs are down tight, and don't bend or jam the conductors. When all is to your satisfaction, tighten the connector sleeve to secure the cable in the box. Then attach the cable to the side of the house with an appropriate number of clips, leaving the top portion hanging free for later connection (Fig. 10-4). Note: in some areas the serving power company will bring small cable assemblies directly to the top of the meter box and make the connections themselves.

If the meter is mounted on the distribution pole, make the cable long enough to reach the top of the transformer, plus enough for the drip loop and connections. You might be able to

298

enlist the aid of the lineman in securing the cable to the pole, especially at the upper points near the transformer and transmission lines. If there is any danger of physical damage to the cable, either on the side of the house or on the pole, some sort of physical protection in the way of conduit or EMT may be required.

If mechanical protection, a service raceway, or a service mast is necessary in your installation, then your next step after mounting the meter box will be to take care of this. Cut the necessary conduit or EMT to reach to a point about a foot below the point of attachment of the service drop. If the meter is polemounted, the raceway should probably extend to a point just below the bottom of the transformer, depending upon power company regulations. Mount a raintight connector into the top hub of the meter box for EMT, or thread rigid conduit directly into the hub. Place a weatherhead at the top end, secure the raceway in place with clips, and feed the conductors

Fig. 10-4. Service conductor loop to meter box.

Fig. 10-5. Service entrance using outside service mast.

or cable through the raceway. The connections are made as previously explained. Don't forget to make the exposed conductors at the top of the mast long enough for the drip loop and connections. In the case of a pole-mounted raceway, this may be as much as 8 or 10 feet, depending upon the distance from the weatherhead to the connection points on the transformer.

Service Mast

An outside service mast (Fig. 10-5) is attached to the finish siding of the house and extends up through the roof eave or overhang. This is best installed before the finish roofing is put on. Cut the raceway to size and drop it down through a properly located hole in the roof into a connector in the top of the meter box, or the service equipment if the meter is polemounted. After the roofing paper has been applied but prior to shingling, slip a roof jack or flashing piece down over the top of the raceway and cement it into place with a liberal dose of roofing compound or some similar sealing material to prevent leaks. If the mast must be guyed, pick solid anchor points that will interfere with the roofing as little as possible. Then you can clamp the insulators for the service drop lines to the mast if necessary, run the service-entrance conductors or cable, and put on the weatherhead.

An inside service mast is installed the same way, except that the raceway runs on an interior wall and is almost always set into the top of the main disconnect or main entrance panel rather than a meter box. For this reason it is usually installed after the service equipment is postioned. In some instances, with either type you may find it easier to make up the whole assembly, conductors and all, on the ground and then slide it into place as a unit.

Main Entrance Panels

The next part of the installation usually involves mounting the service equipment in place. This consists of the main disconnecting means, the main overcurrent protection for the system and may include distribution means as well. They may all be separate units, but the main disconnect and main overcurrent protection are frequently combined in one enclosure. If for some reason they are not, they must be directly adjacent to one another. Note that both must be as close as possible to the point of entrance of the service-entrance conductors into the building. The exact distance is not spelled out, but anything more than 12—15 feet will probably be suspect at inspection time.

The three most common main-entrance combinations are:

1—Main switch and fuses are together, located either inside or outside in a weatherproof enclosure, feeding one or more distribution panels inside.

2—Main circuit breaker serves as both disconnect and overcurrent protection, located either inside or outside in a suitable enclosure, feeding one or more inside load centers.

3—Main entrance panels located inside, including main breaker or main fuses in a pull-out holder with the circuit distribution busses in one enclosure.

A fourth possibility is a split-bus main entrance panel with less than seven breakers or pull-out fuse holders and with the

Fig. 10-6. Service conductor loop from meter box to service equipment, using cable.

circuit busses in one interior enclosure. If desirable, either type of main entrance panel can be placed outside in a weatherproof enclosure, but this is seldom done. A full-sized switch or circuit breaker may also be required at the head end of a long service drop or lateral. This cannot be considered as either a main disconnect or main overcurrent protection, but only a supplement.

Whatever the arrangement, the equipment should be located at a convenient point, fully accessible, and with ample working room around it. Enclosures should be mounted at eye level or thereabouts, so that no stretching or stools or stepladders are necessary to work on any part of the equipment. Be sure to leave plenty of clearance for all incoming circuit lines. Heavy feeder cables will not bend sharply and take up a lot of room; conduit takes even more. In the case of a main entrance or distribution panel, you will have to keep all of the knockout openings clear from framing members and such. Mount equipment enclosures solidly with heavy screws.

Service-Entrance Conductors

Once the service equipment is in place, you can run the service-entrance conductors to the meter box (Fig. 10-6). If an inside mast system is used, follow the procedure noted earlier. If the meter is located on an outside wall, the next step is to bore a hole through the wall at the appropriate point and slip the cable without damaging the jacket, but no larger than necessary.

Clip the cable to the outside wall and run it into the bottom hub of the meter box through a weatherproof connector, strip the conductor ends and secure them under the lugs. Keep the color codes of the conductors lined up—black to black and red to red, with the white or bare conductor going to the neutral lug.

Then feed the other end of the cable into a convenient knockout opening on the main disconnect or main entrance panel, through a cable connector. Bend the conductors into place with care, without making the bends too sharp. Strip the ends and secure them to the lugs. Again, follow the color coding, keeping the conductors in their same relative position. When the cable is in its final lie, cover the outside of the cable hole with a sill plate and caulk around it throughly.

If you are using service raceway in the installation, you'll probably want to put this in place first. If the meter box and the main entrance panel or disconnect are exactly back to back—one on the inside of the wall and the other on the outside—and the right knockouts are properly lined up, you may be able to connect the two with a short nipple and bushings. Then you simply slip short lengths of conductor into place and make the necessary connections as previously outlined. Failing that, you will have to drop from the bottom of the meter box to a right-angle conduit body, then through the side of the house to another conduit body or a pull box, or whatever combination is necessary, and so to the main service equipment (Fig. 10-7). Once the raceway is installed, supported and secured, then you can run the conductors and make connections.

With only one exception, which we will cover later, no splices of any sort are allowed in the service-entrance conductors. (This does not include the connections made at the meter box.) They must make a continuous run from service equipment to service drop or to the point of connection with the utility power source. Nor can small parallel conductors be run to take the place of large conductors in a service-entrance line. (Or at least this is the practical effect since the smallest conductor allowable for *parallel* service-entrance conductors is #1/0, two of which run together as one conductor would have a capacity of 350 amps, far above the requirements of a residence.)

Fig. 10-7. Methods of running service conductor loops from meter box to service equipment, using raceway.

Service Lateral

With an underground service (refer back to Fig. 3-2), there are two possibilities for running the service lateral. One is to use a direct-burial type of cable, the other is to run the conductors in a buried raceway. Buried cables must be put down to a minimum depth of two feet, and they must be protected by a raceway where they emerge from the ground. EMT is generally not approved for this use, but either nonmetallic raceway or rigid conduit usually is; circumstances vary.

If the meter is pole-mounted, drop a length of raceway straight down from the bottom of the meter box to a point about a foot from the bottom of the cable trench, which at this particular point should be about three feet deep to allow for a gentle transition loop. The natural tendency here is to install a 90° sweep ell to get the cable headed in the proper direction, but in cold-weather country this is not a good idea. If the raceway is left straight down and the cable is set in a wide curve coming out of the raceway, frost damage from the cable being pressed against the pipe will not likely occur. In either case, a bushing should be threaded onto the free end of the raceway to prevent abrasion.

Follow the same procedure at the other end of the cable where it rises to the meter box at the house, or to the service equipment. If the meter is not pole mounted, the raceway must extend up the pole for at least eight feet above ground level and be capped with a weatherhead. The inspecting authority may also require that when conduit is used as a protective raceway it be further corrosion-proofed by an application of asphaltum or some other material.

The cable itself may also require some special attention, depending upon burial conditions and upon the judgement of the inspecting authority. You might have to floor the cable trench with a layer of sand and then cover the cable with another layer, especially if the soil is rocky and frost heaving is a consideration. Or, some sort of protective covering such as heavy, treated planks may have to be laid on top of the cable. If the cable will be covered later at any point by a concrete pad for a patio, porch floor, parking pad or whatever, the cable must be run through a protective raceway that extends on each side of the pad. Backfill material used to fill the trench after the cable is laid must be relatively fine and free from large rocks, chunks of concrete or paving material, or anything that might cause damage to the cable insulation.

Raceways

A buried continuous raceway can be either of rigid conduit or nonmetallic material such as PVC Schedule 80 pipe. Much of what has been said before applies equally here. Conduit must sometimes be corrosion-proofed, especially in areas where the chemical action of the soil is harmful. The minimum burial depth for conduit is 6 inches; for nonmetallic raceway, 18 inches. Bedding in sand or gravel or some sort of mechanical protection may or may not be required, depending upon your locale.

In some areas, expansion joints of flexible raceway must be inserted in the line if it is subject to widely varying temperatures from one point in the raceway to another. This could occur, for instance, in a severe winter climate area where a protective raceway might rise out of the ground at a temperature of 20°F, into an early morning outside temperature of −25°F, and then enter a house with an ambient temperature of 68°F. Such wide temperature imbalances can produce enormous pressures on the raceway, with consequent potential damage to conductors, enclosures, and the raceway itself because of expansion and contraction.

The usual procedure in laying an underground raceway is to mount either the meter box or the service-equipment enclosure first, then lay the line toward the location of whichever enclosure is left unmounted. Once the raceway is in place and secured, the second enclosure is mounted atop the free end of the raceway and anchored into place. Unlike plumbing lines, no union fittings are allowed, so the process must be a continuous one of threading fittings and lengths of pipe together from one end to the other.

When the raceway and the enclosures at each end are completed, the conductors can be run. This is usually a two-man job, unless the run is quite short. One man runs a snake through the raceway, and the conductors are taped to the free end. Then the first man draws the snake and conductors back through the raceway, while the second feeds them carefully into the raceway from the opposite end. This, in fact, is the method generally used to run conductors in most raceways, regardless of their size or purpose.

Sometimes construction methods dictate a slightly different procedure. For instance, if the house is to be built on a grade-level poured concrete slab and is to have an underground service, the raceway (or at least a part of it) is sometimes installed before the concrete is poured. A stub of conduit is left sticking up in the proper location and well above the level of the finished concrete, and then plugged or capped to keep debris out. When the building is sufficiently well along, the raceway is completed, conductors run, and enclosures mounted.

The simplest way to get a raceway or cable through a poured concrete or masonry wall, whether it be a foundation or solid or veneered finish walls, is to provide an opening when the wall is built (Fig. 10-8). A sleeve, usually a piece of pipe of adequate length and diameter, is used for this purpose. In a poured concrete wall, the sleeve is placed in the form, then the concrete is poured around it. In other types of walls, the sleeve

Fig. 10-8. Sleeving cable or raceway through masonry.

is positioned as the wall is laid up, and mortared into place right along with the bricks or blocks or stone.

If a cable is to be run through the sleeve, each end of the sleeve must be threaded and fitted with an insulating bushing to protect the cable insulation. If raceway is to run through the sleeve, it can be left as is. In either case, each end of the sleeve should be thoroughly sealed off with duct seal, or perhaps a silicone construction sealant, to prevent the entrance of moisture and particles.

The only time a set of service-entrance conductors can be spliced, except for clamped or bolted connections made in a meter box, is in a service lateral. In some cases it is necessary to bring these conductors into the building, either through a raceway or a sleeve, directly into a *pull box* rather than extending them on to the service equipment unbroken. The pull box is usually located immediately on the inside of the building and must be of a type approved for this purpose. Another section of raceway extends from the box to the service equipment, and splices can be made in the service-entrance conductors located in this enclosure. But since every splice is a potential source of trouble, the line is best left unbroken if possible.

Wherever a service raceway enters an enclosure, it is secured with a double locknut tightened against the surfaces of the enclosure, one outside and one inside. A bonding bushing must be added to the end of the raceway if it is metallic. A length of copper wire is then run from the bushing to an approved lug or connector attached to the enclosure. The purpose of this is to insure electrical continuity between the raceway and the enclosures. You can find the proper conductor size for supply-side bonding jumpers in the table in Table 7-9. Where metallic conduit is used, all of the fittings should be made up *wrenchtight*—run down as tight as possible with standard pipe wrenches This will insure continuity throughout the span of the raceway.

Once the conductors are run in the raceway and connected and settled into their final positions in the enclosures, the remaining open space at the mouths of the raceways should be filled and sealed off. This prevents air and moisture flow through the raceway and into the enclosures, and may help in the prevention of condensation. Various semi-sticky and plastic compounds are made for this purpose, or you can use a heavy caulking of putty-like consistancy. Silicone caulking compound also works well since it sets solidly into place and remains flexible for an indefinite period, yet can be removed if necessary.

System Ground

The one remaining aspect of the service-entrance installation is perhaps the most important: the system ground. A good service ground, assuming that the remainder of the system is in good working order, is the ultimate protection for yourself, your family, and your property. Without a properly established system ground, you are courting potential disaster from possible high-voltage line surges, lightning strikes, atmospheric static electricity, "hot" enclosures, appliance frames (or even pipes, and plumbing fixtures), and slow-operating or malfunctioning overcurrent protection devices.

System grounds are often all but ignored, unfortunately, probably because they are little understood and don't really seem to do much. But don't be misled; their importance cannot be overstated. This holds true of the entire grounding

GROUNDED
LEG

METER

SERVICE
EQUIPMENT

GROUND
RODS

DEVICES

METALLIC
WATER
PIPING

BOXES
AND
CABINETS

WELL
CASING

LIGHTING
FIXTURES

UNDERGROUND
METALLIC
PIPING

APPLIANCES

UFER GROUND

MOTORS

UNDERGROUND
TANKS

DUCT
WORK

METALLIC
PIPING

MESH OR
REBAR
NETWORK

TV, FM,
AND AM
AERIALS

LIGHTNING
PROTECTION
SYSTEMS

GROUNDING ELECTRODE
CONDUCTOR

EQUIPMENT GROUNDING
CONDUCTOR (BOND)

Fig. 10-9. Elements of grounding/bonding system.

and bonding network throughout the whole electrical installation, though the system ground is the heart of the network (Fig. 10-9).

There are three principle points to remember:

1—An adequate path to the earth, in the form of an electrical conductor, must be provided to safely carry over-voltage and fault currents away for dissipation into the earth.

2—The resistance to ground of this conductor must be as low as possible in order to allow quick and effective dissipation. In specific terms, the maximum allowable resistance for residential services (and most others as well) is 25 ohms. Though spoken of as resistance, you should understand that we are actually talking about impedence, and the object is to keep it as low as possible.

3—The grounding electrode conductor, as it is called, must be continuous and unbroken with no splices from its connection point on the supply side of the electrical system at the service equipment to the grounding electrode. The reason for connecting on the supply side is so that if the load side of the system is disconnected, the supply lines will remain protected by a ground.

There are a number of methods of providing yourself with a grounding electrode of 25 ohms or less resistance. Perhaps the most commonly used for rural residences is a simple ground rod. This should be driven for its full length into soil that has been *undisturbed*—no loose fill or backfill. Better results are usually obtained if the rod is not too close to foundation walls; a couple feet of clearance should be sufficient, one or two more if convenient. Then test the resistance of the rod with a ground tester. If the reading is over 25 ohms, you will have to drive another rod about a foot away from the first and connect the two. In an exceptionally dry location, three or four interconnected rods may be necessary.

Any ground rod should be driven to a depth of at least eight feet. If this is impossible because of underlying rock strata, bury the rod, or rods, at least four feet deep. If this too is impossible, you may have to work out another system, such as a shallow-buried grid.

Generally speaking, the wetter the soil the better the ground. In dry areas you can give the soil a little help by adding moisture-attracting chemicals like rock salt to the soil. Drive each rod deep and leave a cup-shaped hollow about a foot in diameter at the top. Pour a good quantity of moisture-attracting chemical solution into the depression. After a short time the ground resistance should lower considerably. The only problem with this method, of course, is that constant attention and testing is needed to make sure that the resistance stays low.

You don't necessarily have to use a ground rod. In fact, you may have a ready-made grounding electrode already on the premises. If your water supply enters the building in metallic pipe, and there is at least 10 feet of pipe buried outside the house, this may make a suitable electrode. Note if a metallic piping system does exist, whether part of a private well or a municipal water system, this must be used in preference to a ground rod, provided that there is no chance that an insulating coupling will be installed at a later date, or that the system might be disconnected or later replaced with nonmetallic piping. If any of those conditions might occur, then the ground rod or some other means must be used with the metallic piping system bonded to it.

There are other possibilities, too. A well casing, especially a deep one, makes an excellent grounding electrode. Buried steel objects such as septic tanks also work well, and so do reasonably extensive (10 feet minimum) underground piping systems. Because of various restrictions, however, *gas pipes* can seldom be used—and they should never be considered without the specific permission of the governing electrical inspecting authority and the gas supplier. Metal frames of buildings, provided that they are themselves well grounded and proven to be so by tests with a ground tester, make excellent grounding electrodes.

If none of these possibilities are usable in your case, you can make up your own grounding electrode in a number of ways. One method is to bury a sheet of copper with a minimum surface area of two square feet and a thickness of at least 0.06 inches. The grounding electrode conductor should be suitably attached, and for this you might use a connection lug bolted to the copper sheet. But under no circumstances can any solder connections be used. You might use a variation of the rod electrode by burying pipe or conduit in the ground, in whatever quantity will produce the required 25 ohms or less resistance. The pipe must be of at least 3/4-inch trade size with a galvanized outer coating and suitable corrosion proofing.

Another system that works well is called the *Ufer Ground*. This makes use of a run of bare copper conductor at least 20 feet long and no smaller than #4 AWG. The conductor is placed almost at the bottom of an area where concrete is to be poured, such as foundation footings or walls, parking pad, or house slab. It should be carefully positioned and supported so that a thin layer of concrete, two inches maximum, will be under the wire. A long loop is left free of the concrete for later attachment to the service equipment. A variation on this is to bond the conductor with tie wires or pressure connectors to lengths of steel reinforcing bars of 3/8 inch or greater diameter, driven three or four feet into the ground at intervals with their tops imbedded in the concrete.

Poured concrete foundations and footings usually contain a considerable linear footage of reinforcing bars, generally called *rebar*. If the rebar is at least 3/8 inch diameter, located at least 2 1/2 feet below grade level, and there is at least 50 feet of electrically continuous rod, you can tie to this with approved clamps outside the concrete or by metal-fusing within the concrete. Obviously, grounding electrodes of this sort must be planned well ahead and installed during the earliest stages of construction.

Some of the details regarding grounding electrode conductors were covered earlier in Chapter 7. With the proper size and type selected as outlined, all that remains is for you to secure the conductor under a suitable lug on the neutral bus of the main entrance panel, main disconnect, or a connector provided for the purpose in the meter box. Run the conductor, properly supported and secured and in a raceway if necessary, to the grounding electrode. Make the final connection there with an approved connector or clamp or other device.

At this point you should have complete electrical con-tinuity from the grounding electrode to the ground lug at the meter box or main disconnect, or to the neutral bus of the main entrance panel, and to the enclosures of each. In addition, there should be a bond from the grounding electrode conductor to the cold water piping system within the house, provided that it is metallic, with connections made at the service main disconnect. This requirement holds whenever any metallic piping system within the building is not used as the grounding electrode. The bonding conductor size can be found in Table 10-1.

DISTRIBUTION PANELS

The next operation is to mount in place any subpanels, distribution centers, or load centers. This is a simple process that consists of attaching the enclosures to their mounting surfaces, bars, or wood supports.

Flush panels should be adjusted so that the cover will lie flat on the finish wall surface with a minimal gap between it and the enclosure rim.

If the load center is equipped with a bonding strap between the neutral bus and the enclosure itself, remove it, since it will not be used. The neutral bus must be isolated from the metal box, just as the circuit busses are.

Table 10-1. NEC Bonding or Grounding Jumper Sizes

Ampere Rating or Setting of Automatic Overcurrent Device in Circuit Ahead of Equipment, Conduit, etc., Not Exceeding	AWG Wire Size	
	Copper Wire No.	Aluminum or Copper-Clad Aluminum Wire No.
15	14	12
20	12	10
30	10	8
40	10	8
60	10	8
100	8	6
200	6	4

BRANCH CIRCUITS

With the distribution panels in place, you can now go ahead with roughing in the branch circuits. This too is a simple procedure, though there are several points to keep in mind as you proceed. Some wiremen prefer to set all the device and junction boxes in place first, then run the cables to them. Others would rather work the other way around, and in some cases there is no choice. We will outline the procedure in one fashion, and you can then rearrange the steps to suit yourself or the conditions.

So with your layout plans in hand and a carpenter's crayon or some other marker, go through the entire building and mark the locations of the various outlets on the floor or studding or joists and rafters, wherever convenient. You may want to mark in the box heights above the floor and any other handy information at the same time. With these plainly obvious notations spotted about, you can pick out the appropriate courses for the branch circuit conductors with a minimum of confusion and traipsing back and forth to the plans.

Boring and Cutting Holes

Begin to run the branch circuit cables, using either armored cable or nonmetallic cable, from the distribution panels to the outlets. This usually requires a combination of boring holes and driving staples, for which the NEC lists a number of rules governing the process, as well as a number of recommendations for doing an effective and workmanlike job.

The first consideration in boring holes is to make sure that there is no chance that the hole might weaken the member. A 2 1/2-inch hole through a load-bearing stud, for instance, would ruin its effectiveness. In every case, the holes should be drilled as close to the center of the member as is practicable. A 2-by-8 is actually about 7 1/2 inches wide, so the hole should be drilled at the 3 3/4-inch mark if possible. Bored holes should be drilled at least two inches from the nearest edge of the member when centering is not possible.

Now obviously this requirement is not possible in the standard 2-by-4 studs, which actually measure about 1 1/2 by 3 1/2 inches. In this case, you can bore the holes centered no closer than 1 1/2 inches from the nearest edge. If the holes must be closer than 1 1/2 inches from the nearest edge. If the holes must be closer than 1 1/2 inches from the edge, then you have two options. First, you can line each hole with a *metal sleeve* at least 1/16 inch in thickness. If you are running nonmetallic cable through the sleeves, then each end of the sleeve should be fitted with an insulating bushing to protect the cable jacket from abrasion and cuts. Secondly, you can nail a 1/16-inch thick steel plate on the face of the stud in line with the hole to protect the cable from inadvertant nail penetration when the finish covering is applied. Depending upon the covering to be used, this may also entail recessing the steel plates into the studs so that no bumps will show up on the finish surface.

Provided that there is no chance of weakening the framing members, you can cut notches in their faces to hold the cables. A couple of quick passes with a portable power saw with the blade set to about a 3/4-inch depth, followed by a rap with a hammer on the cut piece to knock it free will do the job. The notches must be protected by *steel plates* of 1/16-inch thickness, just as with the shallow-set holes mentioned above. Where the cables are to be run through metal studs or other framing members, you must provide bushings, chase nipples, or grommets at each hole if the cable is nonmetallic.

The tendency when boring holes through a series of studs or other framing members is to hold the drill cocked at a slight angle, either from carelessness or because the drill and bit will not fit in between the members to allow a straight start. Another common failing is not lining one hole up with the next, with the result that the installed cable wanders about like a worm. Both of these practices should be avoided wherever possible and practicable, in the interests of good workmanship and ease of installation, but primarily for the safety of the cable jackets. Cockeyed and out-of-line holes invariably when the cable does not feed well through the holes and the scrape and abrade the outer insulation of the cable, especially wireman gets irritated and gives an extra hard yank. Also, it is a good idea to clean away any jagged splinters left on the back side of the hole when the drill bit bursts through.

Securing Cables

Securing cables in place can be done with staples or special cable clips, but never with bent nails. NM cables must be secured at least every 4 1/2 feet, and also at a point 12 inches from each outlet. The same is true of metal-clad cables. Note, however, that both types can be fished into voids where walls, ceilings or floors have been covered or finished, and need not be further secured in this instance. If you are using plastic device boxes, the cable must be secured at a point 8 inches from the outlet, rather than 12 inches.

Running Cables

Cables can be run in floor, wall, or ceiling spaces in just about any fashion you choose when they eventually will be concealed and so protected. You may wish to feed everything from below, or from above, or through the walls; it makes little difference. They can also be run in masonry voids, depending upon whether or not there is any chance of dampness entering and also upon the specific type of cable being used. You can secure cables to the faces of joists or rafters that will later be strapped out and covered with a finish ceiling, for instance, or whatever method seems to be most reasonable from a practical standpoint.

If the cables are run in exposed locations, however, several precautions must be heeded. If nonmetallic, they must always be protected by conduit or EMT, sleeve-fashion, at any point where they might be subject to physical damage. Wooden guard strips that effectively protect the cable can also be used. Where NM passes through a floor, it should be run in a metal sleeve extending at least six inches above the floor surface. In unfinished basements, nonmetallic and metal-clad cables consisting of assemblies no smaller than two #6 or three #8 conductors can be secured to the bottom edges of the joists if the cables are run at an angle to the joists. All smaller sizes must be run through bored holes. Any size, however, can be attached to the joist sides and run parallel with them. Any size can also be secured to running boards nailed across the joist bottom edges and at an angle to them. Cables should not be run parallel with joists along the bottom edges.

When feeding cable, especially with long runs and when going around sharp corners, work carefully. It takes surprisingly little pressure to cut an NM jacket or to separate the coils of armored cable. Have a care when you are driving staples, too. That last shot of the hammer, just to make sure, is the one that usually does the eamage. The cable should be tight and secure, but the jacket must never be crimped or crushed, nor must armored cable be flattened or the coils separated. If a staple starts in at an angle, pull it out and start over because the edge of the staple will surely do some damage. Keep the staples or clips flat.

Bends in the cable are another common source of difficulty, either electrically or in the opinion of the inspector, or both. The NEC specifies that most cables shall not be bent to a radius less than five times the diameter of the cable (see Table 3-3). NM can easily be bent to a smaller radius than this, so you must take pains to make gentle corners. About 2 1/2 inches is approximately correct for most brands of #12/2 Romex, for instance. Armored cable will bend or flex to a certain point with little effort. It should never be pushed beyond this point, or it will crimp and break open. The five-times-radius rule applies here, too, as a maximum bend.

As you run the branch circuit cables from point to point, try to arrange them so that they don't interfere with or will later compress the thermal insulation batts to the point where the insulation will be less effective than it should be. Also keep in mind any items that might later be installed in or recessed into the cavities between framing members, such as ductwork, plumbing, or fixtures that could interfere with the wiring, and avoid such complications.

Outlet Terminations

Leave plenty of free cable where you snip it off at each outlet. You must have at least six inches of free conductor wihhin the outlet box to make connections with later, plus a few inches of extra slack at each outlet in case repairs have to be made at some future date. Switch and controller loops are run in the same fashion as those connecting the various receptacles.

There are occasions when the interior finish covering is installed before the outlet boxes, as when a box is mounted with plaster clips into and flush with the finished wall surface midway between two studs, instead of on an adjustable bar hanger or wooden mounting strips. In this case, be sure to position the cable so that it can readily be found in the wall cavity when the cutout for the box is made. At the same time, place the cable so that it is not likely to be damaged in the cutting process. It is also a good idea to indicate with a mark on the floor exactly where the wire is located, so that you will not end up fishing around in the wrong cavity for it. The best method is to bore a small hole in the finish covering as it is put up, at the approximate center of the box opening-to-be, and pull the cable through.

MOUNTING BOXES

The mounting of boxes is a simple enough chore, but must be done with a certain amount of care, especially where they are to be eventually concealed and thus made inaccessible by the structure or the finish of the building. Those that are not right and go unnoticed until the finish work is complete require a whale of a lot of work to correct so that they present a decent appearance. In addition, as you may have already guessed, the NEC has a few things to say about the subject.

General Rules

You will encounter no particular problems in mounting boxes for switches, receptacles, lighting fixtures, miscellaneous devices, or simply for connection purposes, on wood framing members or other wooden supports, or in plaster or wall-board, or in or on masonry construction. Whether surface or flush mounted, they are considered to be a permanent part of the structure and you can attach them about any way you wish. The principle requirement is that they be mounted securely and ruggedly. You can use wood or sheet-metal screws if you wish; these work well where the box is surface mounted through the back. You can also use nails; broad-headed roofing or shingle nails about an inch long work well. Some boxes are made so that 12- or 16-penny nails can be run in a hole one side, through the box, and out through a hole on the other side for side-mounting on a framing member.

Flush mounting in plaster or wallboard walls without other support requires the use of plaster clips (two per box). These are designed primarily for wall boxes, but can also be used with a few other types.

Mounting boxes within masonry, other than poured concrete, is a matter of placing the box in the proper spot and mortaring it into position as the wall is laid up. Poured concrete installations use a special type of box, set in the form and poured into place. In most masonry work the cable or raceway is installed along with the box.

Surface mounting of boxes on masonry is done with lead shields, anchors, or some similar hardware. The anchors are first set in properly positioned and drilled holes. Then the box is secured to them with whatever type of fastener is required by the particular kind of anchor. The only restriction here is that "wooden plugs" rammed into drilled holes cannot be used. Lead wool can be, however, or studs shot from a stud-setting gun, or masonry nails provided that they will hold securely, or even in some cases epoxy glue.

There are also a few restrictions concerning the boxes themselves. For instance, nonmetallic boxes cannot be used with a metallic raceway system, but can be used with nonmetallic raceway of certain types, or with nonmetallic cable. Metallic boxes can be used with any type of raceway or cable. If metallic boxes are to be mounted on metal, such as plaster lath or metal framing studs, they must be either insulated from the mounting point or grounded to the metal. Note that this includes the metallic surface of insulation, such as the popular aluminum-foil type vapor barrier. Any ordinary box may be mounted in a damp location, provided that you take precaution that no moisture can accumulate in the box, like condensation dripping down the conductors. If the location is wet, or likely to be so at any time, you must use a weatherproof box.

Setting Boxes in Place

Setting concealed boxes needs to be done with care to avoid difficulities later. If the finish wall is noncombustible, such as concrete or tile, then the front surface of the box may be set back no further than 1/4 inch from the finished surface. If the material is combustible, however, which is most often the case in a residence, the front edges of the box may be set exactly flush with the finish surface, or may protrude a bit. But if it extends more than about 1/16 of an inch, you will have difficulty in making the finish device covers sit flush and flat. This means that when installing boxes before the finish surface is applied, you must remove the mounting ears, if any, and then position the box so that it protrudes beyond the mounting surface exactly the same distance as the thickness of the covering. For a 1/2-inch sheetrock wall, this would mean mounting the boxes so that they extend 1/2 inch out beyond the studs. For sheetrock and panelling combined, the thickness might be 11/16 inch. The problem is that you have very little leeway, so the box must be accurately mounted.

If the boxes are mounted after the finish covering is applied, leave the mounting ears in place. This will position the front edges of the box flush with the finish surface. If you are setting them into wallboard or plaster, or a thin panelling with or without a wallboard backing, you will have to secure them with plaster clips. If the covering is 3/4-inch wood, you can drive #4 flat-head wood screws through the holes in the mounting ears.

Another problem lies in making doubly sure that boxes mounted before the finish covering is installed are both plumb and square. Some boxes have projections on them that hit against the mounting surface and keep them from lying flat. It is also possible that the framing member to which the box is mounted may have or later take a twist. Sometimes the box ends up at a tilt, either side-to-side or up-and-down, so that one corner or edge sticks out a bit further than the others. The net result is that they end up askew, with the covers bumped up in one spot and digging into the finished surface in another. If you have a good eye for plumb and square lines, fine. If not, use a small bubble level and a good rule. You'll save yourself a lot of anguish later.

Protecting Boxes

There are times when protecting the boxes after they are set in place saves a lot of later aggravation and extra work. This is especially true when any spray applications of materials are to be made, such as foam insulation, texture plaster on ceilings or walls, or any similar operation where the material is likely to be oversprayed into the boxes. Not quite so bad but still discouraging is the slopover from stucco and plastering jobs, Sheetrock joint taping, or spray painting.

To save scraping and digging at boxes and the tapped device mounting screw holes, you can cover each box with a plastic sandwich bag when you install it, and then cut the plastic away later. Or, wads of newspaper, rags, paper toweling, or scrap fiberglass insulation can be stuffed into the boxes to do the job. The screw holes you can cover with bits of masking tape.

Securing and Labelling Cables

Nonmetallic cables brought into device boxes, and some junction boxes as well, can be secured in place with the cable clamps provided for the purpose in the box. Simply pop out an appropriate knockout slug, run the cable into place, and screw the clamp down. If there are unused clamps in the box, remove them to give yourself a bit more room. Remember that any unused cable openings where the knockout has been removed must be closed again with suitable plugs or covers. Where no clamps are provided, use standard 1/2-inch (or larger for large cable) Romex or cable connectors inserted in knockouts and secured with locknuts. Bushings are not required here. The clamp on the connector should never be run down on the individual conductors but always on the cable jacket with an extra bit of the jacket protruding into the box. This actually takes the place of a protective bushing. Metallic cable uses a special type of connector designed for the purpose, along with fiber bushings that are inserted into the sheath around the conductors.

Another idea that saves time later is to tag each circuit as you install the cable both in the outlet box and in the distribution panel. The small string tags such as some stores use for pricing merchandise work nicely. The conductors in the outlet boxes would be tagged "Circuit 1," "Circuit 2," and so forth. Those in the distribution box should be more specific, such as "Plugs, south living room wall, and dining light," or "attic and crawl space," or whatever. The tags in a junction box should also be specific, as "Circuit 4 feed," "power to kitchen light," "Circuit 4 out," and so forth. When the time comes to troubleshoot or figure out what specifically is where, tags help a lot.

Running the branch circuits into the load centers or main entrance panel is simply a matter of punching out a suitable knockout, affixing a suitable connector, and inserting the cable (Fig. 10-10). When using NM, some wiremen prefer to clamp the connector to the cable first, then insert the cable and connector into the knockout hole and run the locknut up, since there is less chance of damaging the cable jacket with this procedure. In any event, leave a length of free cable inside the enclosure about equal to the length plus the width of the enclosure. This will give you plenty of conductor to work with when you make connections.

The closure of unused openings in panel boxes or any similar enclosures applies the same as with outlet boxes. So does the front-edge placement of flush-mounted enclosures with respect to the finished surface of the wall.

Tap conductors are most easily put in place after the outlet and junction boxes are set. These may run from a junction box to a fixture or fixture outlet box, or perhaps to some other piece of equipment such as a pump motor or a water heater. Whatever the case, run the necessary conductors into a suitable length of flex, no longer than six feet, then secure the cable to the junction box and to the fixture or equipment connection box with flex cable fittings. Pulling in the conductors after the flex is installed is a miserable chore. Leave a minimum of six inches of free conductor at each end for connections.

Low-Voltage Switching

Where a low-voltage switching system is used and the relays are ganged together in central relay panels, you will

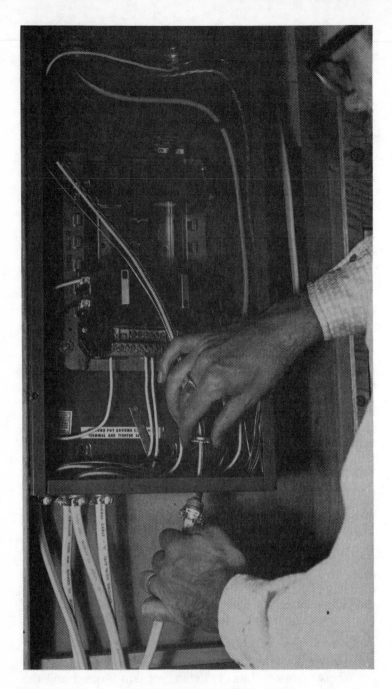

Fig. 10-10. Making up a distribution panel.

run the necessary lighting branch circuits into the high-voltage section of the relay panel. Then an individual line is run from each relay to its load outlet box. In all respects, the cables are installed just the same as any other branch circuit lines. Again, leave plenty of free conductor in the panel for connections.

The regulations for the installation of low-voltage wiring are not quite so rigorous as those for light and power wiring. Nonetheless, you should take considerable care so as to avoid problems later. The conductors are normally quite small, usually #18, and they are delicate by comparison to general wiring cables. Circuitry in this category would include low-voltage switching control, low-voltage thermostats, doorbells, buzzers and chimes, certain signal circuits, communications networks, sound systems, antenna rotating devices, and the like. There is such a great variety of circuit possibilities and installation procedures that the best policy is to follow the instructions made up by the manufacturer of the particular items involved.

Low-voltage conductors are usually plastic insulated. They may be encabled, or the individual conductors may be twisted around one another in a continuous spiral. Cables are easiest to work with, and they can be run through holes and notches, tucked behind moldings or mopboards, stapled, and otherwise handled in the same manner as other nonmetallic cables. Saddle staples are often used because they are less likely to damage the insulation. The supply source is generally a transformer, designed so that it may be mounted upon a standard junction box or with a collar arrangement for mounting in a standard knockout hole. The primary (115-volt) leads are connected within the box, but the low-voltage secondary connections may be left out in the open since they pose no shock hazard.

Most low-voltage devices are designed to fit into standard wall boxes, so you can use these at all outlet points just as for the 115-volt branch circuits. The box wire-fill rule does not apply, and sufficient room for the conductors in a box is seldom any problem since the conductors are small. They may be fastened just sufficiently to hold them in place. They do not need support at any particular intervals or points just so long as they are mechanically secure and safe.

You can run low-voltage wiring in raceways or through sleeves for protection from physical damage. They can be run in raceways from outlet to outlet, though this is seldom necessary. But they should not be run along with the general wiring conductors. If both types come into the same panel, device box, junction box, or other enclosure, the high-voltage portion of the enclosure must be separated from the low-voltage section by a protective partition.

Some manufacturers of low-voltage switching equipment also make a special flat cable assembly for use with their products. Ordinary thermostat wire can be used as well, as it is for doorbells, signal circuits, and sundry other uses. Thermostat wire has the added advantage of being available almost anywhere, while some of the special cables are not.

CONDUIT AND EMT

Wiring systems, or parts thereof, cannot always be installed using cables, but instead the individual conductors must be run in conduit or EMT. Working with raceways is slower, less flexible, more expensive as a rule, and takes a bit of getting used to. On the other hand, raceways also make a safe, neat, and sturdy installation that any homeowner could be proud to call his own.

As mentioned earlier, rigid metallic conduit or EMT is used in conjunction with either threaded or slip-on fittings, respectively, and each type can also be bent to fit into place, using 90° bends, double 45° bends, or whatever else will do the job. The general scheme should be planned in detail and measured out with care as the work progresses. Each leg of pipe, from outlet to outlet to panel or junction, should be completed before any conductors are pulled in the lines. In some cases it is advisable or perhaps necessary to leave the conductors out entirely until the finish work and final connections can be started. The pipeline itself would then mark the completion of the rough-in stage of the installation.

Bending Conduit

Any metallic raceway system is usually a judicious combination of fittings and bends wherever changes of direction are needed. Bends can be made either hydraulically or manually. The smaller trade sizes you can easily bend by hand with a hickey. Note that EMT and conduit take benders of slightly different configurations—so don't mix them up. Hickeys are also made for a specific size of raceway, so using a 3/4-inch bender on 1/2-inch pipe will result in a poor or unusable bend. Instead of stretching properly around the saddle, the pipe will kink and crinkle.

You will probably find that a certain amount of trial and error is necessary when you begin to learn how to bend pipe. You might therefore want to start off with some scrap pieces. Bending a single length so that it fits exactly into place requires experience, but the process will come to you after a while. The basic instructions for use are provided with the bender (there are slight variations depending upon brand). You can probably get some further instruction from your supplier.

The procedure for bending pipes is essentially the same with all types, however. Slip the conduit or EMT into the saddle flange of the bender and through the retaining finger, then place the pipe on a flat, level surface, bender handle pointing up. Place one foot on the pad or peddle of the bender head and push down and slightly forward, while pulling back and downward on the bender handle as though you were prying up a boulder with a crowbar. The back section of pipe will remain flat on the floor, while the forward section will curl smoothly upward to the desired angle. The trick, of course, is to get the bend at the exactly right spot in the length of pipe and to bend it to the correct angle.

Making two or more bends in the same length of pipe is no particular problem, except that you have to consider which bend, or perhaps which part of which bend, should be made first, so that you can make the subsequent bends without the first ones interfering with the process. Tight saddles and offsets sometimes cause a bit of fiddling around, but rather complicated configurations can be turned out with patience and practice. Bends can also be taken too far, and there are some rules to observe.

Bending Rules

In the first place, bends should not be carried past the 90° point because the material can't stand the strain; a kink is sure to result. Also, a series of bends in a length of pipe—that is, between fitting and fitting, fitting and outlet, or outlet and outlet—can not exceed 360° (Fig. 10-11). In other words, you might have four 90° bends in a line between two boxes, but you can't have five. You might, for example, exit a box with an offset (two 45° bends back to back), turn a 90° corner, go straight for a bit, turn another 90° corner, and enter another box with a second offset, for a total of 360°. But a saddle added in the straight section would not be allowable, for this would give you another 180° degrees of bend.

In general, the greater the number of bends in a given run of pipe between outlets, the harder it will be to pull the conductors through the line. However, conduit bodies of the openable variety can be inserted in the pipe line to make that task a good deal easier. Still, the fewer direction changes you can get away with, the better, since the more effort you exert

Fig. 10-11. Allowable conduit bends. (A) 540° total—over the maximum. (B) 270° total, below the maximum. (C) 360° total, the maximum allowable.

in pulling the conductors, the greater the possibility of damage to them, including stretching.

Installation

EMT and conduit is usually supported by means of two-nail straps, single-nail clips, or drive clips that are made with a drive stud as a part of the clip. Either type of raceway must be supported at least every 10 feet, and again at some point within 3 feet of each outlet or junction box. Any outlet boxes should be self-supporting, not dangled in place by means of the raceway connectors.

You can cut conduit or EMT with either a pipe cutter or a hacksaw, but make sure that the cut is straight. The rolled edge that a pipe cutter leaves has to be completely reamed away so that the interior edge of the raceway presents a beveled surface free from burrs or sharp edges. Raceway cut with a hacksaw should be filed as necessary on the outside edges, then reamed to remove any burrs and sharp inside edges.

EMT is secured to the outlet box with a connector slipped over and tightened onto the tubing (Fig. 10-12). Insert the threaded end of the connector into the knockout hole in the box until the flange is snug. Then run a locknut down on the threads to lock the assembly into the box. With conduit, you first run a locknut onto the threads, then insert the pipe into the knockout hole and run another locknut on, tightening one against the other. All of these connections, as well as any along the pipeline at fittings, should be carefully checked to make sure that they are tight, tight, tight. Electrical continuity here must be perfect in order to insure a good equipment grounding circuit.

At each outlet point, whether for load attachment or for interconnection purposes, the raceway must be fitted with a bushing to protect the conductor insulation. If the conductors are lighter than #4, you may use either metal or insulated bushings. Where the conductors are # 4 or larger, you must use insulated bushings, which are generally made of plastic. You can also use fiber sleeves that slip into the throat of the raceway with any size conductor.

EMT can be run either open or concealed, indoors or out. In damp or wet locations, fittings and enclosures approved for the conditions must be used, and in some instances special corrosion-proofing standards must be met, depending upon the particular conditions of the installation. EMT can also be installed in masonry construction, but rarely is it embedded in poured concrete and then only with approval from the inspecting authority. Because of corrosion and chemical reaction, EMT almost never can be used underground, and never buried directly in cinders. Where cinders are used as fill or as an aggregate in concrete, the raceway must be protected by at least two inches of noncinder concrete, or buried at least 18 inches away from any cinder material. And all of this presupposes that direct burial of the EMT is approved in the first place. In addition, EMT cannot be used where there is any possibility of severe mechanical damage.

Fig. 10-12. Securing conduit in enclosures.

Much the same situation holds true for rigid metallic conduit, except that it can be used in some corrosive situations, though additional corrosion-proofing may be required. Conduit can be used in cinder fill and for direct burial in the ground or embedding in concrete under many, but not all, circumstances. Again, some further precautions against corrosion and also mechanical damage from frost or rough fill may have to be taken. There are times when expansion joints should be used with most types of raceway to prevent damage to fittings and the conduit itself, as noted earlier in the discussion of service raceways. And if the conduit is made of aluminum rather than a ferrous material, direct burial or embedding is almost never allowed, and yet there are some instances when this material will stand up better than a ferrous pipe.

Nonmetallic rigid conduits can be used for a variety of purposes both above and below ground and under a wide range of conditions. They are especially useful where corrosive elements are present, but cannot be used in hazardous locations or where subject to possible physical damage. Before using nonmetallic rigid, check with the local inspecting authority to make sure that the installation will be approved; in many areas it is not acceptable.

All types of raceway can usually be installed in the form of *stubs*. In this situation, to take an example, an outlet box might be mounted low in a concrete or masonry wall, or half-wall of a foundation. A length of conduit or EMT runs from the box up through the masonry to a convenient point ending within the wood-framed portion of the building, perhaps in a bay between two floor joists. The top of the stub is capped with a connector and a bushing, but no box is installed there. The cable, either metallic or nonmetallic, which is being used to wire the remainder of the house in normal cable fashion, is then run intact down the stub to the outlet box where connections are made in the regular manner. In this case, the pipe is a protective and convenience item more than a raceway.

OUTSIDE BRANCH CIRCUITS

Once all of the branch circuit cables or raceways (with or without conductors) are run within the main building and the boxes set or not depending upon your construction requirements or personal preferences, you can turn your attention to the outside branch circuits. Virtually all of the practices and principles that we have already discussed apply equally to outside wiring, together with a few additional points.

Outdoor wiring can be done either overhead or underground, but in the interests of esthetics and convenience, most is done underground. Under no circumstances should any conductors be left resting on the ground surface. You can bury cables directly in the earth, using UF or a similar approved cable type, or you can run conductors in rigid metallic or nonmetallic conduit raceways suitable for the burial conditions. Neither conductors nor cables should be exposed where they leave the ground, but must remain enclosed all the way to the point of use. Burial depths, as well as all other requirements, are the same as those noted for service laterals.

Oddly enough, it is permissible to tap or splice underground cable assemblies directly in the earth without benefit of a box of any type. This is, however, a practice to be avoided whenever possible. It is not permissible to bury any junction or pull box because this effectively renders the box inaccessible. Any box of any sort must be completely accessible, whether inside the premises or out, without any digging or removal of any permanent part of the structure. A box mounted behind a removable panel or door on the hollow post of a patio lamp would be accessible, and allowable. So would boxes mounted above removable-panel drop ceilings, in crawl spaces, low attics, and the like. But this can be a judgmental decision, so if you are unsure about a box in your installation, check first with the inspecting authority before you proceed with the work.

FEEDER CIRCUITS

All that has been said of branch circuits applies just as well to feeder circuits. Feeder conductors, which are usually heavier in size, are handled exactly the same and installed under the same rules and precepts of good workmanship as branch circuits.

Outside feeders are sometimes run overhead to outbuildings, and in this case you would install them in the same fashion as discussed under service entrances. Individual spans between supports can not be more than 50 feet long, or use smaller than #10 conductor size. Feeders can also be direct-buried, or run in buried raceways, in the same fashion as service laterals.

The roughing-in process is exactly the same for outbuildings as for the main structure, with all rules and principles applying equally. Each outbuilding must have its own disconnecting means, and the feeder to it is treated the same as a service entrance. It is perfectly acceptable, and often advisable, to provide a separate grounding electrode connected to the subpanel in each outbuilding, just as at the service entrance equipment. In short, each outbuilding is treated as a smaller version of the dwelling.

Chapter 11
Finish and Test

The best time to go ahead with the electrical finish work is after everything else is done and the other tradesmen have packed up and gone on to the next job. At this point the painting is completed and dry, the wallpaper has been hung. Cabinetry and trim carpentry is complete, no more sawdust is swirling around the place. Appliances are waiting to be connected, equipment is installed and ready to go. No more trucks or heavy machinery are rumbling around outside to interfere, finish grades have been run. So there is nobody around to bother you as you concentrate on the final work. That's the best situation, but it almost never happens that way, especially if you happen to be doing all of your own work.

However, let's assume that you have a clear field to work in and you can proceed without interruption of any serious nature. As before, we'll follow one pattern of explanation, and you can rearrange matters to suit yourself.

RECEPTACLES

About the easiest place to start is with the duplex receptacles, and the simplest way to install them is to begin with the handiest one and follow through around the entire building until you come to the last one.

Wiring

To install a duplex in a cable-wired system, first peel the cable jacket away from the conductors, back to the point of box entry, and then strip about three-quarters of an inch of insulation from the ends of the conductors. The next step is to make the outlet box a permanent part of the continuous equipment bonding circuit, and there are several approaches that can be used (Fig. 11-1).

One way is to secure the incoming equipment grounding conductor (generally bare) to the box as in Fig. 11-1A, either with a grounding screw driven into one of the tapped holes provided in the box for that purpose, or with a grounding clip forced over the edge of the box. If you use the clips, position them on whichever side of the box the neutral or "white" side of the duplex will face, making sure that they do not interfere

Fig. 11-1. Circuit continuity of the equipment grounding conductor should be maintained, including the enclosure, even if a device is removed or an outgoing circuit is removed. Note in example (H) that if the upper incoming branch is removed, the enclosure is no longer grounded, so there is still a live circuit in the box.

with the duplex itself. Clearances are a bit slim, but a little experimentation will show you the best spots to place them.

Then, using a scrap piece of the same type of bare conductor, make a short pigtail and attach that to the box with another grounding screw or clip. Turn a 180° curl in the free end of the pigtail with your diagonals (a slight nick at the outer end of the wire made by the jaws will do no harm here) or with needlenose pliers. Hook the wire around the grounding screw (green) in the duplex and tighten the screw down. Place the hook of the wire so that the clockwise tightening direction of the screw is toward the end of the wire. Then the tightening action will draw the hook inward instead of spreading the conductor away from the screw. Next, do the same with each of the main conductors, placing the black wire on the gold-colored terminal screw and the white on the silver or chrome screw.

Always maintain this color code throughout the entire wiring system. Green is used for equipment ground. White or natural gray is used for neutral and is attached to the silver terminal. And black, or some other color such as red, is used for "hot" and is attached to the dark or bronze terminal. Various colors are sometimes used in larger installations for the "hot" wire when it is desirable to color-code different branch circuits.

Figure 11-2A shows the standard connection scheme for parallel-connected receptacles and also for joining a duplex receptacle to a single receptacle. A duplex receptacle also has conductor bars that can be broken out to separate the two receptacles. When wiring each receptacle of the duplex from separate branch circuits, break off both the hot and neutral bars and wire according to Fig. 11-2B. If a split circuit is desired, as when switching only one receptacle, break off only the hot bar and wire according to Fig. 11-2C. Note that Fig. 11-2C has two hot lines with a common neutral and common equipment grounding conductor, whereas Fig. 11-2B has independent hot and neutral lines for each receptacle.

When there are two sets of conductors in the box, you can secure one of the equipment grounding conductors to the box with a clip or screw, then do the same with the second, leaving a pigtail on this one which you then attach to the duplex. The hot and neutral conductors are attached as above. When there are three or more sets of conductors in the box, group the equipment grounding connectors together, cut two pigtails and join them all with an approved connector. One of the pigtails should be then secured to the box with clip or screw, and the second one to the duplex grounding screw. However you

332

A

GND
B
W

2W/WG

GND
B
W

2W/WG

B

BREAKOUT

GND
B
W

GND
B
W

2W/WG

BREAKOUT

2W/WG

C

BREAKOUT

GND
B
R
W

3W/WG   (3 WIRES WITH GROUND)

GND—GROUND   R—RED
B—BLACK   W—WHITE

Fig. 11-2. Receptacle connections. (A) parallel duplexes. (B) duplex with double feed from two separate circuits. (C) split circuit feeding duplex halves.

arrange the equipment ground circuit, continuity must not be broken when the duplex is removed from the box and disconnected.

Mounting

If the box is mounted perfectly flush with the wall surface, break off the support ears on the duplex by bending them back and forth so that the duplex frame will lie flat against the box. This must also be done when you are working with surface-mounted boxes, or the finished cover will not go on properly. If the box is set back a bit from the finish surface, leave the ears in place so that the duplex will lie flat against the surface.

Then, forming the conductors with your fingers into natural loops, push the duplex carefully into position in the box (Fig. 11-3). If you need to bend a conductor loop into a different position and can't reach it with your finger, *don't use a*

Fig. 11-3. Receptacles must be carefully set into their boxes and the conductors bent gently to avoid jamming and danger.

screwdriver. Use a flat-topped eraser on a pencil or anything else that has no sharp edges to damage insulation. When the duplex is fully in place, the conductors should not be jammed up or twisted badly, and the device should have no tendency to spring back out of the box. Run the device holding screws into place, and the job is done.

Switched Receptacles

There are several wiring possibilities with switched receptacles or duplexes. In the event that a single receptacle or both halves of a duplex are to be controlled by a switch, follow the connection diagram in Fig. 11-4A if the branch circuit feed runs first to the switch and then to the receptacle. If the feed runs first to the receptacle and then to the switch, use the connections shown in Fig. 11-4B. Note that in both cases the entire duplex is controlled by the switch. If you want half of the duplex live at all times and the other half controlled by a switch, follow the connection diagram in Fig. 11-4C.

Single and multiple receptacles are wired in the same fashion as described above. Combination devices, which may include switch, outlet, and pilot light (or some similar arrangement), have slightly different wiring patterns depending upon the design of the device, the manufacturer, and the combination of uses involved. Follow the instructions enclosed with the device.

Conduit and Metallic Cable

The situation is just a bit different if the system is wired with metallic cable or with a metallic raceway, or any other

GND-GROUND   B-BLACK   W-WHITE

Fig. 11-4. Switched duplex connections. (A) switch feed, both halves of duplex controlled. (B) duplex feed, both halves of duplex controlled. (C) duplex feed, one half of duplex controlled.

type of cable that contains an approved grounding strip. In this case, no bare or green equipment grounding conductor is required because the built-in grounding strip or the raceway itself serves as the grounding conductor. All that is necessary is to make sure that approved fittings are used, the circuit is complete, and bonding jumpers are used to bridge around any point where the continuity of the grounding circuit might be weak or broken, such as an expansion joint in the raceway.

Install the receptacle as described earlier, using a grounding pigtail from the green screw of the duplex to a grounding screw or clip on the box. Locknuts, clamp screws, or mounting screws should never be used to take the place of grounding screws or clips.

Covers

The final step in finishing off the convenience outlets is to mount the protective/decorative covers over them. Line the device frame up so that it is straight, set the cover on, and run down the mount screw. Covers made of breakable material such as porcelain, wood, or plastic should be tightened with caution. If the duplex is a little bit deep in the wall, the cover can break between the two receptacle openings. And if the box is not set squarely, the cover may break at some other point.

Metallic covers must be made a part of the equipment grounding system, though as a rule this is taken care of by contact through the mounting screws. If you have any doubts as to whether a metal cover is adequately grounded, arrange a bonding jumper of some sort to take care of the problem.

Any cover you use must fit square, flush, and flat, especially at its mounting surface. This will give more than ample protection to the box and its contents.

SWITCHES

The installation procedure for snap switches is much the same as for outlets, except of course the connections are different. The box must be made a part of the equipment grounding circuit, and the conductors are stripped and attached with the same method.

Wiring

The switching device itself usually has no equipment grounding terminal. It relies on the contact made between the mounting screws and the box to ground the device frame. If you are using metal switch covers, it is especially important to make sure that good contact is made here because the only ground the cover will have is through the cover mounting screw and the device frame.

There are two basic possibilities in switch connections. In one case the power feed comes to the switch outlet first, and then goes on to the load (Fig. 11-5A). In the other, the power feed goes directly to the load, and a switch loop or hot leg is brought back to the switch (Fig. 11-5B). In the former situation, a hot and a neutral pass through the switch box; in the latter, one conductor is hot all the time, and the other is hot when the switch is in the *on* position. Despite the fact that one conductor is probably colored white and the other black, you have in effect two blacks.

Whenever you have this situation, where a white conductor is hot part of the time, paint a couple of inches of the end of the conductor black. Model airplane dope or India ink will do the job, but whatever you use, make it *permanent*. The reason for this is to avoid any possible confusion—a cardinal rule of switch connections is that a neutral (white) is never switched. (There are some exceptions to this rule, but none that would be likely to apply in residential wiring.)

There is one situation where this white wire rule does not appear to hold true, though upon analysis it does. This occurs

Fig. 11-5. Single-pole switch connections. (A) line feed, switch ahead of load. (B) load feed, load ahead of switch.

337

when low-voltage switching is used. Each switch assembly, which uses short pigtail leads instead of screw terminals, has one white pigtail, along with a red and a black. But here this is not a neutral, but actually a common line serving one side of each coil in the relay and the common or center pole of each switch. Though hot all the time, it uses the designation *white = common supply* sometimes found in low-voltage applications and not like those of the general wiring field.

Connection diagrams for various switching combinations are shown in Fig. 11-6 through 11-8, in addition to the switched duplexes shown earlier in Fig. 11-4. Note that in three- and four-way systems, as well as low-voltage systems, the conductors are color coded so that they can be wired to the proper terminals to perform the right circuit function without confusion.

Mounting

Switches are mounted in their boxes in the same manner as duplexes. They must be properly aligned and placed with the conductors tucked into gentle loops and out of the way. The standard position for the toggle handle is up for *on* and down for *off*, but if you choose to mount them sideways, or in a three- or four-way system, the toggle position means nothing.

Once the switches are mounted in the boxes, put on the covers or faceplates, and that job is complete. There are dozens of different types and brands of faceplates on the market, by the way, and many of them are not of approved construction. Unless they are, particularly with respect to switch covers, the inspecting authority may choose not to pass them. Look for the UL seal of approval of the Underwriters Laboratories. If there is any doubt with the plates or any other electrical equipment, check with the inspector.

LIGHTING FIXTURES

Installing the lighting fixtures is not quite as straight-forward a job as the devices, principally because there is such a tremendous variety of fixtures and mounting methods. So for this reason we will first discuss the mounting requirements.

Mounting

As far as the mechanical part of the task is concerned, either follow the directions included or use whatever hickeys, clamps, threaded rod, jam nuts, bar hangers, or other hardware seems necessary. Most smaller fixtures are simply mounted with a pair of screws, or studs and cap nuts, run either directly into mounting ears on the fixture box or into a fixture bar mounted on the box (Fig. 11-9). Other types use a center nipple and locknuts (Fig. 11-10). Recessed fixtures may mount by means of screws through the housing or on bar hangers attached to the structural framework of the building. Fluorescent strips can be mounted directly on the surface with screws through the housing, or hung on chains or stems fixed to a sturdy mounting point.

Whatever the process, mount all fixtures solidly, and make sure that if they are suspended from a fixture box, that box is also ruggedly attached. This is especially important with heavy chandelier-type fixtures and with pull-downs. If the fixture weighs more than 50 pounds, it must be supported independently from the fixture box. If the fixture exceeds 6 pounds in weight, or exceeds 16 inches in its largest dimension, it cannot be supported by the screw shell (socket) of the fixture.

Take care to maintain a minimum clearance of 1/2 inch from any combustible materials around a recessed fixture enclosure. Also, unless the fixture is marked suitable for the purpose, no thermal insulation of any sort should be within two feet of the top or three inches from the sides of the fixture enclosure.

Fluorescent fixtures have a heat problem, too, so don't mount them directly on any type of fiberboard surface, such as ceiling tile, unless it has a density of more than 20 pounds per square foot. Some exceptions can be made to this if the fiberboard has been chemically fireproofed and can pass certain tests; check with the inspecting authority on this. You can direct-mount to heavier fiberboards or to plywood or solid wood.

All fixture outlet boxes must be covered, either by the base, pan, or enclosure of the fixture itself, or with a part of the fixture that serves this purpose in a decorative fashion, such as a trim canopy.

On adverse result of covering a lamp is the entrapment of heat. For this reason, further steps must be taken when the fixture is mounted on combustible material. If there is exposed combustible material between the perimeter of the outlet box and the perimeter of the fixture canopy or pan, this area must be covered with a noncombustible material sandwiched between the canopy or pan and the finish surface, as shown in Fig. 11-11. The NEC does not spell out what this material might be. Aluminum foil, for instance will not burn under normal circumstances, but it is a great heat conductor and so is not as effective as it might be. About the best material is asbestos

C

GND
B
W

2W/WG

GND
B
R
W

3W/WG

GND
B
W

2W/WG

D

GND
B
W

2W/WG

GND
B
R
W

3W/WG

W-B
B
R
GND

3W/WG

B-BLACK
R-RED

W-WHITE
GND-GROUND

W-B-WHITE CODED BLACK

Fig. 11-6. Three-way switch connections. (A) load feed, load ahead of switches. (B) load feed, load between switches. (C) end switch feed, both switches ahead of load. (D) end switch feed, load between switches.

341

GND-GROUND W-WHITE B-BLACK R-RED W-B-WHITE CODED BLACK

11-7. Four way switch connections. (A) load feed, load ahead of switches. (B) end switch feed, load between switches.

342

R-RED W-WHITE B-BLACK

Fig. 11-8. Low-voltage control switch connections. (A) single switch, relay feed. (B) parallel switches for multiple control, relay feed.

Fig. 11-9. This indoor fixture pan mounts to the box with center stud and jam nut. Fixture body then attaches to pan with two cap screws.

Fig. 11-10. This outdoor weatherproof fixture mounts to a fixture bar on the outlet box with center stud and cap nut.

Fig. 11-11. A layer of noncombustible material must be placed between a fixture pan or canopy, mounted against a combustible surface.

sheet or paper, which comes in rolls and can be bought by the linear foot. You can cut it with scissors, trim it with a utility knife, and fasten it in place with staples or glue.

Wiring

The wiring of fixtures is simple enough, but there are a few points to keep in mind. The connections are straightforward; one hot line from the switch and one neutral from the feed go to the two fixture leads (Fig. 11-12). The leads or terminals may or may not be color coded. If they are, all well and good, but if not, no matter, because most fixture terminals are well isolated. The exception lies with certain lampholder screw-shells, where the neutral must be connected to the shell. In all cases the tap conductors or pigtails, which sometimes are a bit fragile, must be protected from any possibility of abrasion or other mechanical damage, and there never should be any splices or taps made within fixture arms, stems, or other hidden points. Note too that no circuit conductors should be run within the fixture enclosures themselves, except on a very few occasions where the fixture is also approved as a raceway, or in some fixtures such as fluorescent strips that are designed for end-to-end assembly.

You will connect many fixtures by tap leads provided with the fixture assembly. With others you will have to supply your own taps, making your choice from Table 11-1, depending upon what conditions the fixture will operate under. Whatever the

Table 11-1. NEC List of Commonly Used Fixture Conductors

TYPE	INSULATION	SIZE	MAX. OP. TEMP.
RHF-1	Heat-Resistant Rubber	18	75°C 167°F
FFH-2	Heat-Resistant Latex Rubber	18-16	75°C 167°F
TF	Thermoplastic	18-16	60°C 140°F
TFN	Thermoplastic	18-16	90°C 194°F
AF	Impregnated Asbestos	18-14	150°C 302°F
SF-2	Silicone Rubber	18-14	150°C 302°F
PF	Fluorinated Ethylene Propylene	18-14	200°C 392°F

A

LAMP

SOCKET

METALLIC PORTION

BOX

2W/WG

B GND W
OR
W GND B

B

LAMP

SOCKET

METALLIC PORTION

BOX

2W/WG

W GND B

W-WHITE B-BLACK GND-GROUND

Fig. 11-12. Lighting fixture connections. (A) unmarked terminals. (B) polarized or identified terminals.

situation, all exposed conductive parts of a fixture must be grounded to the equipment grounding circuit. The only exception to this is when a fixture is mounted on some nonconducting surface at least eight feet vertically or five feet horizontally from a conducting and grounded surface. This would mean such things as sinks, tubs, showers, appliances, concrete floors, and tile walls that might have moisture on them. Regardless of this exception, your best bet is to make sure of a positive ground at all lighting fixtures for added safety.

The grounding equipment conductor of the power feed line should be attached to the fixture outlet box, just as with other devices. If the fixture is mounted directly to the box, the mechanical contact of the mounting means may be sufficient to ground the fixture, or you may put in a bonding jumper from box to fixture (sometimes necessary anyway). If there is no direct mechanical connection, then you must provide a grounding pigtail. Where the fixture is connected by means of tap conductors in a length of flex, use approved fittings and the flex will serve as the grounding conductor. However you do the job, the ground should be permanent and effective, with continuity of the equipment ground in the branch circuit ensured when the fixture is removed.

APPLIANCES AND OTHER EQUIPMENT

Connecting the appliances and miscellaneous equipment is similar to wiring up the devices; the same practices are followed.

Appliances

All appliances must have a disconnecting means, which is often merely an attachment plug and receptacle combination, especially with the portable appliances. Where the appliance is wired directly into the branch circuit, the circuit breaker protecting the circuit can be used as the disconnect when it is within easy reach of the user. In single family dwellings, the main service disconnect can serve as the disconnecting means. Attachment plug and receptacle arrangements are frequently used for ease of service, too, such as with ranges or even built-in ovens and cooktops.

Whether you are wiring to a receptacle or directly to the appliance or equipment, observe the color coding. Circuits for 230 volts will have two hot conductors, red and black, plus the white neutral, plus the grounding equipment conductor in most cases (Fig. 11-13). As with the other elements of the system, the equipment grounding circuit is essential and must be connected to the outlet boxes and to the frames of the appliances. With 115-volt plug-in equipment, this is accomplished through the grounding prong on the three-prong plug. The same is true of some 230-volt circuits where a

GND-GROUND   B-BLACK
W-WHITE   R-RED

Fig. 11-13. 230-volt circuit connections.

four-prong plug, one of which is a grounding prong, is used. When wiring direct, use the bare or green conductor, as usual, as the equipment grounding conductor. And as previously mentioned, metallic raceway or an approved grounding tape in a metallic cable assembly can also be used for the equipment ground.

There is one special instance that applies only to ranges, built-in ovens, counter cooktops, and clothes dryers. If such equipment is supplied by a 120/240-volt single-phase three-wire source, or by a 120/208 single-phase three-wire service supplied from a three-phase four-wire system, the neutral conductor and the equipment grounding conductor may be one and the same, provided that the conductor is no lighter than #10. No separate grounding conductor need be run to the appliance, and the frame of the appliance is bonded to the neutral conductor (Fig. 11-14A). Type SE cable with a bare neutral is sometimes used for this purpose, but such cable can only originate from the main service equipment, never from a load center or subpanel. This eliminates the likelihood of the bare neutral touching the isolated equipment grounding circuit which is also present, thus rendering it ineffective. If the appliance is supplied from a subpanel, a cable with three insulated conductors must be used (Fig. 11-14B).

This also raises another potential problem. It is possible that the range or dryer, especially if old or repaired, or installed and reinstalled a few times, has a frame that has been bonded at some internal point to the neutral. If a four-wire cable (Fig. 11-14C) with one equipment grounding conductor is then used to connect the range or dryer, the neutral and the grounding equipment circuit will become tied together. However, this should not happen except at the service equipment itself. You can easily find out if the frame and the neutral are bonded together by attaching the lead of your continuity tester to the neutral terminal or lead of the applicance and then scratching the probe of the tester against

A

B

C

W-WHITE   B-BLACK
GR-GREEN   R-RED

Fig. 11-14. Three possible cooking appliance connections. (A) typical three-wire appliance. (B) three-wire service with neutral used as ground. (C) four-wire service with separate ground.

the bare metal of the frame. If the tester reads positive and there is continuity, the two are bonded and you will have to separate them.

The actual process of making connections is about the same as for those items previously discussed. Simply strip the conductors, follow the color codes, allow the conductors plenty of the free room in the outlet boxes, arrange them in gentle bends, and avoid any abrasion or excessive compression. About the only difference is that instead of using Wire-nuts, you must use approved split lugs or similar connectors on the larger wire sizes, and then insulate the connections with splice tape and electrical tape to a thickness at least equal to the original insulation. More information on specific appliances can be found in Chapter 13.

Miscellaneous Equipment

About all that should be left now, except for the distribution end of the circuits, is miscellaneous odds and ends. Thermostats and other controllers are switches and are wired up as such, breaking only the ungrounded conductors in the circuit. Illuminated devices or motor-driven ones such as time clocks will require the attachment of a neutral as well.

Following the manufacturers' instructions plus the generalities of good wiring procedure and any points in the NEC that are applicable, you can go ahead with the finish installation of these items. Above all, don't forget the equipment grounding details.

MAIN ENTRANCE PANEL

The next step is to make up the main entrance panel. If you have not done so, strip and attach the service-entrance conductors to their respective connecting lugs, red and black to the distribution busses and the neutral to the neutral bus. The grounding electrode conductor should also be attached to this bus, but under its own lug—not sandwiched with the incoming neutral.

Wiring

If the service-entrance conductors are made of aluminum, make sure that the connecting lugs are approved for use with both copper and aluminum conductors. This information should be stamped on the panel, perhaps as "AL-CU approved." Copper and aluminum conductors should not be used under the same unapproved pressure connector or terminal because of oxidation, difference in coefficient of expansion, and the strong likelihood of electrolysis. Where they must be used together in this manner, certain approved procedures should be followed, such as the use of special compounds to prevent oxidation and electrolysis. It is easier in the long run to use only approved pressure connectors and terminals.

Your entrance panel may have the fuse blocks and distribution terminals already installed, but in most cases the panel will not contain the breakers or fuses needed for overcurrent protection. At this point you can go ahead and install all of the circuit breakers or fuse blocks. This merely involves snapping them into place. Some units are anchored to the busses with screws that must be threaded in with care to avoid crossthreading. Others mount by pressure clips or prongs.

Now strip the jackets off the cables, clear back to the connectors. If the cable is armored, this would have been done prior to installing the circuits in the panel. With a raceway system, of course, the conductors are not as a rule encabled anyway. Strip a short bit of insulation off the end of each conductor.

At this point, it is a good idea to make a quick test to make sure that there are no shorts in the system. Connect one lead of your continuity tester to the black wire of the first circuit, and the other to the white. There should be no positive signal, indicating that there is no electrical continuity or contact between the two conductors. If you get a positive signal, now is the time to find out why. Shift the lead from the neutral to the equipment grounding wire; again there should be no positive signal, no contact between those two conductors. Shift the leads to the neutral and the equipment bonding conductor; again there should be no contact. Similarly, there should be no reading between the red and black conductors of a three-wire circuit.

With all of the circuits clear, at least by the preliminary tests, you can connect the circuit conductors into the panel (Fig. 11-15A). Tie each black wire to an appropriate fuse or circuit breaker, leaving the tip of the conductor straight for breakers and fuse blocks that have a groove under the terminal screw, or looping the conductor end for terminal screws lacking a wire groove on the mount plate. The three-wire circuits have the black conductor tied to one half of the two-pole breaker or fuse block, with the red going to the other. All of the white wires go the neutral bus, each under a separate lug or terminal screw. Each equipment bonding conductor also goes into the neutral bus, which serves as the connection point to the grounding electrode conductor. If there are not enough terminals to handle all grounds and neutrals individually, pair each circuit neutral with its own grounding conductor under the same lug. Make sure all of the connections are tight, but not so tight that screw threads jam or the conductors are badly cut or mashed. Also make sure that the bonding strap is attached to both the main panel enclosure and the neutral bus.

As you wire in each circuit, jot down in a list the numbers of each circuit relative to the panel board markings and note the function of each (refrigerator, bedroom wall plugs, etc.). There is a blank label attached to each panel door that is supposed to be for this purpose, but it is usually impossibly tiny. When all of the circuits are made, type up a complete list of the circuits and paste it to the panel door with rubber cement. You will then have a much more satisfactory listing both for your own purposes and for the benefit of those who might later have to figure the system out.

Leave plenty of slack in each conductor so that every one of them will easily reach to any connecting point in the panel, in case there is need to shift some of the conductors around at a later time. Also, you need enough length so that each conductor can be tucked away around the back edges of the box and well out of the way of the bus bars, overcurrent protection devices, and the cover of the enclosure. With everything tidied up to your satisfaction, pop out the appropriate knockout slugs to expose the fuses or breakers in the cover and secure it into place. Leave all of the breakers in the *off* position, or leave the fuses out, until you are ready to energize the system.

Distribution Centers

Load centers, distribution panels, and subpanels are handled in exactly the same fashion as a main entrance panel, with one paramount difference—equipment grounding conductors are never tied to the neutral bus, under any circumstances. Make all of the hot feeder and branch circuit conductors into the panel just as you would in a main entrance panel. But you must provide an approved terminal block or strip for the equipment grounding conductors, mounting it in some convenient location within the enclosure. All equipment grounding conductors are run to this block. If the block, because of its design, is electrically isolated from the enclosure, provide a bonding jumper between it and the enclosure.

In addition, and not like the service entrance, the feeder cable must have four conductors: two hots, a neutral, and a

352

GND-GROUND W-WHITE B-BLACK R-RED

Fig. 11-15. Equipment grounding circuit connections. (A) service equipment distribution panel. (B) load center of other distribution panel, not a part of the service equipment.

353

Fig. 11-16. Making up connections in a low-voltage control relay cabinet.

separate equipment grounding conductor (bare or green). This conductor is also connected to the equipment grounding block. The result is a continuous equipment grounding system or network that ultimately feeds back to the neutral bus and the grounding electrode conductor at the main service equipment, but has no other contact with the neutral system at any other point (Fig. 11-15B).

Low-Voltage System

Wiring the low-voltage relay panel (Fig. 11-16) follows the same precepts outlined before. Connect the black branch circuit feed conductor to one black conductor from the high-voltage side of each relay; which one you use doesn't matter. Instead of trying to lump them all together, it is easier to use short interconnecting jumpers from relay to relay.

Join all of the neutrals going out to the fixtures with the incoming branch circuit neutral, again using jumpers where necessary. The equipment grounding conductor of the branch circuit feed is connected to all of the outgoing equipment grounding conductors traveling to the fixtures or loads, and also to the enclosure.

Fig. 11-17. Low-voltage control system connections.

Then connect the mass of little pigtail leads in the low-voltage compartment of the panel according to the diagram shown in Fig. 11-17 or the manufacturer's wiring diagram, using small Wire-nuts and jumpers between the relays. If each low-voltage circuit is tagged and the relays are labeled, your job of keeping everything straight will be much easier.

TURNING ON THE POWER

With everything installed, checked out, closed up and otherwise squared away, and any possible contact points between the new permanent system and the old temporary construction wiring cleared out of the way, all that is left for you to do is give the power company a call. When the man arrives with a meter tucked under his arm, the first thing he will ask you is if you're the party that called to have a new system hotted up. The second thing he'll ask you is if everything is dead and off, meaning that he wants to make absolutely certain that there is no power feedback from a separate source and that the main disconnect or utility service switch, if any, is in the *off* position. He doesn't want to cram the meter into the socket and have it blown back out at him.

Fig. 11-18. Checking a duplex receptacle with a circuit analyzer.

So, with the meter in place and sealed and the meterman gone, the moment of truth arrives. Cruise around the whole system and make sure that everything is turned off and that everything is unplugged, insofar as is possible. Check to see that all branch circuit breakers are in the *off* position (or fuses out). Then plug in the main fuses or flip the main circuit breaker. Unless something mighty peculiar has happened, they'll hold and not pop off or blow.

Now start activating the branch circuits one by one. If you have done your work correctly, they will hold too. Then you can go around the premises and try all of the various loads. If trouble occurs along the line, shut down that circuit and proceed with troubleshooting.

And last but far from least, go around to every 115-volt receptacle in the system and plug in a circuit analyzer to make sure that there are no opens or reversals and that the receptacles are properly wired (Fig. 11-18).

Congratulations! You have just wired your home.

Part 4:
Special Topics

Chapter 12
Modernization
and Updating

Modernizing is a catch-all term when applied to the general wiring field, and it can mean almost anything. Working with any wiring system that has been in existence for more than a few months is called "old work" by wiremen, and it can range from replacing a convenience outlet or a lighting fixture to putting in a whole new system, called a "rewire." It can also mean starting from scratch in an old, old home that has never had the benefit of a single electric wire. And it can mean bringing a middle-aged home up to more modern standards and capacity. Or, unfortunately, it can also mean redoing a two- or three-year-old home where corners were cut by the contractor and no provision was made for future expansion.

Whatever the circumstances, the modernizing process is seldom an easy one, for any number of reasons. The older the building and the deeper you dig, the more difficult it becomes. The owner or the electrician seldom knows what has gone before. Shabby, shoddy, or downright dangerous materials and workmanship may exist, hidden beneath the finish surfaces. Various articles of the NEC or local codes may have been skimped or altogether ignored. Older houses can be a can of worms that seems without end if they have been wired and rewired, added to both structurally and electrically, and have systems patched up and tapped onto by backyard mechanics with home-brew methods. Such houses usually present any number of surprises, along with considerable aggravation.

Then too, many homes were wired long before any codes, including the NEC, were much more than little-used guidelines. At that time, codes were not particularly complete, nor was the state of the electrical art in general very far advanced. Electrical demand was quite small in the average home. Material was sometimes of dubious quality, especially by comparison with today's standards, and it was rather limited in scope. And of course, age itself takes a toll; there are hundreds of homes about the country today that are still functioning with wiring systems that are three-quarters of a century old, that ceased to be either safe or adequate 30 years ago or more.

There are several ways to find out if your electrical system is up to snuff, some obvious and some not. Systems can be added to, extended, shifted around to use some old and some new parts, or the whole works can be torn out and replaced in any old house. Sometimes this can be accomplished with relative simplicity, sometimes not. And occasionally, all that is necessary is a bit of facelifting, though usually one thing leads to another and the job gets more complicated than expected. Covering every circumstance would be an absolute impossibility, but with a few pointers and generalities, plus a certain amount of thought and ingenuity, you should be able to tackle the task of modernization without any more difficulty than the job entails anyway.

SWITCHES AND RECEPTACLES

Probably the simplest part of modernization is replacement of the devices. Switches and receptacles do wear out in time, what with contact prongs that will no longer hold a plug in place and switch contacts that are burned or corroded and tumbler mechanisms falling apart.

Switches

You can generally make direct replacements for switches. Just pull the old worn and noisy one out and reconnect a new one; many very old wall boxes will accept new switches with no modification. Before you connect the new switch, however, check the insulation on the conductors. If it is dry, cracked, crumbly, or brittle, it has outlived its usefulness and is now both ineffective and highly dangerous. This means new conductors, and more about that a bit later.

If the outlet box is of an odd size, you will have to either replace it or modify it to accept the new switch. Either process may require care and a lot of fiddling around, especially if the box was installed so as to outlast the house.

Surface-mounted switches can be replaced easily, but make sure that the new switch body covers any holes in the

EXISTING DEVICE
BOX, IF SUITABLE

NEW 3-WIRE
GROUNDED DUPLEX
RECEPTACLE

EXISTING
2-WIRE CIRCUIT

COPPER GROUND
CONDUCTOR

GROUND CLAMP

COLD WATER PIPE
OR OTHER NEARBY
AND SUITABLE
GROUND

Fig. 12-1. Replacement of existing two-wire receptacle with grounding receptacle and ground connector.

mounting surface. Use an adaptor plate if necessary, preferably made of thin wood. Or, remove the surface switch and install a wall box and a new flush-mounted switch. The old rotary type of switch is still available today for direct replacement.

Receptacles

Installing new receptacles is usually not a difficult chore. Again, the old wall boxes will often accept the new device. Since the grounding type receptacle is a relatively recent development and its use throughout a home has been required for only a short time, you will undoubtedly find that any duplex receptacles you replace are of the nongrounding type for use with two-prong plugs. This brings up a judgmental situation, both for you and for the wiring inspector. There can be no question that an electrical system that includes an equipment grounding circuit is better and safer than one without. On the other hand, no home more than a few years old will have a grounded or bonded system, and elderly homes may have no effective grounding whatsoever, even at the service entrance.

The NEC does recognize this fact. It states that convenience outlets *shall* be replaced with grounding type outlets, and the grounding terminal *shall* be connected to a nearby cold water pipe (Fig. 12-1). It makes no difference if you plan to use only two-prong plugs on this receptacle—it *must* be a three-prong grounding type.

Now, this ruling gives rise to other considerations. First, you must find a way to get a copper grounding conductor, either bare or green insulated, to a nearby cold water pipe. Then, the water pipe itself must also be grounded, by virtue of being bonded to the service entrance grounding rod, or other electrode, or if a part of that piping system is buried underground for a distance of at least 10 feet, or both. This pipe must demonstrably be grounded and provide a good electrical path to ground with low impedance.

If it is impractical to reach a cold water pipe, or the pipe is not close enough (a judgmental decision), or does not exist, or does exist but is not or cannot be adequately grounded, then you may replace the old receptacle with a new nongrounding type. *Under no circumstances should you install a grounding type receptacle which in fact has no ground.* This could easily give a user a false sense of security and lead to a dangerous situation.

If the receptacle being replaced is outdoors, the new receptacle must be of the type that has a built-in ground fault interrupter (GFI), but only if you are able to adequately ground the receptacle.

As with switches, any frayed, frazzled, or tuckered-out conductors will have to be replaced.

FIXTURES

Replacing fixtures is sometimes a straightforward process of removing the old and reconnecting the new. More often, though, one problem or another arises to complicate matters. Probably you will have to rearrange the means of mounting, at least with a new combination of hardware and perhaps to the extent of replacing the outlet box as well.

Wiring

One common method of wiring old houses was to run practically everything through the fixture boxes, so you may be confronted with a maze of conductors rammed into a small box. There may be soldered taps and splices, wrapped with an unrecognizable gummy (or maybe rock-like) substance that used to be friction or rubber tape back in the good old days. Worse, the conductor insulation may be fried to a fare-thee-well, and even the conductors themselves could be annealed from years of heat dissipated from the fixture.

This situation may present a major problem and is one that you should never try to just get by with. Something will surely go wrong, sooner or later. If the box is overfilled by modern standards, then you must replace it or add extension rings so that the conductors have adequate room (see box fill tables in Chapter 7). Disconnect and remove each lead,

Fig. 12-2. Installation of new junction box and tap conductors at existing fixture outlet.

making notes as you go along as to how they were connected. Run them through new connectors of the appropriate type in the new box or extension and then reconnect.

The old conductors may be snapped up tight so that there is no slack at all to work with; another common practice from 'way back when. In this case, you may be able to add junction boxes at various points and splice in some extra wire, assuming that everything is accessible. You cannot, however, make connections in junction boxes and then throw them back into the ceiling or wall voids where they are inaccessible. Your only recourse is to run new cables. The same holds true if the conductor insulation is in bad shape and you can't gain enough slack to prune them back to the point where they are sound and usable.

If you find cooked conductors in a fixture outlet, or wires that are still usable but show some signs of taking a lot of heat, don't compound the problem by putting another and perhaps bigger heat-producer right back again in the same fashion. Depending upon the size of the fixture, you may be able to set a new junction box a short distance from the old, so that the new fixture will cover both. Then run taps of high-temperature fixture conductors from the new to the old.

If the old fixture box is in an accessible location, such as an open attic floor, you can move it to a different spot and substitute a new fixture box and tap conductors (Fig. 12-2). Failing that, you may have to work up an entirely new arrangement. And just as in new work, any combustible surface exposed between the rim of the outlet box and the outer rim of the fixture base or canopy must be covered with a noncombustible material.

Special Lighting Fixtures

Old surface lighting fixtures are frequently replaced with modern recessed types, and this too can present problems. One is to insure ahead of time that there is sufficient room at the fixture location to make the installation. The clearances from combustible surfaces (Fig. 12-3) used in new installations also apply to remodeling, and so do the insulation requirements. Another potential problem, especially in old houses, is adequate mounting of the enclosures, and keeping old plaster or other building components from crumbling away during the installation process. One advantage is that recessed fixtures, because of their size and construction, will often allow you to cut off the old tatty conductor ends and still have enough slack left to make the new connections without adding cable.

One common desire in upgrading is to replace a fixture that has its own pull-chain or canopy switch with one controlled by a remote wall switch. The first problem here is to get a switch loop from the fixture outlet to the proposed switch location, and the techniques for doing so will be discussed a bit later on. The second problem is that the fixture outlet box will invariably be too small to take another pair of conductors. This means installing either extension rings or a new fixture box, or perhaps additional junction boxes, or even new cables all around, as explained previously.

Fig. 12-3. Clearance dimensions for recessed lighting fixtures.

All newly installed lighting fixtures must have any exposed conductive parts or surfaces grounded to the equipment grounding conductor if one is provided and if the fixture is directly attached or wired to the fixture outlet box. Note that this excludes fixtures that simply plug in. However, if no equipment grounding conductor is included with the feeding circuit, either a separate wire or a metallic raceway or sheath, then the fixtures must be made of nonconductive materials or have no exposed conductive parts. But there is an out. If the fixture is mounted on a nonconductive surface and is at least five feet horizontally or eight feet vertically from any conducting surface, then it need not be grounded. Remember that the surface may only be *potentially* conductive, such as a tile wall or floor that might become damp, or ductwork, pipes, earth, concrete floors even though covered, appliances, sinks and tubs, and so forth. This means that you may have to choose your new fixtures, as well as their locations, with care.

EXTENDING CIRCUITS

Extending existing circuits is another method of modernizing. Usually this is done to provide one or two more convenience outlets on a given circuit, or perhaps to add a lighting fixture. When the load involved is small, like an extra table lamp or power for a radio or stereo, or where the job is undertaken simply to provide more convenience for powering loads already in use rather than additional new loads, there is not much to worry about. But when loads are added that total more than just 200 or 300 watts, care should be taken to avoid overloading the circuit and perhaps the system. Many circuits can already be operating, unknown to the user, under marginal conditions and nearly maximum loads.

There are a number of ways to add to a circuit. One is to sneak a cable through an open floor, ceiling, or wall cavity to the new location. If the extension is from an existing outlet such as a receptacle, you might have to enlarge the box to accomodate the extra conductors. Use a deep wall box if possible, or gang two wall boxes together and insert them in a double-sized opening. You can cover the box with a two-gang plate, one side for the old receptacle and the other side blank. The new items, whether lighting fixtures or receptacles, must be grounded if possible, as in the discussion on fixture and receptacle replacement. If you wish, you can also use cable with a grounding conductor for the extension, grounding the outlet box from which the circuit was extended and replacing the old receptacle with a grounding type. Check, too, to make sure that you do not exceed the maximum number of allowable receptacles on one circuit.

Another possibility is to make what is called an *underplaster extension*. This can be done only in brick, masonry, and concrete block construction. The conductors must be enclosed in rigid conduit, EMT, flex, type AC cable, type MI cable, or a special raceway approved for the purpose. The cable or raceway is set on the surface of the masonry and then buried in finish plaster. There might be cases where a variation of this could be used by embedding an approved type of cable in a framed wall chase and then plastering it over, but the cable or raceway would have to be protected by a strip of 1/16-inch metal for its entire length.

You can also extend a circuit with special surface-mounted material (Fig. 12-4). This is called a *nonmetallic extension* and can only be used to extend existing 15- or 20-amp branch circuits. The equipment cannot be concealed or run through partitions, but must be exposed and completely within the room in which it originates. It also must be used in finished areas, never in unfinished basements or attics. There are various designs and brands of this type of surface wiring equipment, and when installed correctly the finished appearance is not obtrusive or objectionable. Some types start within an existing outlet box; others actually plug into an existing receptacle. They are supported by small nails or screws, placed every 8 inches minimum, and no more than 12 inches from the supplying outlet. You can run them about anywhere or in any direction, just so that the location is finished and dry, neither on the floor nor less than two inches from the floor, and not in contact with any conductive surface

Fig. 12-4. Extension of existing circuit with nonmetallic surface extension equipment.

except the supplying receptacle itself, or its cover. Any taps or attachments must be made with equipment designed for that particular system, and bends should be made with great care.

Extending circuits that are already surface mounted, either in cable or raceway, presents no particular problems as a rule. All you need do is use the same or similar materials and keep going in whatever fashion suits your requirements, provided that the circuit will stand any extra load you might plan to impose upon it.

You might also want to extend one or more circuits to the outside of the building, perhaps for additional lighting or for convenience outlets. All such additions follow the basics of installing service-entrance conductor, as regards clearances, burial, support, insulation, mechanical protection, and the like. Devices must be contained in weatherproof boxes, fixtures approved for damp or wet locations, and receptacles backed by GFI circuit breakers and a good ground.

GROUNDING SYSTEM

When you make modifications or additions to an existing wiring system, whether the system is ancient or only a few years old, check out the grounding system. Only the newest homes will have a complete equipment grounding system, except for a few locales where local regulations have made that process mandatory for some time. Older homes probably will have only a service entrance ground. Some may have none at all, or a grounding system that time, corrosion, or mechanical damage have rendered worthless.

Inspection

An eyeball check is the first step. There should be a grounding electrode conductor leading out of the meter box or the main service equipment. You can measure the conductor diameter and check with Table 7-9 to determine whether it is large enough. In some cases it is.

Follow the conductor to its point of attachment at the grounding electrode and check the attaching hardware. There should be a good solid connection, bare metal to bare metal, with no sign of deterioration. If the hardware is damaged or eaten away, or if it consists of a plain metal strap, remove the old connector, clean up the contacting surfaces, and put in a new and approved type of clamp.

If the grounding electrode conductor is too small or is in bad shape, replace that as well. You can use as large a piece of copper as seems practicable, for the point is to provide ample ampacity for any fault current that might be carried and to keep the impedance as low as possible.

The chances are that you will not be able to tell visually whether the grounding electrode itself is effective or not. You can check the bonding system though. There should be solid connections, with the proper size copper jumpers (see Table 10-1) between all of the following grounding elements that exist on the property: service-entrance grounding electrode, aerial ground rods, lightning-protection ground rods, buried water-piping system, inside cold-water piping system, metal heat-ducting system, metallic sewer pipe and soil stack system, and interior gas piping as allowable.

In addition, you may wish to bond such items as buried metal septic tanks or well casings, or perhaps an irrigating piping system. A bonding jumper must also be run from one side to the other of a water or gas meter in cases where a grounding electrode conductor leads back to the service equipment and is connected on the service side of the meter, rather than on the supply or street side. Where the bonding of gas pipes or meters comes into the picture, discuss the situation first with the gas supplier and get their recommendations and requirements.

A metallic water-piping system, 10 feet or more of which is buried, is the preferred grounding electrode and must be used unless it does not effectively exist. Where another type of grounding electrode is used, the interior metallic water-piping system must be bonded to the grounding electrode at the service equipment, not at some random point within the building. Other bonding jumpers may be used to interconnect other elements you might like to tie into the grounding circuit. These can be placed according to necessity or expediency, but the conductors must be sized according to Table 10-1; you can't just tie on a handy piece of lamp cord.

The first thing to check, then, is whether or not the grounding electrode, the water-piping system bond, the main grounding electrode conductor, and the neutral bus of the service entrance equipment and distribution panels are all properly and effectively tied together. You can then prowl around the premises to locate other bonding jumpers that may have been installed at one time, checking to see if they are effective and complete. If the jumpers are too small or damaged, or connected with straps instead of approved clamps or pressure connectors, redo them. If you plan to install a larger service entrance, all of the bonding jumpers will probably have to be increased in size. Here again, soldered connections should never under any circumstances be used because under a heavy fault current they will simply melt. With the old bonds repaired, you can then add other bonding jumpers that you or the electrical inspector deem necessary or advisable.

Testing

You can't tell by looking just which parts of a piping or ducting or any other kind of metallic network are effectively bonded and which may not be a part of the circuit. You can check, though. All you need is an electrical continuity tester and a conductor long enough to reach from the attaching point at the grounding electrode to the farthest spot in the house where there is something to be checked. The conductor can be a reel of wire of about any size, or a series of extension cords hooked together.

Secure one end of the conductor to the grounding electrode—you can do this with a clamp or a pair of Vise-grips or in any other way that will ensure a tight electrical connection—and attach one lead of the continuity tester to the other end of the conductor. Trailing the conductor behind you, travel through the house, touching the other lead of the tester to any metallic surface you want to check. If you get a positive signal, that part of the system is bonded to ground. If you do not, scrape a bright spot on the metal and try again. If there is still no signal, probably that section of pipe or duct is electrically discontinuous with the grounding circuit because of a loose joint, an insulating coupling, never having been bonded in the first place, or some other problem.

By working your way through the network, testing section by section, you can discover where and what the problem is and correct it. This usually involves tightening up joints or perhaps installing more bonding jumpers around some electrically noncontinuous parts. Movable or removable parts, like pipe unions, should be bonded around in the same manner as meters or expansion joints.

It is absolutely essential that every point on a grounding circuit be thoroughly effective, and that the circuit and the path to earth have as low an impedance as possible. In the event of an electrical fault or a lightning strike, current will follow the course of least resistance to earth, and if there is a weak link in the system, especially if coupled with a high impedance at the grounding electrode, this is where trouble and probably damage will occur. Heavy fault currents can be incredibly powerful and at times explosive.

A case comes to mind that illustrates both the strange ways of electrical force and the need for careful installation of equipment and inspection of grounding systems. The incident took place in a large house about a hundred years old that was being partially remodeled. A relatively modern main entrance equipped with circuit breakers had been installed sometime earlier at the front end of an unfinished cellar. A new subpanel was being put in at the back of the cellar to feed a new kitchen under construction directly above. The grounding electrode was the water piping system, and an extensive cast-iron soil pipe system was bonded to the ground. A nonmetallic cable feeder from the main entrance panel was stapled for about 70 feet along supporting beams and across the cellar to serve the new subpanel.

With the installation complete, the wireman threw a circuit breaker on at the main service panel to energize the new subpanel. The cellar lights dimmed slightly, and there was a peculiar loud sputtering sound from some point far away in another part of the cellar. Then the lights grew dim, and a reflection of eerie flickering blue light could be seen. The effect was much like an electric arc welder at work. After some six or seven seconds—a very long time to wonder what in the world is happening—the main breaker tripped and everything went off.

A quick check revealed no apparent or obvious problems with the new installation, nor could the source of the blue light or sputtering noise be found. By this time the circuit breaker had cooled down. The wireman tried again, with exactly the same strange result. A more complete search ended behind a stack of cordwood, where a line of 4-inch waste pipe slanted down to join with a larger line. One length of pipe had developed a long crack in the bell, right at a joint between two sections, and forced apart the poured lead seal in the joint. This was the weak link in the grounding system. High heat developed from a fault current in the new installation as the current arced over the bad joint, completely melting the lead so that it ran out into a pool on the floor. The connection of the grounding electrode conductor to the water supply pipe was also found to be faulty, but the problem surfaced, dramatically, at the weakest point in the grounding system.

The fault was finally traced to a spot midway along the new feeder cable, which was secured by means of long-legged staples. A leg of one staple had struck a nailhead in the beam, turning itself inward and completely around. As the other leg was driven into the beam, the first drove back down into the cable, directly shorting two of the conductors. Under circumstances just slightly different, the results of this odd combination of malfunctions could have been tragic.

The final proof of the worth and effectiveness of a grounding system lies in making actual tests. This is done only with a ground tester, and these instruments unfortunately are both expensive and not in widespread use. You will probably have to search for a bit to find someone who has one and knows how to use it. But by contacting your power company, electrical supplies dealer, or local electrical contractors, you can surely arrange to have the tests made. The path to ground at the grounding electrode must be no more than 25 ohms, and the less the better. You can also have the bonding system checked out to make sure of a low impedance there as well.

Probability says that a good grounding system will last for years. But the possibility exists that it will not. So checking the impedance every year or two is strongly recommended, even though that recommendation is seldom followed. In localities where the soil is nearly always moist, where the average humidity is fairly high, and where rainfall is at the least moderate and consistent, grounding electrodes will usually keep a good path to ground. In arid areas, however, this may not be true, so the ground path should be checked frequently.

ELECTRICAL SYSTEM

How do you know when part, or all, of an electrical system needs replacing? Sometimes the matter is forcibly brought to your attention. One particular circuit, for instance, may repeatedly blow fuses or trip breakers. This could happen a number of times in a short period, or perhaps only every once in a while. Either way, that circuit is in trouble for one reason or another, and consequently so are you. Or, a sudden short circuit, either in the line itself or in an attached load, may overheat the conductors to the point where they are unserviceable and have to be replaced. Someone may get a shock when touching a metal surface, eventually leading to a bit of rewiring. Or perhaps the addition of household electrical equipment will dictate one change, which may then lead to further work.

Inspection

Often, a system may be operating marginally, unknown to the user. Thoroughly checking a system is an excellent idea, even if the home is fairly new and everything seems to be in pretty good shape, and it's not a difficult job. Visual inspection comes first, and you can start with the grounding network as just described. Check exposed conductors in the attic and basement, looking for signs that the insulation has been gnawed by mice or squirrels, or for any other mechanical damage.

With the current turned off, locate a slack piece of cable and flex it gently. If the insulation is intact and still flexible, the chances are good that the same is true throughout the entire house. But if it is brittle and cracked (Fig. 12-5) or otherwise damaged, this could be the case elsewhere in the building. More extensive investigation is then in order.

If the wiring is armored or in raceway, look for signs of rust, corrosion, loose fittings, and potentially poor or nonexistent grounds. Again, if you see this at the few points you are able to inspect, the problem is very likely prevalent.

Pull a receptacle or two out of their boxes at random and check the condition of the insulation. If it is brittle and crumbly here, it may be throughout the system. You might find a similar, or worse, condition at the fixture outlets, but this may be from the heat of the fixtures rather than a general problem. Also open a few junction boxes to see how ram-jammed full they are, and note the insulation condition here, too.

Then pull the cover off the main distribution panel and see how that looks. In theory, there should be only one hot

Fig. 12-5. NM cable cracked open from installation under improper conditions.

conductor under the terminal at each branch circuit overcurrent protection device. Once in a while you may find two, wired that way more for convenience than to try to squeeze a little more out of the system. If there is a jumble of wires with two or more tied to each circuit point, the panel is too small, though not necessarily overloaded electrically.

Take a look at the service drop and service-entrance conductors, too. If they look frayed and frazzled or have suffered some mechanical damage, they have been there too long.

Testing

If your system passes the visual test and you don't notice anything that seems out of the ordinary and everything appears in good shape, you can make some electrical tests.

First, run down each individual circuit and see what loads are imposed on them. You will probably have to use a trial-and-error elimination method. Turn off all circuits except one, then go around the house with a tester and see what is alive. Make notes as you go along, and then draw out a circuit tree like the one in Fig. 7-1. Note the size of each circuit conductor, which may or may not be the same ampacity as the overcurrent protection device in the circuit. If you are unsure, disconnect one of the conductors and check it with a wire gauge. Then tally up the individual loads, plugging in any demand or other factors as necessary, to arrive at a total load. Check this figure against the ampacity of the conductor, assuming a new circuit (see Chapter 6). Don't forget to count the number of receptacles to see if the total exceeds the allowable maximum. Any excess will have to be reduced to obtain the proper maximum.

When you have made these calculations for each circuit, the resulting figures noted on the circuit tree will tell you which circuits are overloaded. It will also tell you which have almost no load, as often happens, and you may be able to simply rearrange and rebalance the loads so that each circuit shares equally in the total load. (There are times when this relatively simple procedure can save doing an entire wiring job.) Then you can add up the totals for each circuit, plug in any necessary demand and diversification factors, and see whether or not the feeders and distribution panels, if any, and the main service equipment and service-entrance conductors are of sufficient capacity to take care of your needs. If they fall short or are marginal, new service equipment is in order. If you find overloaded circuits, you may have to install new ones to pick up the overload, provided that the service equipment can handle the additions. You will need both physical circuit space and current-carrying capacity.

Another more satisfactory method of checking an electrical system for its capacity is to make a series of current readings while the system is in normal operation. You can do this by disconnecting each circuit's hot line, attaching it to one terminal of an ammeter, and then connecting a jumper lead from the second ammeter terminal to the circuit terminal. Any current drawn by the circuit will register on the series-connected ammeter. This is a lengthy and cumbersome process, though, and the job can be done much more easily with a snap-around meter such as the Amprobe or Amp-clamp (Fig. 9-11). With this instrument, movable jaws clamp around the live conductor and the small current induced in them is translated to a true reading on the dial. You can quickly take a series of readings on individual circuits, and on the total current draw of the system as well, at different times over a period of days. By averaging out the readings and noting the peak currents, you can easily determine which circuits are approaching or exceed the danger level and whether or not the service entrance equipment is of adequate capacity. If the answer is no or maybe at any point in your tests, you had better think about doing some rewiring or rearranging.

Another indicator of problems within the system is the way your lights act. If they dim or flicker when an additional load is thrown on the line, you may be reaching the capacity of the circuit, or there may be a substantial voltage drop on the line. There is also the possibility that the supply voltage coming to the house is sometimes lower than it should be. You can check your incoming voltage with a voltmeter across the two main legs of the service entrance, making your checks at various times of the day and night. If the problem seems to be on the supply side, contact your power company, show them your figures, and ask them to take their own readings and try to correct the problem.

To check your branch circuits, you will have to take voltage readings first with the circuit unloaded, then with the load activated. This entails first diagramming the circuit so you know which loads are where, then determining which outlets are the farthest, in terms of conductor footage, from the source. Then read the voltage at the head end of the circuit; let's say that you measure 115 volts. Turn the circuit off, attach the leads of the voltmeter across the terminals of the farthest load, making sure that everything is clear and no shorts can occur while the outlet box is opened up, and turn the circuit back on. With the load turned on, the reading might drop perhaps to 109 volts. Other loads on the circuit may drop the voltage even further, depending upon conductor size and the location and nature of the loads. But even at 109 volts, you would have a 5% voltage drop, which 2% more than the allowable maximum for branch circuits under the NEC. This drop is too much for proper operation of the loads on the circuit. Again, this means either rearranging the existing circuits if possible, or adding new circuits and dividing the loads up, or rewiring the system.

Operating loads in a consistant under-voltage condition, aside from the dangers to life and property involved, can be an expensive proposition. Efficiency is lowered, operating costs rise, and early failure of the load equipment is inevitable, so repair and replacement costs rise rapidly. This is especially true where expensive motor-driven equipment is concerned. The costs of rewiring an inadequate system would probably be returned to you in savings over a relatively short period of time.

ELECTRICAL SAFETY

As each year goes by, the state of the art in the general wiring field, as in most others, advances and hopefully improves. New equipment and material is introduced to the market, old material and equipment is upgraded. Systems are made safer, more trouble-free, more efficient. New regulations come into effect in the National Electric Code and in local codes. The result is that you could make a perfectly safe bet that not one residential wiring system in a hundred that was installed more than four or five years ago would pass inspection if installed today with the same materials and methods. This gives rise to the question of whether any old wiring system is safe.

Assuming that a particular system was installed using quality materials and with good workmanship, and that it is not now overloaded, deteriorated, or otherwise inadequate for the demands placed upon it, then the system is probably "safe" regardless of age. But the question is really one of degree. A new system, all other things being equal, is undoubtedly safer than an old system in terms of potential hazard to life and property.

You cannot be compelled, under most circumstances, to completely rewire an old installation simply because it does not come up to today's standards and code regulations, unless or until that system fails or becomes totally inadequate. Otherwise, practically every house would have to be redone every two or three years. But there is much to be said for bringing parts of an old system more into line with new standards on a periodic basis, just to increase personal safety and convenience if nothing else. You do this with your automobile, your sporting equipment, and your furnishings—why not your electrical system?

For example, one particularly hazardous point in the residence has been, and still is, the ungrounded two-slot outdoor receptacle. Many a person has received a dismaying electrical jolt while standing on the wet ground and working with an ungrounded power tool. The 1971 NEC made mandatory the use of a new device, the ground fault interrupt (GFI) circuit breaker, to protect all outdoor receptacles, thus greatly minimizing this potential hazard. By equipping your outdoor receptacles with this same device, you can provide yourself with additional safety at low cost, and bring your electrical system more into line with present standards. In 1975, the NEC extended this GFI requirement to all bathroom receptacles, where essentially the same hazard exists. This would be another way to upgrade your system. A good ground must be provided in both cases, of course.

You can easily find out where your system is deficient by reading through the latest NEC and any local codes that might apply, and then comparing with what you have. A discussion with your supplier, the electrical inspector, and perhaps an electrician might help you to decide what modernizing steps you should consider, and how those steps might best be taken.

BUILDING CONSTRUCTION

Adding substantial new loads to a system, running new circuits, extensive rewiring of old circuits, or the installation of a new service, usually involves a lot of digging into the innards of the building. This means that knowing how the building was put together will be considerable help to you during the rewiring process.

Modern Homes

Nearly all of today's homes, whether precut, tract, or architect-designed and custom-built, are constructed on a modular basis, the root of which is the 4-foot by 4-foot panel or section. Multiples of two are widely used in overall dimensions, room dimensions, and so forth. Framing members are placed on 16-inch or 2-foot centers, with standard thermal insulation made to fit exactly between them. Door sizes are standardized for the most part, window sizes to a somewhat lesser degree. Floor and ceiling tiles come in standard sizes, as do brick and masonry block.

377

In addition, certain methods of construction and principles of building are widely followed throughout the country, so that there is always a certain predictability about any given kind of home regardless of location. You can be reasonably sure that a frame house in California will be built in a similar fashion to one in Vermont, for instance. Masonry veneer houses have similar characteristics, as do stuccoed homes. Even solid masonry, solid log, sprayed foam, and geodesic structures each have common characteristics of construction. So when you start to work in a fairly modern frame house, one built since the Second World War, you are fairly well assured that the wall studs will be mostly 16 inches from center to center, floor joists probably the same, the depth of the walls will be about 3 1/2 inches, there will be a sole at the bottom of each wall, a single or double plate at the top, and so on.

This modular concept, however, was a product of the late 1800's and was not in much use until well after the turn of the century. More widespread acceptance was aided by the standardization of materials and hardware by the building trades industry, so the idea came on strong in the late 1930's and after.

Older Homes

Houses built a hundred and more years ago were put together differently. In fact, each one was built differently than the next, or so it seems when trying to work in them. Often as not the final dimensions or the manner of construction was determined by the size and type of whatever building material came to hand. Materials were not necessarily sized to any particular standard, but rather to fit the job in progress. A house might actually be six inches higher at one roof corner than another, and beams of different sizes on odd centerings were the rule rather than the exception. Sawdust was sometimes poured into wall cavities for insulation, some walls were bricked solid under the plaster. Overhangs and mismatches from floor to floor were common, as were exposed corner posts and inside Y frames for strength.

The point is that before you start doing any major rewiring, especially if extensive remodeling is not also part of the project, you should try to determine how your house was put together internally. There never was a wireman doing "old work" who didn't wish for Superman's X-ray eyesight so he could figure out how to get a cable from here to there without tearing the house down in the process. Poke around in the attic or basement or crawl space—any unfinished area that you can squeeze into—to get some idea of how the framing members are positioned, whether they follow a pattern that is predictable, the thickness of the flooring, how the upstairs partitions line up with the downstairs ones, and that sort of thing. Look for ways to drill cable holes so that they won't appear in the middle of a finished surface. Check the thermal insulation and look for horizontal fireblocking in the wall spaces or solid bridging between joists to see how and where you might be able to fish cables through.

If you are not familiar with general construction principles so that you can visualize the entire framing of a house, just the skeleton with no covering, then a session at the library might help. Architectural and construction books are loaded with drawings of all kinds, including the old methods, and these may give you some clues as to how you might proceed. Remember, though, that you must expect the unexpected, and the older the building, the more unexpected's there are likely to be.

OLD WIRING SYSTEMS

There are two general types of old home wiring that deserve special mention. One is an armored cable known as *BX*, about which much has been written and a lot more said—largely uncomplimentary. The second is called *knob and tube*.

BX Cable

BX was sometimes improperly installed and often has its connections crossed up as various mechanics worked on the system down through the years, with the result that the steel casing became "hot." This problem is so common that electricians always approach BX with wariness; even with the circuit turned off, the cable might be hot from touching another one.

Ground faults in BX cable caused the steel to heat up and started fires. No grounding conductor was included in the cable, as it is with the newer type *AC*, so the jacket impedance was high, causing grounding problems. The special fittings supposed to be used with BX were often ignored, or disappeared as the system was repaired or revised.

If you have a BX system, treat it carefully and inspect it thoroughly. You might be well advised to replace the cable completely.

Knob and Tube

Knob and tube was widely used in the early days of electrical wiring. It was a good system and still remains in functioning order in many homes today. The name comes from the fact that the circuit conductors were run individually, either concealed or in the open, supported above a mounting surface by porcelain knob insulators called *stand-offs*, or through porcelain tubes inserted in holes bored through framing members. Taps were made in the open without benefit of boxes. The conductors were enclosed in short lengths of flexible tubing, made of what looked like heavy canvas and called *loom*, wherever they entered the device boxes.

This type of wiring can now be used only in extensions of existing installations, but only under certain conditions and following certain regulations. Article 324 of the NEC has complete details on restrictions. Knob-and-tube wiring can also be left intact and tied into the distribution panel of a more modern system under some conditions. But more often than not, from a standpoint either of convenience or local regulations, this type of wiring system is completely eliminated whenever extensive rewiring is undertaken.

REWIRING

Trying to tell you exactly how to go about rewiring your house would be a waste of time. Even you won't know exactly how to do some parts of the job until you actually tackle them; a certain amount of trial and error is inevitable, especially in early homes. The rules for redoing old work is the same for installing new systems as far as grounding, bonding, capacity and ampacity, circuit loading, connections, rating and derating, specific use of materials and equipment, and general wiring principles and practices are concerned. Branch circuit, feeder, and service entrance calculations are made just as though you were making the installation in a new home, whether the work involves one circuit or a whole system. The problems involved in rewiring are seldom electrical, but most often physical and mechanical, actually getting the conductors and boxes into place with a minimum of work and building repair.

Perhaps the easiest way to describe how to proceed would be to outline the installation of two different circuits. The first is the rewiring of an existing circuit, retaining the old outlet locations. This would also cover lengthy extensions from existing circuits and some switch loops as well. The second is the installation of an entirely new circuit with new outlets. The same would apply to new switch loops or to new feeder circuits.

Rewiring an Existing Circuit

We will assume that in rewiring the old circuit that no new devices or loads will be directly connected and no new outlets are needed, and that any additional loads that might be thrown on the circuit will not overload the distribution panel. And since most of the older branch circuits were run with #14 wire having a 15-amp capacity, we can also assume that you will replace the conductors with #12 for a bit of additional capacity and safety margin.

Electrical Work. First, remove the fuse or turn off the circuit breaker for the circuit to be rewired. Arm yourself with a flashlight, turn off the main disconnect, and remove the cover of the main entrance or distribution panel. Locate the hot wire of the circuit and its associated neutral and pull them away from the rest of the group. Next, turn the main disconnect back on—after you get all of your fingers and screwdrivers out of the way.

Now place one lead of a voltmeter or a circuit tester on the neutral bus and the other on the hot wire of the circuit to be replaced, making sure that it is dead. The reason for making this check is that sometimes the circuits in old houses get scrambled around so that they have two feeds, and just because the circuit is turned off doesn't necessarily mean that it is dead. If the circuit is still live, turn off the rest of the circuits until you find the one that is backfeeding.

Turn the main disconnect off again, remove the circuit conductors from the panel, replace the panel cover for the time being, and turn the power back on. If there was a backfeeding circuit, leave that one off too, but place a piece of tape over the fuse socket or breaker handle so that someone won't turn it back on while you're not looking.

Devices. Remove all of the devices on the dead circuit from their boxes, take down the fixtures, and disconnect any other loads. Check each one with a circuit tester before you tear into it, just to make sure that it is indeed dead and you have not picked one on another circuit by mistake. A direct short circuit can ruin a five-dollar pair of diagonals and raise havoc with your heartbeat. Clip the conductors off close to the connection point, leaving as much slack in the outlet box as you can. Unless the receptacles are the grounding type and in good shape, consign them to the nearest trash bucket. Switches likewise, unless they are fairly new and their actions are solid and contacts good. After all, replacements are only a buck or two. The lighting fixtures may be all right as is, just plain junk, or suitable for repairing, rewiring, or remaking into a unit of entirely different appearance; it's your choice.

Outlet Boxes. All concealed outlet boxes will have to be removed. Loosen up the cable clamps and work the boxes out of their holes carefully to preserve the surrounding finished surface. Feed the conductors backward through the box holes without losing them into the cavity. If the wires show a tendency to disappear, tie them to a scrap of wood propped across the face of the box hole. Surface-mounted outlet boxes or those which are accessible from both front and back, like a ceiling fixture box with an open attic floor above, can be left in place if they are big enough to handle the number of conductors you plan to bring to that point, and if they will accept the proper cable fittings or clamps used with the new cable.

At this point, it is best to double-check and make sure that you have disconnected every element of the circuit. If you have a backfeed problem, go back along the circuit and make sure that none of the clipped-off conductors are touching. Turn the backfeeding circuit on again, and with a circuit tester make the rounds of each pair of conductors. Only one pair should be live. Cap these conductors with wire nuts for identification.

Now turn the backfeeding circuit off, figure out from which of the live outlets in the electrical system the conductors originate, and disconnect them. The best policy is to remove those conductors completely if you can. If they are caught or stapled within the structure, you may have no choice but to clip them off short and shove them far back into the cavity. But if you do that, make double sure that those conductors are still dead when you turn the backfeeding circuit on again.

Running Cable. Next you can figure out the best general course to take with the new cable; that may or may not be the same as the original path. If you can, go from outlet to outlet to outlet, rather than to one outlet and then spraying out to several more from there. The cable, incidentally, should include an equipment grounding conductor even though the old one most likely did not.

With the course mapped out the fun begins, and you will probably need a helper to feed or pull the cable with you. One old trick is to pull a new cable in as the old one is pulled out (Fig. 12-6). Strip the ends of both conductors of the old cable, likewise with the ends of the new. Form hooks in all of them, hook them together and crimp the hooks flat. Wrap the joint with tape until you have a smooth covering that won't snag, but not so thick that the joint can't hinge or bend a little. Then, if the old cable is not hung up or stapled down at some hidden point, you should be able to slip the new one right into place. If this proves impossible, you will have to clip the conductors off, stuff the remains back into the opening out of the way, and start fresh.

Fig. 12-6. Joining old and new cables, to pull the new in as the old is pulled out.

There are any number of possibilities for getting a cable through. (Figure 12-7 shows a few.) Exactly what you must do depends upon how the building is put together at the place where you are working. If there is an open and unfinished attic above or cellar below, you may be able to drill or bore into the cavity where the outlet is located and feed the cable up, or down, into place as in Fig. 12-7A and B. If the outlet is near the floor, you may be able to work a drill bit on an extension into the opening on a steep angle and bore down through the sole and into the open. Failing that, you might be able to remove the mopboard or molding and drill down through, then pull the wire up, cut a notch or a slot in the wall, re-feed the cable to the outlet, and cover the work area with the mopboard. If there is a wide molding at the ceiling line, you might be able to use the same trick to bring a new cable down from above.

To feed two or more outlets in a line, you might have to drop a line into the basement from one and come back up to the next, and so on, or perhaps go up into the attic and then drop back down. There might also be room enough behind a mopboard to tuck the cable into a gap between the wall and floor, which often exists in old houses, or you may be able to cut a channel to serve the same purpose, notching the cable back into the wall cavity above the sole at convenient points. If you do this, make sure that the cable is well protected and that you don't cleave a conductor with a nail when you replace the trim. If the building is of post and beam construction, you may have to box out around one or the other with molding or planks, building a new trim section to enclose the wiring.

Fig. 12-7. Various ways of fishing cable through voids in existing structures.

Fishing. Fishing or double-fishing is often necessary to get across long spans or past obstacles (Fig. 12-8). For instance, to get a cable up through a wall and across an enclosed ceiling to a fixture outlet could mean first cutting a small hole in the wall just below the plate (Fig. 12-8A). Bore a hole up through the plate and run a snake into the ceiling cavity, with a crook in the snake so that it points out toward the fixture outlet. Have your partner run another snake into the fixture outlet opening and across above the ceiling (the two snakes must be in the same cavity) and try to hook the snakes together by twisting and rattling them about. This sounds difficult, and often is, but it is the method most widely used by all electricians for pulling concealed cables in old work. Three principal requirements are necessary: time, patience, and persistence. A little round of cussing sometimes helps, too.

Once the two snakes are hooked, one man keeps a constant tension by pulling back slightly so that the hooks don't slip apart, while the other pulls both snakes through until the hooks are in the open. Then the cable is hooked and taped to the snake that passes through the cavity, and the cable is pulled back. You now have a cable from the fixture outlet dangling out of a hole in the wall near the ceiling. The next step is to run a snake up through the wall, let's say from a switch outlet, to the hole. Tape the snake to the cable and pull it back through, using extra care when the last few inches of cable double into a loop and disappear through the hole. Don't let the cable bend too tight or damage the insulation with the last tug.

This same system can be used any time you are working blind from hole to hole. You can also sometimes push a length

Fig. 12-8. Double-fishing to pull in new cable.

Fig. 12-9. Routing cable around an existing wall blocking or firestop.

of cable with the end bent into a loop, into an outlet hole and then snag it from the opposite direction with a snake, and so pull the whole works back through. Both holes usually need to be big enough to stick your hand into, though, rather than a small-bore hole.

Another method of fishing wires that sometimes works better than a snake actually makes use of genuine fishing gear. Use a length of heavy nylon fishing line of 10-pound test or more, or squidding line, with a fairly heavy sinker attached to one end. You can even leave the line on the reel for extra convenience! Drop the sinker and line down through a drilled hole, wall cavity, old chimney or pipe shaft, or whatever, and pick it up at the other end, either with a snake or your hand. A length of light-fixture chain will do the same job.

Obstructions. Many houses are built with blocking or firestops, with chunks of dimension stock nailed horizontally between the wall studs in the outside walls or with solid bridging blocks between rafters or joists. Here you have two choices, both of which lead to later repair work (Fig. 12-9). You can cut a small hole above and below the blocking, chisel a notch out of the member, feed the cable out through one hole and in through the other around the blocking, and then plaster over the cable. Or, you can cut a somewhat larger hole on one side or the other of the blocking, drill a hole through the member, and fish the wire through the hole. Then cut the plaster out a bit wider than the hole in the plaster lath, insert a piece of backing—a piece of mesh or perforated sheet metal works well—and plaster over. If the covering is Sheetrock, mortise out a groove around the edges of the hole for the backing to rest on, and then plaster over.

Installing New Devices. Once all of the conductors are in place at the outlets, the worst is over. Replace all of the boxes, using new ones where necessary because of size, damage, or incompatibility of the old ones, and secure the cables into the boxes. Fished cables do not have to be stapled as do those installed in new work; they remain unsupported except at the box clamp itself.

Install the devices elsewhere, just as though the work were part of a new system. Make sure that the boxes are correctly set and the gaps between box edges and mounting surfaces are plastered in tight, and that the combustible surfaces under fixtures are covered with a noncombustible material. Mount the various cover plates and canopies, and check the circuit for shorts by testing each conductor to ground with a continuity checker.

Powering Up. Turn off the main disconnect, remove the main panel cover, and install and connect the circuit conductors. In this example, you replaced a 15-amp circuit with a 20, so though you don't have to, you may want to change the overcurrent protective device to a 20-amp rating.

If fuses are used, install a new 20-amp type S fuse. Chances are that the panel is equipped with standard fuses, so you may want to replace all of them with type S, which will mean that the various sizes cannot be interchanged as is possible with standard fuses. This done, slap the panel cover on, turn on the power, and check out the devices and loads in the usual manner.

Installing a New Circuit

Installing a new additional circuit is easier in some respects than replacing an existing circuit. The first step is to design the circuit itself, as outlined in Chapters 6 and 7. The second is to make sure that the service entrance panel, or subpanel, is capable of handling an additional circuit, both physically and electrically. To do this, you may have to undertake the calculations and tests outlined earlier in this chapter. You probably have already decided on the approximate locations of the outlets on the circuit, so if the answer to your calculations is a yes, then you can mark the specific locations, remembering that at this point they may still be tentative. That's because upon close inspection, you may find that you can't get from there to here with your cable. And that's the next step.

Routing. Look around for the easiest routes to run your cable and try to make certain ahead of time that you will be able to reach all of your proposed locations. Again, wiring point-to-point is best, but you can always fall back upon a conveniently located junction box (remember that it must be accessible) from which you can feed two or three outlets if absolutely necessary.

Locate the studs, joists, or other framing members. Look for blocking or other obstacles. Check for problems with thermal insulation or getting around door and window openings and past foundations, sills, bearing beams, and such. Look for natural paths, such as abandoned pipe or chimney shafts, alongside existing ducts or vents (but not in them), hollow partitions that line up from one floor to another, closet corners, and under book or cabinet shelves where the cable can be later covered with a trim piece.

Cutting Plaster. With the outlet locations and the cable route spotted as closely as possible, you can cut the opening for the boxes. A word of caution if you are working with old plaster—take your time! Plaster cracks and crumbles easily, and you can wind up with a major repair job before you realize what has happened. Hair plaster, a mixture of lime plaster and horse hair, was commonly used in old houses and is especially difficult. The old wood lath used as a plaster backer is tough and springy and sometimes not attached too well to the framing members. When you outline the hole to be cut, position it so that the center lath is completely cut away, but only part of the upper and lower laths is cut into (Fig. 12-10). This will give you less trouble and a more solid mount. Poking about with a screwdriver—or better yet a scratch awl—will show you where the laths are. There is usually about a quarter-inch gap between each one.

Drill holes at the corners of the box outline with a 1/2-inch bit, then cut across top and bottom. When you make the side cuts, the chunk of top or bottom lath will come out first, allowing you to exert some stabilizing pressure with your fingers on the middle lath to keep it from chattering back and forth as you cut. If you don't hold the lath, the sawing action will loosen the plaster and probably break the wall out from stud to stud. Beware of splinters, too; most of that old stock was rough-cut and is covered with dirt and mouse-carried

PLASTER APPLIED
OVER WOOD LATHS

PLASTER
LATH

NOTCH TO
CLEAR DEVICE
SCREWS

CUT ONLY
CENTER LATH
COMPLETELY

STUD

Fig. 12-10. Cutting wall box into old plaster lath.

germs. A strip of masking tape around the perimeter of the hole may help to keep large chips of plaster from breaking away as you make the cuts.

Installing Boxes. If there are any surface-mounted device or junction boxes in the circuit, these can be mounted right away; the concealed boxes will wait until later. Run the cable to the surface boxes, and then to the concealed location openings in whatever manner suits the occasion. This process is little different than the one described in the rewiring of an existing circuit. Drill holes, chop out little sections of walls and ceilings—whatever is necessary to get the job done. After the cable has been pulled to the concealed locations, install the outlet boxes.

You may not be able to mount the boxes with ears flush against the finish surface if you are troubled with crumbly plaster. Instead, dig a bit of plaster away from the lath at the mounting points, reset the ears so that the face of the box is flush with the finish surface when the ears are against the lath,

Fig. 12-11. Mounting wall box, inset into old plaster.

and secure the box in place (Fig. 12-11). Screws do not always work, and if loose or in a soft material, they are useless. Use plaster clips in this case, as well as for mounting boxes in wallboard. The conductors should be pulled into the box and clamped before the box is fastened down.

Finishing Up. Install the various receptacles, lighting fixtures, and loads as explained previously. Then apply the finish covers and plates. Test the circuit, take the conductors into a circuit slot at the panel, put in the overcurrent protection device, and run the final test.

INCREASING SERVICE

It may be that the existing wiring system is in good enough shape but there is not enough there for convenience. Perhaps there is no more room or capacity available at the service or distribution panel. Or, you may be planning an extensive addition to the house, or more and bigger appliances, which will necessitate additional wiring. These or other reasons will mean leaving the existing branch circuits pretty much intact, but replacing the service entrance equipment with one of greater capacity.

Planning

Again, planning is the first step, and you can refer back to Chapters 6 and 7 for more complete information. You will need first to figure out what your present load is, both circuit-by-circuit and total, and decide which if any of the existing circuits needs to be rearranged, and if so, how. Then you should plan the additional circuits and make allowance for later expansion. This will lead you to a total connected-load figure for immediate usage, plus some extra and as yet only estimated future load capability.

Contact the serving utility and have someone come to the house to look the situation over, telling him what you have in mind. Find out if the utility insists upon installing the new service drop in exactly the same location as the old, or if it can be moved a couple of feet one way or another. If you don't have to use the same location, this means that you can put in a new meter box, service-entrance conductors, main disconnect, ground, and main entrance panel in a new spot. Then you do what electricians call a *changeover* or a *cutover*. With this method, while the power company removes the old service drop and installs the new one, you can move the circuits from the old box into the new and be without power for only an hour or so, rather than a day or so.

Installation in New Location

Assuming that a new location is agreeable to all parties, you can go ahead and install the new service equipment as outlined in Chapter 10, much as though the house were a new one. If the new equipment is close to the old, allow yourself plenty of working room. Check over all of the existing circuits where they approach the old panel to see how much available slack the cables have, how they might be rerouted into the new box, and which ones will have to be extended in order to reach.

Prepare the new panel completely, putting in the fuse blocks or circuit breakers that you will need and the connectors for the incoming cables. It is also helpful if you tag all the existing circuits so that you will know what they are and where they will go when the time comes. Time is a commodity that you will not have much of when the power company arrives to make the cutover, so you'd best be as ready as you can.

Mount junction boxes at handy spots next to the circuits that have to be extended. Run a length of cable from each back to the new entrance panel and to an appropriate circuit slot. Then at cutover time you can simply slip each old circuit off, slip it into the junction box, and Wire-nut the connections white to white, black to black, and ground to ground, assuming that there is an equipment grounding circuit in the existing system.

When the linemen arrive to make the cutover, your new service should be complete and ready to go, save perhaps for some unfinished new branch circuits. Anything you might need to work with during the power outage should be right at hand, and a battery-powered working light should be rigged if necessary.

As soon as the power is cut, remove all of the remaining circuits in the old panel, transfer them into the new one, and make up the connections. Then clip off those existing circuits that are to go to junction boxes and splice them to the extension loops that you have already run from the new panel. Make a quick check to be sure that there are no shorts in any of the circuits, turn circuit breakers to *off* or remove the fuses from the new panel, and replace the cover. If the line crew is still at work, this is as good a time as any to tear down the old service entrance equipment, remove the service entrance conductors, meter box, raceways, and so forth.

When the new meter is plugged in and the new service activated, check out each circuit to make sure that it is operating properly. The job is done, and the downtime usually runs only a couple of hours or so.

Installation in Same Location

If the new service must be placed in exactly the same spot as the old, your task may well be a bit more complicated. The length of downtime depends upon the demands of the household for power. If you can send the rest of the family off on a vacation and close the house up for two or three days, that's fine, and it's the easiest solution.

Call the power company and have them remove the meter and disconnect the service drop at the distribution pole. Rig up some battery lamps or move in a portable generator and plug a trouble light into it. Or, run a long extension cord from your friendly neighbor's house for working lights if necessary. Then tag all of the circuits, pull them out of the old panel, and remove all of the service entrance equipment and wiring. Make a direct replacement with new material, set the new panel, and connect and check the circuits. Then call the power company back to set the new service drop and meter.

If a short downtime is essential, part of the work will have to be done with the panel energized. This is a tricky business.

Electricians work in this fashion quite often, but unless you are entirely sure of yourself, you might prefer to opt out. Each case is different, but the general idea is to keep as much of the system live for as long as possible, or to keep essential circuits operational while others are being changed around. This is done with a series of jumpers and temporary lines. For instance, it may be possible to demount the old main entrance panel and suspend it from the ceiling, wiring and all, and mount the new panel in place. Then temporary feeders can be run from the old panel to the new and the circuits changed over one by one, cutting the power off for short periods as necessary. Just before the new service drop is installed all temporary wiring is removed, and during the changeover when the power is off, everything impossible to finish beforehand is completed. If you are faced with this sort of situation and there seems no way to do the job differently, you might be well advised to seek some professional help.

LOW-VOLTAGE SWITCHING SYSTEMS

Often one of the greatest deficiencies in old houses is a lack of lighting fixtures, and an even greater lack of remote switching from convenient locations. Instead of a handy wall switch at the room entrance, one has to stumble through a dark area to operate a switch at the fixture itself. Compounding the problem in many structures is the difficulty of adding new fixtures and remote switches for them because of the construction of the house. Not much can be done about the problem of adding new fixtures—all you can do is pull the new lines in as best you can. But the situation is ideal for a low-voltage switching system.

The easiest way out is to open up each fixture box, change the box if necessary to gain more room, and install a low-voltage relay right in one of the box knockout openings. Most such relays mount directly in a 1/2-inch trade size hole, or in a 3/4-inch hole ringed with a rubber grommet. The grommet will absorb most of the clicking sound made when the relay operates. Then run a length of #18 three-wire cable to the switch location The two-wire power feed can be run eigher to the relay or the switch. This cable is so easy to use and so small that it can be readily fished across voids, drawn through 1/4-inch drilled holes, worked through small openings and cracks, tucked behind moldings and mopboards, run in the crack between two boards, or even notched into a wall and then plastered over.

You don't have to mount the switches in a wall box, although that is the recommended practice. Connections are simply made with tiny Wire-nuts. In addition, by looping a three-wire cable from switch to switch—any way you can get there—you can control one or more fixtures from two or a dozen switches in different places if you wish, with a minimum of effort and expense involved in the installation. And you can also run cables to a multipole master switch, in the bedroom for instance, from each relay that you wish to control. Then you can operate a whole battery of lights all over the house, inside or out, from your bedside. You can install the whole proposition in a few days' time and for not much more than dinner and dancing for two at a fancy nightclub.

ELECTRIC HEATING

Should you have an inclination to convert your old heating system to electric comfort heating, proceed with caution and thoroughly investigate all of the pertinent details. Many old houses fall far short of the requirements necessary, especially in the thermal insulation department, for efficient and effective operation of electric heat. Not only must you be able to install sufficient thermal insulation without completely tearing the building apart, the structure must also be sound and tight. Vapor barriers are a necessity, crackage must be eliminated insofar as is possible, extensive caulking must be done, and thermal or double glazing is a prime consideration. In some old houses the expense of preparing the building to receive electric heat could easily equal the value of the house.

A lesser consideration is that of the electrical system itself. A new service entrance will undoubtedly be a necessity along with a certain amount of new circuitry and of course the installation of the heating unit or units and their controls. All of this can be approached as explained in earlier part of this chapter.

On the plus side, electric comfort heating offers some advantages, not the least of which is its ease and flexibility of installation and operation. In some cases, it may be possible to adapt a heat pump or electric furnace to existing air ductwork or fluid radiation units. Individual ceiling, wall, or baseboard units can be placed anywhere there is enough room; they are particulary useful for heating new additions or newly finished expansion sections of an existing structure. For more complete information on electric heat and its calculation and installation, refer to Chapter 4 as well as to Chapters 6 and 7 for general layout and engineering material.

Chapter 13

Appliances and Equipment

The most frequent additions to existing residential or farmstead electrical wiring systems are appliances of one sort or another, or the installation of new equipment or machinery. These added loads may not, but often do, necessitate the running of new circuits or the installation of a new service entrance for increased capacity and may even instigate a rewiring job.

Whenever any load, especially a heavy one, is added to an existing system, the circuits and the service must be checked to see whether the load will overburden any part of the system. In addition, many of the appliances must be wired with certain conditions and restrictions in mind whether in new construction or in an existing home. Each new appliance must be installed using the maximum current draw as a minimum load, plus any derating or overrating factors that might apply.

In many cases a better idea is to run oversized circuits or feeders that are large enough to handle later replacements involving a still bigger load. Almost invariably a clothes dryer, for instance, is replaced after a few years of service with an electrically (and perhaps physically) larger and more complex unit. If the serving circuit was oversized to begin with, no rewiring is necessary.

If rewiring is required, or new circuits must be run, or the service entrance equipment has to be changed, refer to the applicable information in earlier chapters. The following pages will give you detailed information on installing various specific units, useful for making allowances for future additions, for installations in a new system, or for additions to an old one.

COMPACTOR

The compactor, or trash-masher, is becoming an increasingly popular appliance and is one of the easiest to install. This appliance is designed to operate on 115-volt circuits and is equipped with a short cord and attachment plug assembly for quick connection to a convenience receptacle. The plug is three-pronged and should be used with a grounding receptacle so that the metal frame of the compactor is adequately grounded. If grounding proves to be impractical, you may be able to omit it, depending upon local codes.

A compactor might be considered as either a fixed or a stationary appliance, depending upon whether it is built in under a counter, freestanding, or simply slid into an undercounter opening without attaching it to anything. If considered fixed, the required disconnecting means could be the branch circuit breaker if that is readily accessible. If the serving branch circuit is protected by a fuse, you may have to provide a separate disconnecting means at a convenient location, or the main disconnect may serve in a single-family dwelling. But judgment of the inspecting authority will prevail. If the compactor is considered stationary, either an accessible branch circuit breaker or the attachment plug and receptacle may serve as the disconnecting means.

Most compactors draw between 400 and 500 watts. As they are motor-driven, there is a certain amount of starting current surge, and there is also the possibility of their working under heavy load or jamming, which means high potential current draw. On the other hand, they are used infrequently and only operate for a short period of time. You could run a separate circuit to a compactor, either 15 or 20 amps. You could also wire it to an existing circuit that is not loaded to more than about half of its 80% capacity, provided that no lighting fixtures are on the same circuit. This is an appliance that could be used easily at the same time other loads on the circuit are also being used, so any diversification factor would probably not come into play.

DISPOSAL

The garbage disposal is a fixed appliance, and installation is governed by local codes in many places. Made to operate on a 115-volt branch circuit, most consume about 400—500 watts of power, but some may go as high as 1000 watts. A disposal is permanently installed in a drain line under the sink, and it usually is wired directly from a wall junction box by means of suitable tap conductors run in a length of flexible conduit to a connection box within the unit. Most are operated by a wall switch located near the sink. The disconnecting means can be that same controlling switch; backed up by the main disconnect switch in a single-family residence or the apartment disconnect switch in a two-family residence. Or, it can be the branch circuit breaker if readily accessible.

The preferable means of installation is to provide a short cord and three-pronged attachment plug to connect to a grounded receptacle located close to the disposal. This method is preferred because a disposal is sometimes installed and connected by a plumber, rather than electrician, who may not know the value of the ground connection and make it improperly or not at all. Or, the unit may be removed by a serviceman with the same lack of knowledge and replaced without benefit of ground. With a grounding plug attachment, there is no chance for foul-up.

The disposal is motor-driven and has a start-up current surge plus the possibility of high current draw if it bogs down or jams. Most are protected internally against burnout or overheating of the motor with a resettable protective device. They should be, and are often required to be, supplied by a separate 20-ampere circuit and wired with #12 conductors.

FURNACE

Fuel-fired furnaces are not usually considered as appliances, but in effect they are. Though the NEC has nothing specific to say about them, local codes usually do. In addition, the fuel supplier, such as a gas utility, may have some special requirements. This is the first thing to check when you contemplate a new furnace installation.

Power requirements vary widely. They might include a small control or control circuit transformer, gun or blowers, circulating pumps, and other equipment. Usually a single 20-amp circuit is sufficient, and though sometimes not required to be a separate circuit, they should be wired that way anyway. There is less likelihood of furnace downtime if no other load is on the circuit.

One common method of wiring a furnace circuit is to bring a power feed to a junction box on the ceiling above the firebox inspection door at the front of the furnace (Fig. 13-1). A special *lead-slug safety switch*—the slug melts and cuts the power if a fire starts—is mounted on the junction box. A length of rigid conduit or EMT is run from the overhead junction box to another junction box about two feet off the floor and slightly to

Fig. 13-1. Fuel-fired furnace circuit with safety switches.

one side of the furnace front, and another empty piece of pipe is run from this box to the floor, where it is secured on a flange to act as a support.

A service or safety switch, usually finished with a bright red cover, is connected in the hot feed line at this box. Then the feed continues to the furnace apparatus, be they controls, safety limit switches, transformer or whatever. Another switch may be located at the head of the cellar stairs or some other handy spot so that the circuit can be remotely killed in case of emergency. Circulating pumps, plenum high-limit controls, high-low temperature fan controls, damper motors and other auxiliary equipment are fed from the same circuit in the most convenient manner, but in such a way that when either the fire-control switch or either emergency or service switch is turned off, nothing operates and the entire circuit is dead.

All metallic surfaces must be properly grounded and bonded. And since the ambient operating temperatures are likely to be higher than normal, high-temperature conductors are most desirable even if not specifically required by local regulations. This is considered by the NEC in their material on conductor operating temperatures. Typical ambient temperatures for furnace and boiler surroundings should be considered as at least 40°C (104°F) and can easily run as high as 60°C (140°F) in a poorly ventilated furnace room or closet with a hot-air furnace running. This means that the minimum conductor ratings would be 75°C to 90°C (167°F to 194°F). You can choose specific insulation types from Table 6-3.

CENTRAL AIR CONDITIONING

Central air conditioning is usually installed by professionals who either have a qualified electrician on their crew or contract the work out. But if you are interested in and familiar with control circuit wiring, there is no reason you can't do the job yourself, unless it should happen to void the guarantee on the equipment.

There is a wide range of units, of course, and as far as the total power requirements are concerned, you will have to get the specific numbers for the piece of equipment that you will be using. If you are planning for the future, make an allowance based on the advertised power requirements of several brands of units of the approximate cooling capacity that you will need. Central air conditioning systems are designed to operate on 208/240 volts, and power consumption can run anywhere from about 2500 watts on up to 10,000 watts or more.

The specific manner of installation depends to a great degree upon the particular size and type of the equipment. Obviously, size will determine the capacity of the branch circuit, which must be a separate one. Whether or not the unit uses a hermetic refrigerant motor/compressor makes a difference. Conductors are chosen on the basis of the equipment nameplate rated-load current; or the rated load on the compressor nameplate. The final choice may be from 100% to 125% of this current, depending upon the size and type of the motors involved.

You must provide a disconnecting means, which may be a motor circuit breaker. The rating and interrupting capacity of the disconnect is based upon the load characteristics, but is at least 115% of the rated load and may be more.

The branch circuit must be equipped with an overcurrent device for both short-circuit and ground-fault protection. This too must be sized according to the load characteristics of the equipment, and may run as much as 175% to 225% of the rated load, but sometimes less.

If your interest is in providing sufficient ampacity and circuit space at the main entrance equipment, the first step is to determine the approximate wattage of the size unit you will ultimately need. Work out the approximate current draw $(I = P/E)$ for your source voltage, multiply by 125%, and find the next largest size conductor in Table 6-3. To be on the safe side, allow for the next step up in the table. For instance, if your calculations say to use #6 conductors for an ampacity of 55 amps, go to the next step and figure on #4 for an ampacity of 70 amps.

If the air conditioning is an added load, you should use this as a load figure in determining the size of the service entrance equipment. But if you are presently heating by electricity and the heating load is larger than the air-conditioning load, no increase in capacity is needed because the two will never run at the same time. If the heating load is smaller, then you must provide capacity in the amount of the difference between the two. As far as circuit space is concerned, a two-pole circuit slot is all you need.

REFRIGERATOR AND FREEZER

Domestic refrigerators and freezers are made to operate on 115 volts and are supplied with cords and attachment plugs. A few might be considered as fixed appliances but most are stationary, and the disconnecting means is the attachment plug and receptacle combination. Most units consume only a couple hundred watts of power, though some may go as high as 400 watts. The usual procedure is to plug them in anywhere, without much regard for the rest of the circuit load. But they are motor-driven and thus have a starting surge, so plenty of ampacity margin is desirable.

Separate circuits are a good idea for both appliances, to insure greater safety for their expensive contents if nothing else. If another load on the refrigerator circuit trips the circuit out unnoticed, the result could be a major household crisis. The separate circuits can be either 15 or 20 amps. If either appliance is placed on a circuit with other loads, take care that the circuit is only lightly loaded and has no chance of being used for portable appliances of high wattage. You have no control over refrigerators or freezers—they will come on whenever they have a mind to—and if they should happen to start up when the toaster is on, poof!

Most appliances that operate on 115 volts are not required under the NEC to have their frames grounded, but refrigerators and freezers in household use are two important exceptions. Both must be adequately grounded. With new equipment and a fairly new electrical system that includes a grounding equipment conductor and three-slot receptacles, this is no problem. But older units will have two-prong plugs and two-conductor cords, and the existing receptacles may well be nongrounding. In this case you should convert them if possible, and if you cannot, it may be necessary to provide a separate grounding conductor from a nearby cold water pipe to the appliance frame. Your best bet is to discuss this situation with your electrical inspector.

CLOTHES WASHER

Clothes washers are made to operate on a 115-volt branch circuit and are supplied with a cord and attachment plug rather than being direct-wired. This is another of the excepted instances when the appliance must be adequately grounded. If you have an ungrounded electrical system, examine the possibilities of installing a new grounded circuit, or a grounded extension to an existing circuit, using the three-slot grounded receptacles. An old washer can be rewired with a three-conductor appliance cord and plug, with the third (green) conductor attached to the frame of the machine. Failing this, you may have to work out a separate conductor grounding system, according to local codes and the desires of the electrical inspector.

Small washers may use as little as 60 watts of power. The medium range seems to be about 400 watts, but many large units take as much as 900 watts. This means at the least that you should carefully size the specific washer load to an existing circuit, using either a lightly loaded circuit or one which has loads that are not likely to be used concurrently with the washer. This is a motor load, so the starting surge has to be considered. Large machines should be supplied by a separate circuit, or one that has only a small lighting load, for instance, with no chance of other appliances being plugged in along the line. The disconnecting means is usually considered to be the attachment plug.

CLOTHES DRYER

There is a wide range of clothes dryers available on today's market. The smallest are made to run on a 115-volt branch circuit and consume about 1400 to 1500 watts of power.

These must be fed by an individual 20-amp branch circuit. Other types are designed for 208/240-volt operation, fed by an individual three-wire branch circuit. Power consumption may run from about 4400 to 7000 watts for automatic dryers, while the high-speed type may consume as much as 9000 watts. The common practice is to provide conductors matched to the size of the present dryer load, but a better practice is to install a heavy line adequate for the largest types, though with the overcurrent protection device sized to the *present* load. Then later replacement with a larger size dryer can be made with no difficulties.

A dryer is a combination load, involving both a motor load and a heating load. To arrive at the proper conductor size, find the current draw of the motor and multiply that by 125%. The heating load will probably be given in watts; divide that by the source voltage and add the answer in amperes to the motor amperage.

If your voltage source is somewhat different than listed on the motor, you might wish to convert for greater accuracy. For instance, your motor load is 5 amps at 220 volts, but your source is 208 volts. Find the wattage $(P = IE)$; multiply 220 times 5 to obtain 1100 watts. Then divide the wattage by 208 $(I = P/E)$ for an answer of 5.3 amps. Then multiply that by 125%, for a total of 6.6 amps motor load. Assume the heat load wattage to be 5000; divide that by your source voltage of 208, for a current draw of 24 amps. Add the two, and your total load is 30.6 amps. By consulting Table 6-4 you can see that the nearest conductor size to this figure, in the 60°C class, is #8 copper.

An alternative method of figuring is to take the total wattage of the dryer, convert to current draw, multiply by 125%, and pick the appropriate conductor size on that basis. If the total wattage is 8000 and your source voltage is 208, then $I = P/E$ or $8000/208 = 38.46$ amps, times 125% equals approximately 48 amps. From the conductor table, then, the proper 60°C conductor size is #6. Note that in some installations it may be necessary to go to a higher temperature class.

The frames of all dryers must be grounded. This means that a fourth conductor (green or bare) must be included in the cable assembly of the branch circuit. If the circuit is run in a metallic raceway, then the raceway can be used as the equipment grounding conductor. There is one exception though: If the supplying service is 120/240-volt three-wire, or is 120/208-volts derived from a three-phase four-wire supply, the neutral and the equipment grounding conductor can be the same, as long as that conductor is no smaller than #10 copper. This means that the cable can consist of three conductors instead of four, with the neutral (white) also attached to the appliance frame.

An old standby for wiring dryers is type SE cable, which has a stranded bare neutral conductor wrapped around the two hot insulated conductors. There are some fairly new restrictions on the use of this type of cable—it can now be used for this purpose only if the branch circuit originates in the main service equipment, and never from a subpanel or load center. A dryer branch circuit originating from a subpanel must have all insulated conductors, three or four as the situation demands. Also, if your supply source is different than those mentioned above, so that the equipment grounding conductor and the neutral cannot be the same, then the conductors must all be insulated.

A dryer may be wired directly into the system by means of tap conductors, or it may be provided with a heavy-duty molded cord and attachment plug assembly (usually called a pigtail) and a matching receptacle that may be either flush or surface mounted. These come in various standard sizes: 20-amp, 30-amp, 40-amp, and so forth. Whenever possible, the plug and receptacle arrangement should be used for ease of service. Every dryer must have a disconnect, which can be the attachment plug if readily accessible, or the branch circuit breaker. It is also permissible to use the main service disconnect or the branch circuit pull-out fuse holder.

ROOM HEATER

There are dozens of types of electric room heaters, some portable and others built-in. Many operate on 115-volt branch circuits, while others, almost invariably of the built-in type, function on 208/240 volts. The portable variety may consume anywhere from a low of 200–300 watts to a high of about 1500 watts, and some models also are equipped with a small fan for air circulation. The portable types should be used with great care and only on individual or lightly loaded circuits, depending upon the wattage of the unit. A 1500-watt unit, for instance, goes beyond the 80% loading factor of a 15-amp circuit; it has only a 3-amp margin on a 20-amp circuit. Cord and three-prong grounding plugs are usually provided, and the units should always be well grounded.

The wattage rating of the built-in types may be as low as 500 watts, or range up to 6000 or more. Some are equipped with fans and others are not, and they may be controlled by line-voltage thermostats built into the unit or mounted on the wall, or by low-voltage wall thermostats. Conductors are sized on the basis of 125% of the current draw; divide the wattage of the unit by your supply voltage, and multiply by 1.25 to find the answer.

Larger heating units are served by individual branch circuits sized to the load, and the insulation can be in the 60°C class unless a notation to the contrary is stamped on the equipment. They must be properly grounded and have overcurrent protection sized to the circuit conductors. The disconnecting device can be the branch circuit overcurrent device, or the main disconnect if readily accessible to the user. Several units may be wired to one circuit or controlled by one or more thermostats.

WATER PUMP

Water pumps are usually installed in a location where they can be served by surface wiring. They may be wired direct to a junction box by means of tap connectors in a length of flex, or they may be equipped with an attachment plug and receptacle. There is a wide variety made to run on 115 volts, more for 208/240 volts, and some are arranged so that they can be wired either way by simply connecting the proper combination of terminals in the connection box. Power consumption runs from about 450 watts on up, depending upon the size of the pump.

Smaller units can be served by a branch circuit that includes additional loads, though the most advisable situation is a separate circuit. A disconnect or service switch may be required at or near the pump, and sometimes overcurrent protection is provided at the equipment in addition to that at the head end of the branch circuit (Fig. 13-2). This protection is sized closely to the normal demands of the motor, so that if it begins to drag or bog down, as when trying to pump against a frozen water line, the circuit will trip out quickly. Thus, a motor with a normal running current of 7.8 amps might be protected with a time delay fuse rated at 8 amps, which would

Fig. 13-2. Circuit for individually close-fused motor or other equipment or machinery.

trip long before a 20-amp branch circuit protection device might.

Water pumps must be well grounded. The disconnecting means may be attachment plug and receptacle, branch circuit overcurrent protection device, service switch, or a separate disconnect, depending upon local codes, the specific equipment, and the installation conditions.

WASTE AERATORS

This equipment is relatively new to the market and is made under several trade names. Designed to take the place of a standard septic tank, these units break down household waste and sewage by mechanical and bacterial action, and the resulting output is almost completely pure water. They constitute motor loads, and though there are various sizes, none made for residential use draw any great amount of power.

This equipment is made to be buried and so requires an underground branch circuit to supply the power. The usual method is to run #10 or #12 type UF cable, properly buried and nowhere exposed to sunlight. Alternative methods may be required by local codes, such as metallic or perhaps nonmetallic raceway. Installation procedures and methods are the same as for a service lateral. The equipment must be grounded and protected by an overcurrent device sized to the serving branch circuit, with additional protection sized to the motors if desired. Ground-fault current protection would be a good idea.

ROOM AIR CONDITIONER

Room air conditioners, whether wall- or window-mounted and whether fixed, semiportable or entirely portable, are considered as appliances. Most units designed to operate on 115 volts consume power in the range of 800 to 1500 watts, so they should be supplied by individual branch circuits, though some of the smaller portable types can be plugged into existing general-purpose circuits provided that no other heavy loads are involved. The 208/240-volt models may consume from 2000 to 4500 watts, possibly more, and are served by separate three-wire branch circuits. Two or more can be attached to the same circuit, provided that the loading reaches no more than 80% of the conductor ampacity of the circuit.

Most 115-volt types use an attachment plug for connection to the circuit. The 208/240-volt types can be wired direct, but often they too are fitted with attachment plugs for ease of service and removal. The receptacles are similar to those used for dryers and may be either flush or surface mounted. The branch circuit rating should be not less than 125% of the current draw of the unit, and the overcurrent protection device can be sized to the ampacity of the conductors. Proper grounding is required, and the disconnecting means is generally the attachment plug or the branch circuit overcurrent device, or it can be the main disconnect.

DISHWASHER

Automatic dishwashers are made in two types: portable, which roll around on castors, and fixed, for undercounter permanent installation. All run on 115 volts and most consume about 750 watts of power, though a few go higher. For the roll-around type, a specific convenience receptacle close to the sink should be provided where the machine can be plugged in. The branch circuit should really be an individual one, or at the least one that will have no other load attached at the same time the dishwasher is operating. The same is true of the built-in models, except that in many localities this type is required to be on a separate circuit.

Fixed dishwashers are generally supplied with only a connection box for direct wiring into the system with tap conductors run in a length of flex to a junction box usually placed in or on the wall behind the machine. A better system is to provide a short cord and attachment plug and a grounded receptacle. This makes servicing easier and avoids the possibility that some serviceman might not replace the grounding conductor properly. The branch circuit may be rated at 15 or 20 amps, with the overcurrent protection device or the main disconnect serving as the disconnecting means. A separate service disconnect may also be required. Proper grounding is an absolute must.

FREESTANDING RANGE

A freestanding range is the familiar cabinet type with burners and ovens all in one enclosed unit, ready to be set and attached to the electrical system with no further installation necessary. Though a few are made to run on 115 volts, the great majority operate in the 208/240-volt class. The load rating is often expressed kilowatts (kW) where k symbolizes the figure 1000 and W stands for watt. Thus, a 6.75 kW range consumes a maximum 6750 watts of power. Some ranges also include motor loads such as timers, rotisseries, and vent fans, but they are such a minute part of the total load that they need no special consideration.

Ranges can be direct-wired to the system, but most of the models used for domestic service are supplied with attachment plugs and cords (known as pigtails) for use with matching receptacles of either flush or surface design. The usual practice is to run a branch circuit cable to a receptacle mounted on or near the baseboard directly behind the range. Both pigtails and receptacles are made in certain ratings, each of which has a slightly different prong configuration so that they cannot be interchanged. They step up in increments of 10 amps, such as 30, 40, and 50 amps. Some of the largest models are wired direct, especially in ratings where pigtails and receptacles are not readily available or do not exist.

Most of the ranges on today's market are rated from 8 kW to 12 kW for units that include four burners and one oven, and up to 20 kW for double-oven models. Sizing the branch circuit conductors can be done in two ways. The first is simply to compute the maximum current draw from the nameplate rating of the range. Thus, if your voltage supply is 220 and the range is rated at 16 kW, the current draw would be just under 73 amps. This would require a conductor size in the 60°C class of #3 copper.

The second method involves the use of a demand factor, which recognizes the fact that a range almost never will be operated at maximum current load. The demand factors for home ranges are as follows:

- If the range rating is 1.75 kW or less, use the full rating for conductor sizing.
- If the range rating is more than 1.75 kW but less than 8.75 kW, use 80% of the nameplate rating for conductor sizing.
- If the range rating is more than 8.75 kW but less than 12 kW, you may use a flat 8 kW for conductor sizing.
- If the range rating is over 12 kW but less than 27 kW, use 8 kW as a base load. Then add 40% of every 1 kW (or fraction thereof) over 12 kW of the nameplate rating for conductor sizing.

To explain a bit further, if the range rating is 6 kW, you can use 80% or 4.8 kW as the load. If the range is rated at 11 kW, then you may consider 8 kW as the load. If the rated wattage is 16 kW, then you start with 8 kW, and since 16 kW is 4 kW more than 12 kW, you add 40% of 4 kW, or 1.6 kW, to make a total of 9.6 kW.

After computing the current draw from the load wattage, according to your source voltage, choose the proper conductor size from Table 6-3. For instance, the 16 kW range load may be considered as 9.6 kW, which at 220 volts works out to an amperage of 43.6 amps. In the 60°C insulation class, this would mean a conductor size of #6, which has an ampacity of 55 amps. Then you should check to make sure that this size conductor will not be loaded past 80% ampacity; so 80% of 55 equals 44 amps, and the load is just right.

The advantage of using the second method is that usually a smaller size can be used in the branch circuit, with correspondingly less expense and effort. The disadvantages are that there is a possibility, though perhaps remote, of exceeding the 9.6 kW demand load, and also that there is no room for expansion when a new and larger range is installed. The circuit would then have to be rewired.

In the event that two ranges are to be wired to the same circuit or feeder—an unlikely event in most households—the demand factors will change. You can find further information on this in the NEC.

The usual procedure in selecting the overcurrent protection device is to pick a standard rating closest to the ampacity of the branch circuit conductors. In the example used above, though the conductor rating of #6 is 55 amps, you could use a standard 50-amp circuit breaker. In any event the device cannot be rated at more than 150% of the demand load of the range, or more than 100% of the conductor ampacity, or less than the demand load of the range.

The branch circuit breaker or fuse pullout, or the main service disconnect, acts as the disconnecting means, provided that it is readily accessible to the user. The attachment plug can be considered as the disconnect only if it is readily reachable, or is accessible from the front of the unit by removing a drawer to expose it.

The frames of all ranges must be grounded. This is either taken care of by the fourth (bare or green) conductor in the cable assembly, the metallic raceway carrying the circuit conductors, or in some special cases a separate grounding wire. The range is another appliance that carries the same exception to the general grounding rules as the dryer, however, so that if conditions are right, the neutral and the ground may be the same.

DROP-IN RANGE

Drop-in ranges are somewhat similar to freestanding ranges in that they are complete in a self-enclosed cabinet. Most models include four burners and a single oven, and most are of the same approximate dimensions. The principle difference is that they are designed to be solidly installed in a specific framework in the kitchen base cabinetry. Some operate on 115 volts, but most are designed for 208/240 volts, with nameplate load ratings of up to 12 kW.

A separate circuit is usually provided, and the conductor sizing, demand factors, grounding, grounded neutral exception, and overcurrent protection requirements are the same as for the freestanding range, as just discussed. Though usually wired direct, with tap conductors in flexible conduit run to a junction box in an accessible location near the unit, they may also be provided with attachment plugs and receptacles for ease of removal and service. The drop-in range is a fixed appliance, and the disconnecting means is the branch circuit breaker or pull-out fuse block, or the main service disconnect in a single-family dwelling.

COUNTER COOKTOP

The counter cooktop is a shallow, self-enclosed unit housing only ring burners, anywhere from two to five of them, and their control units and accessory items. Some operate on 115 volts, but most are designed for 208/240-volt service. Nameplate load ratings go from 5 kW to 10 kW for the most part.

A separate circuit should be provided if the cooktop is not being used in conjunction with other cooking appliances. But if a wall oven, for instance, is to be installed as well, the two can be run on the same circuit. In arriving at the proper conductor size, no demand factor applies if the cooktop is wired singly. Use the nameplate rated wattage, divide that by your source voltage, and choose a conductor size that will carry the current without being loaded to more than 80% of its ampacity. If one or two wall ovens are to be wired to the same circuit, add the nameplate load ratings of all the units. You can then size the conductor to the full rated combination load, or you may use the demand factors for ranges, as discussed earlier under *Freestanding Range*.

The counter cooktop is a fixed appliance, whether connected by plug and receptacle or not, and the disconnecting means is the overcurrent protection device or the main service disconnect. Grounding, grounded neutral, and overcurrent protection considerations are the same as for free-standing or drop-in ranges. Tap conductors may have to be of a high-temperature insulation class in many instances.

WALL OVEN

Wall ovens are self-enclosed units designed to be built into the kitchen cabinetry and are considered to be fixed appliances. They are available for use on either 115 volts or 208/240 volts. Single units carry a nameplate load rating of 5 kW to 6 kW for single units, and 6 kW to 9 kW for double units.

An individual circuit should be provided for a wall oven operating by itself, but a double wall oven can be wired on the same circuit as a counter cooktop. Similarly, two separate wall ovens can be wired to the same circuit, or joined with a cooktop. All that is required is that all the units connected to the circuit are in the same room.

Where a single or double unit is supplied by a separate circuit, the conductors should be sized according to the full nameplate load rating with no demand factor applied. When a cooktop and not more than two ovens (or one double oven) are connected to the same circuit, you can size the conductors to the total full nameplate load rating, or you can treat the combination as one range and use the demand factors as explained earlier under *Freestanding Range*.

Like the counter cooktop, wall ovens are usually direct wired, but may be connected with an attachment plug and receptacle arrangement for servicing ease. When direct wired, the tap conductor insulation should be, and may be required to be, of a high-temperature class, perhaps one of the 90°C types. Overcurrent protection, disconnecting means, grounding, and grounded neutral conditions are the same as for free-standing ranges.

SUMP PUMP

Sump pumps are generally used in the home to keep the basement clear of water during the spring flooding. All are designed for 115-volt operation and are usually equipped with an attachment plug and cord, to be plugged into any handy receptacle.

Smaller models, which use about 500 watts of power, may be plugged into an existing circuit. But since this is a motor load, which as a rule will turn on automatically upon demand, the serving circuit should not be a heavily loaded one, probably not more than 50% of ampacity. Manual types of pumps can be run with other loads on the same circuit and turned off when necessary. Large sump pumps will use as much as 1000 watts of power, and if automatic should be served by an individual branch circuit.

These pumps can also clog easily under some circumstances. If this looks like a possibility, install a time-lag motor fuse assembly in a junction box at some point between the pump and the branch circuit outlet, with the fuse sized closely to the normal running current of the motor.

The plug attachment and receptacle can serve as the disconnecting means, but if the pump is permanently installed, you may be required to have a motor-rated service switch or other separate disconnect close to the equipment. Proper grounding is particularly important.

BARBECUE AND SMOKER

Electric barbecues and smokers have become increasingly popular over the past few years. Often as not, they have been improperly installed. These portable appliances are designed to operate on 115 volts and are supplied with a grounding attachment plug and cord. Power consumption may run anywhere from 1000 watts to 1500 watts, which means that they should be served by individual branch circuits. And since these appliances are invariably used outdoors, that circuit is usually placed underground in a rigid metallic conduit raceway or protective stub, and should terminate in a weatherproof outlet box equipped with a weatherproof grounding receptacle.

The installation of the circuit itself, whether done with direct-burial cable or in buried raceway (if local codes so specify), should be treated as though it were a service lateral. The overcurrent device must provide protection for both short circuits and ground faults, so a GFI breaker must be used.

VENT FANS

Ventilation and exhaust fans of one sort or another are usually found in the home, perhaps as part of the heating system, or in bathrooms, range hoods, roof ventilators, photographic darkroom air changers, and the like. Most are designed to be wired directly to the system, except for the portable plug-in cooling fans, and most draw only a small amount of current. Even large attic venting fans usually draw no more than three or four amps. This means that almost invariably they may be connected to any branch circuit, provided the total draw including the fan does not load the circuit to more than 80% of its conductor ampacity.

Permanently installed fans should be grounded in the usual manner wherever possible. In some cases, such as a roof or attic venting system where the ambient temperatures are likely to be above normal, insulation of a high-temperature class must be used on the conductors. Usually a minimum 75°C class is adequate. Switch or controller loops should be of the same temperature class. When figuring loads, use 125% of the nameplate current rating.

FOOD CENTER

Food processing centers are made to be permanently installed in a convenient countertop location in the kitchen. They are used for blending, grinding, and chopping operations of relatively short duration. They operate on 115 volts and consume around 400 watts of power. As motorized loads, they should be figured at 125% of the nameplate rating for total load current draw. They can be readily wired to an existing small appliance circuit or general-purpose circuit that has sufficient load capacity, particularly one that has loads attached that are not likely to be in use at the same time as the food center.

WALL TOASTER

This is a relatively new appliance that is modeled after the standard kitchen toaster, but is permanently installed in a case recessed into the wall and pulls out on a slide for use. Wiring is done by direct connection to the circuit. The operating voltage is 115 volts, and the power consumption may run from 800 watts to 1600 watts. A large unit should be served by an individual 20-amp circuit; a smaller one can be connected to a small appliance circuit that has sufficient remaining load capacity, or where there is ample diversification of existing loads so that there is no chance of overload. Proper grounding is essential.

FIRELOGS

Electric fireplaces and firelogs are available in two different types: those that produce heat and those that do not. The non-heating variety only draws 60 to 100 watts of power, about the same as a light bulb, and so can be attached to almost any circuit. Such models are equipped with a cord and attachment plug. There seems to be a rule that says there is never a receptacle right where you want to position the fireplace, though, so you will probably have to arrange a branch circuit extension (see Chapter 12).

The heating type of electric fireplace is also made to operate on 115 volts, and it can be connected by means of a special high-temperature heating appliance cord and attachment plug, or wired direct. Power consumption is about 1500 watts. This may be considered as a continuous load in the judgement of the inspecting authority since it could easily be operated for more than three hours at a time. For this reason, the supplying branch circuit should be #10 size (30-amp capacity) because the 80% loading factor is exceeded on a 20-amp circuit. This would hold true for any unit where 125% of the rated current, computed on the basis of your own voltage supply, would exceed 16 amps. A 30-amp circuit could be protected with a 20-amp breaker if supplying only the fireplace, or with a 30-amp breaker if other loads are connected to the same circuit. As with all equipment of this sort, grounding is essential.

HUMIDIFIER

Portable humidifiers are designed to operate on 115 volts and are supplied with cord and grounding attachment plug so that they may be plugged into any convenient receptacle. Power consumption is low, from about 250 to 600 watts. They usually constitute a continuous load, so when determining the total load, multiply the nameplate rating by 1.25 and convert to amperage draw. Thus, a 500 watt unit on 110 volts would have a total demand of $500 \times 1.25 = 625$ watts, which divided by 110 equals 5.68 amps. A humidifier can be used on any circuit, provided that the ampacity of the circuit is not exceeded.

EVAPORATIVE COOLER

Evaporative coolers are designed for operation at 115 volts, with a power consumption of 400 to 1000 watts. The same considerations apply as for humidifiers, except that larger models should be connected to individual or lightly loaded branch circuits.

MICROWAVE OVEN

Microwave ovens have been perfected to the point where they are now reliable and coming into widespread use. They are supplied with cord and attachment plugs, operate on 115 volts, and consume from 1 kW to 1.5 kW. They must be used with grounding receptacles and are best served by an individual 20-amp branch circuit, or one that has only additional light loads, or loads that will probably not operate at the same time as the oven. The attachment plug and receptacle arrangement serves as the disconnecting means.

WATER HEATER

Water heaters are subject to a great variety of regulations and different wiring schemes. A few are made for 115 volts use, but most operate on 208/240 volts. Power consumption may run anywhere from about 2500 watts on up. There are two basic styles: those with single elements and those with double elements. The latter are sometimes known as *quick-recovery* or *QR* tanks. Such a unit may have, for instance, one 5500-watt heating element at the bottom of the tank and another near the top, so arranged and thermostatically controlled that the incoming water is heated rapidly and kept hot, especially in the top section near the outlet pipe.

Most manufacturers provide specific information as to the necessary conductor size and overcurrent protection to be used, and the easiest course is to follow them explicitly, together with the connection diagrams. All of the smaller units use no smaller than #12 copper conductors for the branch circuit and taps, while the larger use at least #10. The usual 208/240-volt arrangement uses two hot conductors and an equipment grounding conductor; no neutral is necessary. A water heater is a fixed appliance, connected by tap conductors in flexible conduit to an individual branch circuit outlet box. Often a separate disconnect close to the tank is required.

The specific method of wiring the water heater circuit varies. In some cases it is connected into the load center or main entrance panel just as any other branch circuit is. In other cases it must, or may, be wired through a separate watt-hour meter and time clock provided by the utility company and mounted near the main meter, and a different billing rate is applied. Another system just now coming into use involves a special electronic receiver that picks up a signal transmitted by the power company and turns the equipment on during periods of time when area electrical demand is low. Then too, some types of entrance panels have special connection lugs for water heaters. Everything depends upon your equipment, your local codes, and your power company requirements. Talk to the power company and the inspecting authority; they will provide you with all of the necessary specifics.

SOUND SYSTEMS

Extensive sound systems and well-equipped amateur radio shacks often have substantial power needs, especially if some or all of the gear is of the old tube type or uses a lot of power semiconductors. There is also usually a need for quite a number of convenience receptacles, simply in order to be able to plug everything in without the use of a basketful of cube taps and extension cords.

All outlets should be of the grounding type so that the equipment in use will be properly grounded. Usually a single individual 20-amp circuit is sufficient, though two may be necessary. In the case of a large ham shack, a small subpanel might be the best way out. One limiting factor is sometimes the number of receptacles that can be placed on a circuit. With a permissible loading factor of 180 volt-amperes per receptacle, a 20-ampere circuit operating at 120 volts can supply 13 single receptacles or 6 double-convenience receptacles. At 110 volts, this number reduces to 12 single receptacles. A 15-amp circuit at 120 volts will carry only 9 singles or 4 doubles.

However, this equipment could easily be construed as a *continuous* load—one which may run for three hours or more at a stretch. In that case, the circuits must be derated to 80% of ampacity. This means that a 20-amp circuit will carry 10 single receptacles or 5 doubles at 120 volts, or 9 singles and 4 doubles at 110 volts. The 15-amp circuit has a capability of 8 singles or 4 doubles at 120 volts, or 7 singles or 3 doubles at 110 volts.

All of this holds as true for both individual receptacles and plug strips. This means that you must carefully consider the number and size of circuits, the number of receptacles per circuit, and the potential loading of each circuit. With a large complement of plug-in gear, all factors will be closely interrelated.

Rarely would all elements of a sound system or an amateur radio station be in operation at the same time. This means that although there are no prepared demand factor tables that can be followed, you can eliminate the lesser of two loads from your calculations when it is provable that both of the two loads will not be in operation at the same time. By working through the various potential combinations of equipment that will be operating together and apart, you can arrive at a reasonable total load wattage or current draw on which to base the size of the serving branch or feeder circuits.

Much of this equipment is portable and supplied with cords and attachment plugs, which also serve as a disconnecting means. If the equipment is built in or otherwise installed in a fixed manner, the branch circuit breaker or the main service disconnect can serve as the disconnecting means. In many cases it is desirable, however, to provide a master switch or secondary disconnecting means at or near the equipment, which will open all of the ungrounded (hot) conductors at the line side of the equipment. This is a safety measure that will insure that every piece of the equipment is turned off and there is no chance that one item will be forgotten, soon to overheat and burn out. This type of simple interlocking is especially important for hot-running equipment that is normally cooled by blowers. It is quite possible that an ordinary wall snap switch, especially if not T rated, will be unable to handle the potential load across the contacts if thrown under a full load—the inductive surge could easily burn out the switch contacts. If the possible load is over 12 amps or so, a heavy-duty motor-rated switch would be preferable, and perhaps essential. Where more than one circuit is involved, use a small panel with a main circuit breaker.

OTHER APPLIANCES

There are, of course, a great many more appliances than those listed here. Most of them require only small amounts of power for short periods of time, and provided that the wiring system in the house is in good shape, they can be used without any undue worry.

Medium loads, those of about 500 to 750 watts or so, should be attached with some care to any given circuit, especially if you know that circuit is heavily loaded. With larger loads, from 1000 watts on up, check the circuits carefully before you use them regularly, or perhaps before you purchase the equipment.

The two principles to remember are: (1) make sure that everything you use is properly and effectively grounded whenever possible, and (2) never overload or load to the upper limit on any circuit, but try to keep the maximum at around 80% of ampacity.

STEP-UP TRANSFORMERS

Not every residential service in the country is served with dual voltages. Some are still limited to only 115 volts, though it is but a matter of time before this situation changes. In the meantime, however, there may be good reason, if only a desire on the part of the homeowner, to use an appliance designed for operation on 208/240 volts on his 115-volt service. This can be accomplished by the use of a dry-type step-up transformer (Fig. 13-3).

Fig. 13-3. Circuit for operating 230-volt appliance or equipment on 115-volt circuit.

The leading consideration in an installation of this type is the expense. The transformers are fairly expensive, as is the installation, and operating costs are also somewhat higher. Once this problem is reconciled, the next step is to determine the size of the load to be supplied. As an example, we can use a 6000-watt clothes dryer; the general installation procedures are about the same for any appliance.

Break the load down into two parts: the motor load and the heating load. Find the current draw of the motor at its specified voltage, let's say 220. Multiply this by 1.25. Then determine the current draw of the heating load at the same voltage, and add the two together. Let's say your answer is 28.1 amps. This means that the transformer secondary must be capable of supplying a minimum of 28.1 amps at 220 volts. The primary of the transformer will of course operate on your supplying voltage, which we will assume is 110 volts.

Next you must consult with your electrical supplier, or you may have to contact a transformer manufacturer direct, and find a standard transformer that will meet these requirements. It is also possible to have a transformer custom built to your specifications, but the cost will likely be too high. Transformers are made in many configurations and ratings and with various tap-connection possibilities, so the easiest way out is to tell the supplier what your specific needs are and have him make the final selection for you.

Transformers are rated usually in kVA (kilovolt-amperes), which is a measure of power. A volt-ampere is the product of voltage and amperes, which in direct current is the equivalent of a watt. One kVA is the equivalent of 1000 volt-amperes. A transformer for your dryer would have to be capable of handling slightly more than 6 kVA (220 volts times 28.1 amps, divided by 1000), so you would have to step up to the next largest size transformer available.

The next problem is to determine the size of the branch circuit needed to supply the transformer primary. In theory, since this transformer exactly doubles the voltage from primary to secondary, the current draw at the primary would be double, or 56.2 amps. But this assumes that the transformer is 100% efficient, which is a practical impossibility. Depending upon the design of the transformer the actual load might be, for instance, 60 amps. Whatever the maximum rating of the transformer you finally install, it will be noted on the nameplate, and this is the figure to use. The conductors may be oversized if you wish, but the closest ampacity above the nameplate current rating is sufficient.

The final task is to locate the transformer in a convenient spot, spaced at least a foot from any combustible surfaces. Then run a cable from the main entrance panel or load center to the primary leads, and another from the secondary leads to the dryer. Provide the primary circuit with overcurrent protection at the panel, either fuses or a breaker, rated at not more than 125% of the rated primary current of the transformer. If the amperage is an awkward value where there is no standard breaker or fuse size in the range of 100% to 125% of that figure, then you can go to the next standard size that will cover at least 100% of the load, since a smaller one obviously will keep tripping out. You may also install overcurrent protection on the secondary side if you wish. As always, grounding of all exposed metal surfaces or frames is necessary.

Chapter 14

Auxiliary Systems

There are several auxiliary wiring systems, at least one or another of which is found in nearly every house in the country. Some homes have several or possibly all of these systems. With one exception, though, they are generally added at some later date after the house is completed and the task is much more difficult, rather than being considered as a part of the building construction or remodeling.

In the more populated areas, specialists can sometimes be called upon to make these installations, but otherwise the homeowner is left to struggle with the addition of auxiliary systems as best he can. The result is often something less than an easy and efficient installation. But if considered as part and parcel of the residential wiring system, and designed and installed at the same time the wiring is put into a new house or an old one is rewired, the results are far more satisfactory, not to mention less expensive. In addition, prewiring for auxiliary systems, whether or not you feel right now that you want or can afford the completed systems, gives you not only easy flexibility in completing the systems later but also a worthwhile increase in the potential resale value and later salability of the house—all for a modest outlay in cash and effort.

ANNUNCIATORS

Most often referred to as the *doorbell system*, this is the exception mentioned above that is found in nearly every house and is almost always taken into consideration during the installation of the electrical wiring system. Probably the most common arrangement is a simple pushbutton switch (momentary contact) at front and back doors, with a conveniently located bell or buzzer inside the house. The bell may be two-sectioned, producing two different tones to distinguish between front and rear doors. Another common arrangement uses a chime instead, again arranged so that two or more different tones or series of notes will ring for the different doors. Power for the system is supplied through a small low-voltage transformer, usually providing somewhere around 8 to 18 volts, the primary of which is wired into a branch circuit at any handy point.

Though the basic circuitry for most systems is about the same, there are numerous practical possibilities in wiring them. The specific terminal combinations vary with different manufacturers of chimes, so follow the supplied instructions. More complex systems that cover three or more entries will of course require additional buttons, tones, and conductors.

When installing an annunciator system, first mount the transformer on the outside of the main entrance panel, a distribution panel, or on a junction box at some convenient spot (the transformer itself may act as the cover for the box, or it may mount in a knockout hole). With the power turned off, connect the primary of the transformer to the branch circuit conductors. Drill or cut any necessary holes for the pushbutton assemblies and for mounting the bells or chimes. Then run the cables, which may be the so-called *bell wire* or more commonly #18 thermostat cable with the required number of conductors, to the bells or chimes and to the pushbutton locations.

Two-conductor cable is most often used and can be run in just about any fashion that will get you where you want to go. You can route it through drilled holes, fish it in partitions, tuck it behind mopboards or moldings, whatever is easiest. Wherever possible, secure the cable with saddle staples. Treat it rather gently, too; the conductors are small enough and the insulation covering thin enough so that either can be easily broken or damaged. Once the wires are in place, make the final connections, which need not be done in junction boxes, but can be left out in the open. Boxes do make for a neater job, but are seldom used, especially at the pushbuttons. When the connections are made, mount the pushbuttons and the trim plates if there are any, secure the bells or chimes in place, and turn the power on.

SOUND SYSTEMS

There are two distinct types of sound systems found in the home. One is a combination of intercom and piped sound from the radio, which will be treated in the next section of this chapter. The other is what is loosely called *hi-fi* or *stereo*. These systems are primarily used for reproducing recorded music, but may also be so arranged as to carry AM and FM radio signals. There are literally hundreds of equipment combinations ranging from small, self-contained and portable units to shelfloads of complex gear. All, however, have one thing in common: several loudspeakers from which the sound finally emanates—and this is where the additional wiring comes in. The 115-volt power requirements are treated in Chapter 13, under *Sound Systems*.

Planning

There are three basic types of sound systems. One is *monaural* or *monophonic* which has one channel of sound. The *sterophonic* system uses two channels of sound. And the relatively new *quadraphonic* system uses four channels. Thus, loudspeakers are used in singles, pairs, and foursomes to deliver the sound.

The great flexibility of the more powerful sound systems makes it possible, and desirable, to place loudspeakers in parts of the house remote from the equipment itself, such as the bedrooms, kitchen, den, workshop, outside on the patio, or wherever seems reasonable. Each remote speaker is connected at the equipment center to a selector unit that allows speakers to be turned on and off in various combinations and also maintains a proper "impedance balance" between the speakers.

The details of correct loudspeaker placement for best listening results, as well as the proper methods of balancing, impedance loading, and overload protection can be found in a number of manuals on the subject. Our interest here is to *prewire* the loudspeaker locations. This is such a simple job that it should be done in every new house during construction and along with every remodeling and rewiring job where possible. Provisions should be made for a central stereo equipment location, or "music wall." The first step is to choose a logical spot for the equipment, and then decide which rooms are to have loudspeakers, how many, and exactly where (Fig. 14-1).

Wiring

Run a two-conductor cable from the equipment location to each speaker outlet. For short runs of 15 or 20 feet, the

Fig. 14-1. Layout of loudspeaker arrangement for prewiring sound system.

inexpensive #20 or #22 speaker wire is adequate. For longer runs, however, use a heavier cable to avoid line loss; voltage drop applies in this case just as with ordinary residential power wiring. Thermostat cable and "bell wire" in a #18 size is often used, and so is standard two-conductor lamp cord with rubber or plastic insulation. If the runs are quite long, you may have to go to #16 conductors, depending upon the characteristics of the sound equipment to be used. In some cases it may be desirable to use a special shielded coaxial cable (co-ax), available from electronics or audio equipment supply houses.

Like other low-voltage wiring, you can run the cables in whatever way seems suitable and expedient, taking care not to damage them, especially the co-ax, which is not as flexible as the other types. Stapling is not required, though often necessary, and it is a good idea to keep the sound cables well apart from power cables—in no case closer than two inches from conductors not encabled or in a raceway.

If the sound equipment is not to be put in until some later date, merely leave the loose ends of the cables, tagged to show what they are, dangling inside the walls with plenty of slack. Make up a sketch plan with dimensions showing exactly where each cable lies so that a small hole can be bored later and the cable pulled out for use. Keep the sketch with the house plans, property deed, or other household papers.

Speakers

The speakers may be built into the walls or ceilings, or may be surface mounted. Where they are on the surface, you can install a standard wall box in the usual manner, run the cable into it, and finish the box with a blank cover plate that has been drilled and fitted with a phono jack or some other type of jack. Connect the two leads to the jack, provide the speaker with two short leads and a matching plug, and you have a simple detachable arrangement.

Direct connections to the speakers usually require soldering the leads to solder-lug terminals, unless the speaker has been factory-fitted with a cord and plug. Connections at the equipment may call for plugs, or there may be screw terminals. If the cable you use has stranded conductors, tin the ends of the conductors with solder to make a stiff solid wire, then bend each into a hook and attach to the terminal screw. Shielded cable, or co-ax, is generally used with special fittings having soldered connections.

Should you want to make a permanent outdoor speaker installation, other factors come into play. The speakers, of

LEADS TO VOICE COIL

SOLDER LUGS MOUNTED ON LOUDSPEAKER

SCREW TERMINAL STRIP FOR #12-14 CONDUCTORS

Fig. 14-2. Terminal interface for joining large to small conductors.

course, must be of the weatherproof type, unless they are connected with plugs and meant to be taken indoors at all times except during actual use. If the cable is run so that it is exposed at any time to sunlight, you will need to use a type that is designed to withstand the sun's rays without deteriorating. In addition, the cable should be moisture-proof and capable of withstanding temperature extremes—rubber and most plastic cables will not stand up under this use. The simplest solution is to use one of the general wiring cable types, such as SNM, which will hold together under these conditions.

If the cable is to be buried in the ground, use type UF, which is designed with this purpose in mind. The only drawback is that the #14 size conductors are sometimes a bit difficult to attach to the electronics-type terminals usually provided with speakers. But with care it can be done, or you can work out a terminal-strip interface to jump from one type to the other if necessary (Figure 14-2).

INTERCOMS

Intercommunication systems have become quite popular in recent years as the cost has gone down and the quality and flexibility has increased. In large houses they often save a lot of steps, and in any home they are handy as "baby-sitters" and as a link between a sickroom and the rest of the house.

There are a great many different types of intercom sets. The simplest just plug into any handy convenience outlet and use the power lines to carry messages back and forth between sets. The only requirement is that all the sets have to be on the same power system—you probably could not communicate with your neighbor, for instance. Others operate on their own network of conductors and are desgined for communication only between a master station and one or more remotes or substations. With this arrangement only the master station can originate calls. Or, the system may be made up of all master stations, where every station can both receive and originate calls. With either type there may also be "listening posts." The most complex systems (Fig. 14-3) may include AM/FM radio, annunciators, provisions for tape deck or record player, message center and clock, and sooner or later they will probably include closed-circuit TV as well.

The particulars of wiring these systems vary widely with the individual manufacturers and must be taken from the specific instruction sheets. Prewiring is a chancy business until you know exactly what type of equipment you are going to use and what its requirements are. In general, however, the cable used is of the multiple-conductor type. The minimum number of conductors depends upon the design and complexity of the equipment, and the size is usually #18 for low-voltage and communications work. Often it must be of the shielded type. The stations and other equipment may be designed for table or countertop use or built into the wall. The latter is

Fig. 14-3. Complete master unit for radio/intercom center. (Courtesy NuTone Division of Scovill)

direct-connected to terminal strips, and the former by plugs into wall jacks.

Once the locations of the stations and any other equipment are determined, the proper type of cable can be pulled from point to point, or from central location to remotes, as required. Again, this type of cable can be run in any convenient way, but is best kept away from power lines, heat, and moisture. A source of 115-volt power is usually needed at the central equipment location; this can be taken from a nearby branch circuit and run to the equipment in the usual manner. In some cases the power connections are made directly into the equipment, while in others all that is needed is a convenience receptacle outlet to plug into. Make the interwiring connections according to the manufacturer's instructions, either to wall-mounted jacks or directly into the stations and other equipment. The use of junction boxes is not required in most cases, but where they are used, never make both low-voltage and line-voltage connections in the same box unless the box is partioned for such use. When making the line-voltage connection, remember to turn the branch circuit off.

TELEPHONE

It is quite true that if you order a telephone, or extensions, the company will undertake to install them for you at any location you would like, one way or another—and of course for a fee. But if this is done after construction of the house is complete, the result is an obvious telephone wire, stapled, though unobtrusively as possible, across various surfaces. But the installation can be made more easily, usually with less expense and far more flexibility, by prewiring the system for however many telephone locations you think you would like to have in the future. Whether or not you actually use them all is relatively immaterial; with the cables in place, you will at least have some easy options.

Check with your telephone business office first to see what they have available in the way of equipment, and what the usual installation and billing methods are. All equipment must be supplied by the telephone company, including the cable and even most of the hardware; anything you buy elsewhere will be disallowed, with the possible exception of some types of telephone sets that may be allowed in some locales. Certain material is provided at no extra cost, some you must purchase outright from the company, and most is offered on a monthly rental basis. Exactly which is which varies from place to place. Also, all equipment is installed and later maintained by the telephone company, except that in many instances they will provide the homeowner with a coil of wire to pull in as the new home or major remodeling progresses. They may also supply a special underground cable for the homeowner to bury from the distribution pole to the building, or they may insist upon doing the job themselves, usually at no extra charge. In cases where underground telephone service is required to be run in rigid conduit, the homeowner often must provide and bury the pipeline at his own expense, following the telephone company's specifications.

Standard wall and tabletop telephones are available as well as a series of decorative styles. As accessories, you may be able to get long or extra-long (15 to 20 feet) coil-cords, extension bells, extension loudspeakers, cutoff switches, and floor or wall jacks into which a portable set may be plugged. All of these items give you a great deal of flexibility in locating the telephone sets in the home where they will be most convenient. A business representative or engineer from the telephone company will help you lay out a complete installation for your home, including provisions for future expansion. Then, as the construction progresses, either you or they will run the necessary cables, and the finish equipment can be put in as and when needed. The end result, if you have planned carefully, is a concealed telephone system (Fig. 14-4) of maximum convenience and great flexibility, with a savings of time and money for both you and the telephone company.

AERIALS

There are few homes these days that do not have aerials for television reception, and perhaps FM radio as well. Sometimes these two functions can be joined, but in other cases this may be either inadvisable or undesirable. In any event, unsightly tangles of leads can be avoided by the simple practice of concealing the leads—and sometimes the aerials themselves—within the framework of the house. Leads for antenna rotators can be similarly hidden.

Television and FM antennas are most frequently mounted atop the roof of the building, but can also be placed inside an attic, in a basement, in or on a garage, or underneath a high deck. Whatever the location, you can bring the lead-in cable, which may be coaxial cable or foam-filled plastic twin-lead, from the antenna to some handy but hidden spot within the building, such as the attic, basement, or a utility closet. Run the cable through the structure as you would a branch-circuit power cable, concealing it wherever necessary and leaving some slack to work with at an eye-level termination point.

Fig. 14-4. Telephone layout plan for prewiring telephone system.

MASTER BEDROOM

GUEST BEDROOM

UPSTAIRS JACK

PORTABLE DESK SET

JACK

JACK

TELEPHONE LAYOUT
JOHNSON RESIDENCE
NOT TO SCALE

BATH

STAIRS

HALL

KITCHEN

WALL SET
—LONG CORD

DESK SET

DINING ROOM

TELEPHONE ENTRANCE

OUTSIDE BELL

LIVING ROOM

JACK

Then spot every place in the house where you think TV/FM will or may be used. This could include nearly every room, and there might well be two or more locations in a single large room, such as a rec room. Remember too that furnishings are likely to be moved around from time to time, so you could easily end up with quite a number of spots where either a TV set or an FM radio might be used.

At each location, cut a hole in the wall or floor to fit a standard wall box, preferably nonmetallic, or if construction is not yet complete, mount the box in place. Then run separate cables from each individual box back to the point where the antenna lead-in terminates. In all cases, keep the cables away from electrical wiring or plumbing or other metal objects, preferably separated by a joist or stud. And never staple the cables in place—use specially made insulated "standoffs" to hold the cables if they need support. To finish out the boxes, connect the cables to special TV or FM jacks, mount the jacks to the boxes, and put on the trim covers (Fig. 14-5).

How you proceed from this point depends upon the number of sets you anticipate having in operation at any one time, as well as the signal strength in your area. If you have only one TV and the signal is strong in your area, your task is simple. Connect the antenna downlead to the cable going to the particular room or outlet you intend to use. Then equip your set with a short piece of cable ending in a TV or FM plug to match the receptacle.

Fig. 14-5. Typical TV/FM connection point installed in standard wall box and capped with standard duplex finish plate. (Courtesy Jerrold Electronics Corp.)

If you plan to use two or more TV sets at the same time, you will need to use a *splitter*—a small box that is connected between the antenna downlead and the several cables going to the different rooms. Splitters are generally made for two or four TV sets, and they provide an "impedance matching" function that permits two or more TV sets to operate using the same antenna without the possibility of distorting the TV picture with multiple images, or "ghosts."

Signal boosters, or *antenna amplifiers*, are often required in areas located a considerable distance from the transmitting TV towers, usually over 50 miles. This gear is normally tailored to the requirements of the individual installation, but most units incorporate a splitter for use with two TV sets. Antenna amplifiers are used to enhance weak or marginal signals, those appearing washed out and speckled with "snow." Several different models permit the selective enhancement of VHF, UHF, and FM signals in various combinations, and their prices vary accordingly. Antenna amplifiers normally come in two parts: a small amplifier unit is mounted to the antenna mast just below the antenna, while the power supply and splitter unit is located in some convenient central point near a source of 115-volt power, and all connections are made to various rooms from this central location. The power supply feeds low-voltage current through the antenna lead to the amplifier, so no extra conductors are required. Your local TV or electronics shop can advise you as to exactly what equipment would be best for your installation, and they can supply you with all the necessary material and hardware.

An antenna rotator system can be installed the same way, by simply bringing the rotator cable from the motor on the antenna down into the interior of the building to a wall box placed close to the control box location. The motor cable, usually having four or five conductors, terminates at a jack in this box, which should be located close to a convenience receptacle. The rotator control box, which draws little current, plugs into the convenience outlet with a line cord, and then into the motor cable box with another special plug. By parallel wiring the motor cable to other jacks, you can provide several points where the control box can be used, too.

In areas where FM signals are strong, a multiple-element FM antenna may not be necessary. Instead you can use one or more *dipole* antennas either store-bought or made up from television twin-lead (Fig. 14-6). These can be tacked up, at the two ends, on the surface in the basement and attic, with the lead-ins concealed in the walls and run to jack boxes at the

Fig. 14-6. FM dipole antennas can be bought commercially or home-made. They are readily installed almost anywhere for FM reception.

points of use in the living quarters. Or, you may be able to conceal the entire aerial in the walls or ceilings. With enough of them (they are quite inexpensive) spotted about the house, you will never need scrambles of antenna wires draped about your rooms.

SECURITY SYSTEMS

The omnipresent dangers and horrors of fire have always been with us. Vandalism, burglary, and other assorted crimes have been a problem for many years in urban areas, and they are now sharply on the upswing in suburban, exurban, and even some rural areas as well. These facts have given rise to a great variety of security systems designed for home installation, either by professionals or by the homeowner himself. Often the cost of a complete and complex system will return itself within a matter of years by way of reduced insurance premiums alone, with the peace of mind and protection afforded by the system as a free benefit.

There are two basic types of security systems: the fire alarm and the burglar or intrusion alarm. Often the two are

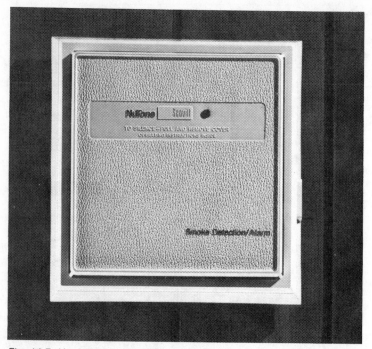

Fig. 14-7. Unobtrusive but highly effective smoke-detector units can be bought in either plug-in or direct-wired models, or for use with central built-in security systems. (Courtesy NuTone Division of Scovill)

combined into one package to serve both purposes. Some systems use individual plug-in units (Fig. 14-7), each of which includes an audible alarm that sounds when the sensor detects smoke, excessive heat, or the disturbance of an ultrasonic field by an intruder. These systems are effective and efficient, require only enough planning to insure that all areas or zones in the building are covered sufficiently by using an adequate number of units, and need only a few convenience outlets to plug them into. In fact, some units do not even need that, since they are direct-wired or battery powered. This type of system is particularly advantageous in existing buildings, especially where the installation of a built-in system might be difficult or expensive.

The best and most effective systems, however, must be planned well and permanently installed, preferably during new construction or remodeling (Fig. 14-8). For the most part, these systems are concealed, except for a few sensors and the main readout and test panel. Circuitry is straightforward, the sensors or detectors are basically simple switches designed for particular purposes, and the wiring is done with two-conductor #18, #20, or sometimes #22 cable. The power supply at the main panel operates from a 115-volt branch circuit, and wiring to the sensor circuits is low voltage, usually 6 or 12 volts. Often a battery pack is included so that in the event of power failure the system will continue to operate.

There are any number of possibilities and combinations of equipment (Fig. 14-9) that can be used in a home security system. A typical installation might be arranged as follows. In the master bedroom is a main readout panel with lights representing four zones of the house: basement, first floor, second floor, and attic. Also in this panel is an audible alarm, a test button, and a battery pack. From the terminals of each of the zone relays, a cable travels to each zone, where it attaches to a line of detectors. For fire detection, these sensors may be the photoelectric type that "see" smoke, or they may be 135°F or 190°F heat sensors. Intrusion detectors might include photoelectric beams, pressure-sensitive mats under the carpeting in hallways or near doors, magnetic contacts at the doors themselves, door-trip switches or contactors, mercury

Fig. 14-8. Installed master control unit of a built-in intruder/fire-alarm system. (Courtesy NuTone Division of Scovill)

Fig. 14-9. Some of the components of a built-in intruder/fire-alarm system. (A) master panel. (B) infrared perimeter detector. (C) inside horn alarm. (D) automatic telephone dialer. (E) surface-mount magnetic switch. (F) surface-mount plunger switch. (G) window foil contact switch. (H) recessed plunger switch. (I) window foil and foil connector blocks. (J) fixed temperature heat detector. (K) recessed magnetic switch. (L) entrance key switch. (Courtesy NuTone Division of Scovill)

tilt switches at transom or casement windows, magnetic tape on the window glass, vibration contacts, or any of several other specialized switches located wherever an intruder might gain entry. Depending upon the type of switch and the design of the system, fire or intrusion opens or closes one or more of the switches, which then activates the zone light and audible alarm at the master panel. The test button is operated manually every once in a while just to make sure that all parts of the system are still operational.

Several accessory items can be added as well. An outside alarm bell is often used, or a flashing or revolving signal light, to attract neighbors or a passer-by. Rechargeable batteries with a built-in battery charger are common. Some units are equipped with an output for connecting to a special-service telephone line that carries a warning signal back to the local fire or police department. Other types have a great many zones and will differentiate between fire and intrusion in any zone, as well as indicating power failure. Any number of remote alarms can be added, even at a neighbor's house, and additional indicator panels can be parallel-wired to the master and placed in other well-used parts of the house, such as the kitchen or family room. Circuits can also be added to pinpoint other problems as well, such as loss of heat in the greenhouse or failure of the cooling unit in a freezer.

Obviously, an extensive security sytem requires planning and the running of a lot of cable. The first step is to decide what you want to cover and where you will need to locate the various detectors, sensors or switches (Fig. 14-10). For intrusion, you might use the proper type of switch at each door and window. For fire detection, you will need a number of detectors of your choice, positioned to afford maximum coverage in square footage. This varies according to the type and design of detector, so follow the maker's recommendations. Then you will have to determine how many zones you want, if any. For each zone, run a two-wire cable from the master indicator panel to the first detector or switch in that zone, and then from device to device until you reach the last one. If the system operates with normally-closed devices, they will be connected in series or "in line." Then any time a device opens, or the line breaks, the alarm will sound regardless of the status of the other devices. In some systems, a return line from the last device on the zone circuit is brought back to the master panel for manual testing.

The running of the cables is done in the usual manner of low-voltage lines, working them through bored holes, partitions, and structural cavities, and stapling on the surface as necessary. Keep the lines away from excessive heat or moisture, and be extra careful not to damage them in any way. Rough handling could weaken a conductor or its insulation and lead to eventual failure just at the wrong time. Connection or junction boxes are seldom used, except for convenience or for those that are an integral part of the equipment. Connections to the devices are usually made to a pair of open-screw terminals and to the master panel at a screw-terminal strip within the cabinet. Warning devices are often relay activated

- **Perimeter**
- **Interior**
- **Fire**
- **Alarm**
- **Control Unit**

① **Heat Detectors** in the middle of every room, including attic, basement and garage.

② **Smoke Detectors** in passageways leading to bedrooms, at top of stairways.

③ **Magnetic or Plunger Perimeter Detectors** at every window and exterior door.

④ **Infrared Detector** to cover area of major traffic flow.

⑤ **Magnetic or Plunger Interior Detectors** at inside doors to basement, valuables closet.

⑥ **Pressure Mats** at exterior doors and stairways.

⑦ **Outside Alarm** on the highest point of the roof, under eaves.

⑧ **Inside Alarm** in central interior location.

⑨ **Control Unit** in Master Bedroom or other convenient location.

⑩ **Telephone Dialer** in Master Bedroom, closet, etc.

⑪ **Emergency Pushbuttons** at each exterior door and in each bedroom.

⑫ **Exit/Entry Controls** at main exterior doors.

NOT SHOWN

⑬ **Foil Tape** at every basement window.

⑭ **Radio Controls** for hard-to-wire locations and at nearby detached buildings such as garage, barn, workshop.

Fig. 14-10. Typical layout plan for intruder/fire alarm system. (Courtesy of NuTone Divisoin of Scovill.

on the low-voltage side of the system, but draw their power to operate from a 115-volt source. In this case, line- and low-voltage conductors and connections must be kept separated, and all line-voltage wiring is treated the same as the rest of the residential wiring system.

Detection, sensing, and switching devices of all kinds are available through your local electrical supplier, or can be purchased by mail from the catalogs of a number of companies that specialize in this type of equipment. Master panels and readouts are available through the same sources, or you might like to design or build your own unit, tailored to your needs. Plans for a number of different setups can be found in various do-it-yourself magazines and project books. The diagram in Fig. 14-11 shows a simple, easily built and installed system for limited coverage in a small home. The same basic circuitry, enlarged but still based on a simple repetitive electromechanical series, is expandable to any number of zones, area sizes, or diverse applications.

TELLTALE PANELS

A telltale panel consists of one or a number of visual or audible information signals located at one central location that will tell at a glance the status or function mode of remotely located equipment. Though common enough in some commercial and many industrial applications, they are seldom found in homes, for no apparent good reason. Yet they can serve useful and helpful purposes, especially in larger homes or farmsteads, and the lover of electronic and electromechanical gadgetry can have a grand time designing and installing telltales.

The basics are quite simple. All that is needed is a sensor at one end of a (usually) two-wire cable, with a warning device, perhaps coupled through a relay, in a central readout panel at the other end. The panel also contains a power supply for dropping from 115 volts to 6 or 12 volts, with perhaps a rectifier to convert to DC and a battery pack to keep the system operating in case of power failure.

There are dozens of possibilities, but a simple panel of six indicators might be set up like the one in Fig. 14-12. Lamp #1

Fig. 14-11. Simple fire alarm circuit, using normally-open heat sensors. Use normally-closed relay contacts with normally-closed heat sensors. Alarm may be rewired through relay to operate on 115 volts.

PILOT LIGHT

GARAGE LIGHT

GARDEN DRY

CELLAR DOOR

PUMPHOUSE HEAT

HEAT CABLES

ROOF VENT

NAME PLATE

TRIM COVER

FLUSH-MOUNT BOX TO HOLD COMPONENTS

Fig. 14-12. Telltale panel for installation in a heavily trafficked area.

shows that you have forgotten to turn the garage light off. When you turned the light on, you also activated a relay that closed the indicator light circuit. Lamp #2 indicates that it is time to water the vegetable garden. The moisture sensing probes buried in the garden "saw" that the soil had become too dry, and closed that indicator circuit. Lamp #3 shows that someone has left the cellar door open. A simple normally-open switch mounted in the door frame closed when the door was opened and completed the circuit. Lamp #4 shows that the temperature in the well pumphouse has dropped below 40°, and the heat has probably gone off. In this case, a thermostatic switch set to close at 40°F has done so, energizing the indicator. Lamp #5 tells you that the electric heating cables on the roof have turned on. This is done by energizing a relay at the same time, which in turn closes that indicating circuit.

Note that this does not tell you whether the cables are actually heating as they should, but only that they are energized. And lamp #6 shows you if the roof ventilating fan is operating. The relay is actuated by turning the fan on.

As you can see, the possibilities are limited only by your imagination and requirements. Figure 14-13 shows you the circuitry involved in typical sensor circuits; all you need do is adapt the circuit to each particular application and make up the appropriate main indicator panel. Where critical situations exist and any malfunction may cause a minor emergency, such as no heat in a brooder house or failure of a heat tape on a water line, an audible warning signal can be added. Also, sub-indicator panels can be parallel-wired to the main panel with multiconductor cable and mounted in other convenient spots in the house.

Wiring telltale panels is primarily a matter of running the necessary cables—thermostat wire works nicely—to the various sensor locations from the main indicator panel. Mount the sensors in whatever way is called for, and make the connections. At the indicator panel you can make direct connections to the power supply, relays, and lamps, or, to make for a more workmanlike job and easier servicing, run all leads to terminal strips. Wherever line-voltage wiring is involved, be sure to keep lines and connection spaces separated, use covered junction boxes, and follow the dictates of the NEC and local codes as they apply to regular residential wiring.

Master control panels (Fig. 14-14) can also be included as a part of telltale panels, providing that any low-voltage wiring is kept separate from the line-voltage equipment. This might

Fig. 14-13. Typical simple sensor circuitry. Use normally-open contacts with normally-open sensors, and normally-closed relay contacts with normally-closed sensors. Alarms can be operated on 115 volts by rewiring through isolated relay contacts.

HIGH-VOLTAGE SECTION

Fig. 14-14. Master control panel, joining low-voltage control functions and telltale circuit functions with other line-voltage control functions, for the electromechanically operated household.

include a bank of switches, either line- or low-voltage and with or without indicator lamps, for various special purposes. Among those included, for instance, might be the controls for motorized draperies, dimmers for banks of decorative lighting, switches for heating cables on the eaves or buried in walkways or drives, switches for yard lights or outbuilding entrance lights, or controls for a motorized swimming pool cover.

LIGHTNING PROTECTION

Lightning is a phenomenally powerful natural force, an awesome manifestation of raw electricity, about which we know a good deal less than we would like to, despite an ageless fascination and decades of study. We do know that it frequently does strike twice in the same place—often dozens of times—the old saying notwithstanding. We also know that strikes may occur from the ground to clouds, or vice-versa, or from cloud to cloud, and that high and unprotected spots are usually the most vulnerable.

A common misconception about lightning rods is that they *attract* lightning. This is untrue. In fact, if the lightning protection system is doing its job, it will greatly *reduce* the likelihood of a strike. The purpose of the pointed spikes of the lightning rods is to continuously discharge or bleed off the accumulating charges between the clouds and the structure being protected, and in this way it discourages the formation of charges that would lead to a lightning strike in the first place. The lighting rods bleed off the charge into the atmosphere through the ionization of the air surrounding their pointed projections, which is why all lightning rods have one or more points. However, the entire lightning protection system should be heavily constructed so that if a lightning strike does occur, the charge will be safely carried through the conductors to the earth. Thus, a lightning protection system really has two jobs: first to discourage lightning strikes, and second to safely carry away any strike if it does occur.

A Potent Force

Studies have shown, among other facts, that the potential between the earth and an overhead cloud mass just before a lightning bolt lets loose may be anywhere from a hundred million to a billion volts, which makes our 115/230-volt household service look pretty puny. And though our 100-watt bulbs draw about 1 ampere of current, a lightning bolt may carry a current of anywhere from 5000 amperes in a scrawny little jab to as much as 150,000 amperes or so, with 25,000-ampere blasts being quite common.

Though thunderstorms or "electrical storms" are quite common throughout the United States and can be severe in some areas, personal injury or death from lightning strikes happens only infrequently. But lightning can be a destructive force, often in a capricious way, and the total loss of property, worktime, and dollars throughout the country is substantial. When lightning strikes a nonconducting material, wood for instance, the material may be shattered, but hardly scorched. Or again, a roaring fire may start, depending upon the electrical characteristics and the duration of the strike. Conductive materials such as copper or aluminum wires may fuse into lumps, weld together, or simply disappear. In fact, lightning can vaporize the conductors of a cable and leave rubber insulation intact, unbroken, and completely empty. If a lightning strike finds its way into a small enclosed space such as a house chimney or a closet, it can literally explode like a bomb, shattering everything within a radius of several feet. It has also been known to do such strange things as ripping a narrow path of shingles away from the full length of a roof, or blowing an electrical panel or a wall telephone set clean through the wall of the building to end up in the yard.

A Little Protection

The upshot of all of this is that there are circumstances and locations where a homeowner might well want to take a hard look at providing his chattels with protection against this tremendous and upredictable force. Like insurance, lightning protection costs money and involves some effort, but there might come a day to make that all worthwhile.

There was a time when lightning protection systems for entire dwellings, sporting the familiar "lightning rods" along the roof peaks, were in widespread use. Fewer installations of this sort are made today because of modern construction techniques and the fact that lightning danger is not as great in urban and surburban areas where homes are packed tightly together. With the recent move toward outlying areas for both primary dwellings and second homes, however, this situation may change. Lightning protection systems are still manufactured, and there are numerous specialty companies that install them.

Whether or not you should have lightning protection, and if so how much, depends upon a number of factors. If thunderstorms never occur in your area, or only happen two or three times a year, there probably is little point in being concerned about them. Where storms may be few but of exceptional ferocity, they are worth worrying about. And in locales where they are common, even though perhaps not intense, the probability of a lightning strike at one time or another in any given location increases greatly.

Where electrical storms do occur, by and large the most susceptible spots are those that are relatively isolated and not closely surrounded by other structures. Wide open spaces like fields or prairie are vulnerable, and so are hilltops or even knolls, especially if they happen to be the highest points in the immediate vicinity. Homes built in mountain areas, especially in the higher chains, and particularly when in unsheltered locations to make the most of the view, are under constant attack. Farm buildings, especially large barns containing machinery and equipment, seem to be prime targets. Rivers,

442

railroad tracks, high tension lines, extensive stock fencing, metal roofs, and similar metallic objects or systems can act as attractants, too.

Installation

A complete lightning protection system is not difficult to install, provided that you do not mind a bit of crawling around on the roof. It must, however, be installed correctly and then inspected at regular intervals thereafter, and repaired or maintained as necessary. A half-baked or deteriorated system is a good deal worse than no system at all, since it might be hit by lightning strikes but not have the full capacity to safely dissipate them.

There are only a few simple parts to the system: air terminals, chimney points, conductor cable, bonding plates, grounding rods, and assorted types of fasteners, clamps, and hardware for various purposes (Fig. 14-15). All of the equipment can usually be installed using nothing beyond ordinary hand tools, and perhaps a power drill for boring holes in masonry if necessary.

The air terminals, or lightning rods, are mounted on the ridgetops of slanted roofs or along the edges of parapets or flat roofs, 25 feet apart if they are of the standard 2-foot lengths, and on special saddles made for the purpose. Chimneys or other nonmetallic projections are equipped with chimney points set so that the tips project at least 10 inches above the element being protected. They are attached with straps or bands around the projection, or with masonry anchors.

Cables

The cable is usually made up of solid copper strands braided together, with a minimum weight of 187.5 pounds per 1000 feet and having a minimum strand size of #17 AWG. Aluminum conductors may also be used, with a minimum weight of 95 pounds per 1000 feet and a minimum strand size of #14 AWG. However, if aluminum is used, it can at no point come in direct contact with the ground, and all of the other elements in the system must also be made of aluminum.

The cable is run along the ridges, parapets, or roof edges in a continuous run and secured solidly to the saddle of each air terminal and to each chimney point. Bends cannot exceed 90 degrees, nor have a radius tighter than 8 inches, since if the bends are too sharp or continuous a lightning surge can become untracked and shoot right off the curve like a runaway freight train. The cable must be attached about every two feet with an appropriate type of clamp, strap, or fastener. The

Fig. 14-15. Basic component parts of a lightning protection system. (A) two chimney points. (B) three types of air terminals. (C) bonding strap. (D) pipe grounding clamp. (E) tee cable connector. (F) reels of round lightning conductor. (G) cable attaching clip. (H) cable clamp. (I) masonry fastener. (J) ground rod clamp. (K) ground rod. (Courtesy Thompson Lightning Protection, Inc.)

Fig. 14-16. Typical residential lightning protecton system. (Courtesy Thompson Lightning Protection, Inc.)

points and interconnecting cables should be arranged so that when a point is struck, current can travel in two directions along the cables. "Dead ends," where a point is located at the end of a cable branch, should be avoided. Where unavoidable, the maximum allowable length is 16 feet.

Then the cable is gently curved away from the ridge and along the rake edge of a slanted roof, or over the edge of a parapet or flat roof, and is secured along the corner of the building to the ground. If the house is rectangular there should be a *downconductor*, as it is called, at two diagonally opposite corners. An ell or other addition calls for at least one more downconductor, and in the case of a large structure, one at each corner is not too much. A sizable barn might also have downconductors midway along each long wall. On any building, enough downconductors should be provided so that none are more than about 100 feet apart.

All metallic projections such as soil stacks, ventilating pipes, and roof ventilators are bonded to the cable at the most convenient points by means of special bonding straps, or lengths of cable and clamps. The cables can be joined with T splicers or straight-line splicers where necessary, though as few joints as possible in the entire network should be your goal. Any joints that are made must be mechanically strong, and the electrical contact area between the two cables being spliced must be equal to a minimum of twice the cross-sectional area of the conductor. Don't use solder, since it forms a weak joint and will melt at the first heavy surge of current through the lines. Figure 14-16 shows a typical home installation.

Grounding

The final step of the installation involves putting in the grounding electrodes. Where deep soil is found, 10-foot (or longer) copper-clad steel rods or galvanized pipe can be driven straight down. If the soil is shallow and underlaid by hardpan or bedrock, or is largely dry gravel or sand, then rod or pipe supplemented by lengths of heavy bare cable, plates of copper or steel, or strips of metal, must be buried a couple of feet deep in trenches directed away from the building. If this is not possible, then an extensive network of bare cables must be laid out surrounding the building in a grid network laid on or just below the surface. In any case, no grounding electrode elements should be closer to the foundations than two feet, to avoid any possible damage.

Unlike the electrical service ground, the lightning protection system resistance to ground is not critical. The lower the better, of course, but if it is 50 ohms or so, no harm is done. Each downconductor must be securely clamped to its own grounding electrode or to the grounding network, as the case may be. Where a water pipe passes into the protected building, a downconductor should be securely attached to the pipe where it enters the building, and then to the grounding electrode or system. Metallic water-piping systems in the immediate area of the building or any substantial lengths of metal pipe in a nonmetallic system, or any well casing within 25 feet of a building that is served by nonmetallic pipe, should be bonded to the grounding electrode or network. Radio or television masts made of metal that are in the immediate area must be bonded directly to one of the main conductors of the system with a bonding conductor of full size. Any other nearby grounding electrodes, such as for telephone or electrical service, must also be bonded to the grounding system.

Even though the dwelling may not have a complete lightning protection system, there are some individual points where protection should be provided, and this can be done with little effort or expense. Any metal mast, whether freestanding or mounted on the building, and especially if it projects above the highest point of the building or higher than surrounding structures, and whether used for television or radio antennas, flying flags or whatever, should be grounded. All this requires is the clamping of a grounding conductor, preferably copper and no smaller than #10 AWG, to the mast, routing it to a driven ground rod in the shortest possible way, and securing it to the rod. All grounding electrodes should then be bonded together for maximum safety and effectiveness. The system should also be bonded through to any existing underground

Fig. 14-17. Mast-mounted TV or radio antenna lightning-discharge units, for use with flat or round lead-in cable. (Courtesy Thompson Lightning Protection, Inc.)

metallic water piping system. (See Chapter 10 for more on grounding and bonding.) In addition, each conductor of each antenna lead-in cable should be provided with an appropriate antenna discharge unit (Fig. 14-17).

Appendices

Glossary

AC—Alternating current.

air heating coefficient—The figure used to express the heat wattage needed to warm interior air volume. Usually expressed as 0.0053 watts per cubic foot of interior air volume of heated space per degree Fahrenheit temperature difference between interior and exterior surfaces.

alternating current—An electric current that reverses its direction at regularly recurring intervals, having a magnitude that varies continuously in a sinusoidal manner.

ampacity—The current-carrying capacity of electric conductors, expressed in amperes.

ampere—The practical mks unit of electric current that is equivalent to a flow of one coulomb per second, or to the steady current produced by one volt applied across a resistance of one ohm.

amplitude—The maximum departure of the value of an alternating current or wave from the average value. The intensity or magnitude of a current. The absolute value of the maximum displacement from zero value during one period of an oscillation.

angstrom—A unit of wavelength of light equal to one ten-billionth of a meter.

angular frequency—A measure of the frequency of an object varying sinusoidally; equal to 2π times the frequency in cycles per second.

apparent power—The power at the two points of entry of a single-phase two-wire circuit that is equal to the product of the effective current in one conductor multiplied by the effective potential difference between the two points of entry.

appliance—Utilization equipment, generally other than industrial, normally built in standardized sizes and types, which is installed or connected as a unit to perform one or more functions such as clothes washing, air conditioning, food mixing, etc.

approved—Acceptable to the authority having jurisdiction.

attachment plug—or plug cap, or plug. A device that, by insertion in a receptacle, establishes connection between the conductors of the attached flexible cord and the conductors connected permanently to the receptacle.

average voltage—A value of AC voltage that is equal to 0.636 times the maximum or peak voltage.

branch circuit—The circuit conductors between the final overcurrent device protecting the circuit and the outlets.

Btu—British thermal unit. A measure of heat energy; equal to the amount of heat energy required to raise the temperature of one pound of water by one degree Fahrenheit.

cable—A stranded conductor (single-conductor cable) or a combination of conductors insulated from one another (multiple-conductor cable).

candle—or candela. A measure of light intensity, now defined as a unit of luminous intensity; equal to one-sixtieth of the luminous intensity of one square centimeter of a black-body surface at the solidification temperature of platinum.

capacitance—The property of an electric nonconductor that permits the storage of energy as a result of electric displacement when opposite surfaces of the nonconductor are maintained at a difference of potential. Also, the measure of this property equal to the ratio of the charge on either surface to the potential difference between the surfaces.

capacitive reactance—The opposition of capacitance to alternating current; equal to the reciprocal of the product of the angular frequency of the current times the capacitance.

capacitor—A device for accumulating and holding a charge of electricity, consisting of two equally charged conducting surfaces having opposite signs and separated by a dielectric. Also referred to as a condenser.

circular mil—A unit used in the measure of wire; the area of a circle having a diameter of one mil.

closed circuit—A circuit without interruption that provides a continuous path through which current can flow.

concentric lay—A method of composing a cable from multiple strands of wire using a central core surrounded by one or more layers of helically laid wires.

conductivity—A measure of the ability of a given substance to conduct electric current; equal to the reciprocal of the resistance of the substance.

conductor—A substance or body capable of conducting or transmitting electricity, heat, or sound, depending on the requirement.

conduit body—A separate portion of a conduit or tubing system that provides access through a removable cover to the interior of the system at a junction of two or more sections of the system or at a terminal point of the system.

continuity—The state or quality of being continuous; uninterrupted connection or flow.

convenience outlet—A utilization point in an electrical system generally taken to contain a duplex receptacle.

cornice lighting—A technique of lighting which uses direct light projected onto walls from ceiling-mounted line-of-light fixtures.

coulomb—The practical mks unit of quantity of electricity; equal to the quantity of charge transferred in one second across a conductor in which there is a constant current of one ampere.

cove lighting—A lighting technique that uses indirect light projected onto ceilings from wall-mounted line-of-light fixtures. Usually used on opposite or all walls of a room.

current—The movement or flow of electricity. Also, the time rate of flow measured in amperes.

cycle—A complete alternation in which a phenomenon attains a maximum and a minimum value, returning to a final value equal to the original one.

DC—Direct current.

degree day—A theoretical measure of the amount of heating needed for one day; usually defined as the difference between 65°F and the actual mean daily temperature for a given day.

demand factor—The ratio of the maximum demand of an electrical system, or part of a system, to the total connected load of a system or that part of a system under consideration.

demand load—The load that is drawn from the source of supply at the receiving terminals, averaged over a suitable and specified interval of time, expressed in kilowatts, amperes, etc.

derate—To change the rating or value, as a given conductor ampacity, to a lower rating in consideration of other affecting factors.

device—An electrical component that carries current, but does not use it, such as a switch or receptacle.

dielectric—A nonconductor of direct electric current.

direct current—An electric current of constant direction, having magnitude that does not vary or varies only slightly.

disconnect—Also disconnecting means. A device, or group of devices, or other means by which the conductors of a circuit can be disconnected from their source of supply.

dust-ignition-proof—Capable of excluding ignitable amounts of dust or amounts that might affect performance or rating, and that will not permit arcs, sparks, or heat otherwise generated or liberated inside of the enclosure to cause ignition of exterior accumulations or atmospheric suspensions of a specified dust on or in the vicinity of the enclosure.

duty cycle—The percentage of time during which a load is in operation over a specified time span.

electromotive force—The energy available for conversion from nonelectric to electric form, per unit of charge passing through the source of the energy. Also, the potential difference between the terminals of a source of electrical energy, expressed in volts.

electron flow—Electric current. The transferance of electrons from one atom to the next in continuous movement along a conductor.

emf—See *electromotive force*.

enclosure—Any box-like unit designed to house or protect electrical equipment, devices, conductor connections, or splices.

explosion-proof—Capable of withstanding an internal explosion of a specified gas or vapor and of preventing the ignition of an external specified gas or vapor by internal sparks, flashes, or explosion.

farad—The mks unit of capacitance, equivalent to one coulomb per volt; equal to the capacitance of a capacitor in which one coulomb is transferred from one of the conductors of the capacitor to the other, per volt of potential difference between the conductors.

feeder—The circuit conductors between the service equipment and the branch circuit overcurrent device.

field—See *magnetic field*.

fixed appliance—An appliance that is fastened or otherwise secured at a specific location.

footcandle—A unit of illumination equivalent to that produced by a source of one candle at a distance of one foot; equal to one lumen incident per square foot.

general background lighting—Overall lighting level in a room or area, exclusive of local lighting.

general light—The total light or light level in a room or area, including all sources.

grounding electrode—A heavy conductor or network of conductors, usually buried in the earth, to provide a conducting connection between an electrical circuit or equipment and earth.

henry—The mks unit of inductance; equal to the inductance of a circuit in which an emf of one volt is produced by a current in the circuit which varies at the rate of one ampere per second.

hertz—The mks unit of frequency, replacing "cycles per second."

home run—The portion of an electrical circuit conductor that extends from the distribution point overcurrent protection device to the first outlet on the circuit.

impedance—The total opposition to alternating current by an electric circuit; equal to the square root of the sum of the squares of the resistance and the reactance of the current, expressed in ohms. Symbolized by the letter Z.

inaccessible location—Not capable of being removed or exposed without disassembling or damaging the building structure, trim or finish, or permanently enclosed by same. Also, as applied to equipment, not readily approachable because of obstructions.

inductance—The property of a circuit by which a change in current induces by electromagnetic induction an emf in the circuit.

inductive reactance—The opposition of inductance to alternating current; equal to the product of the angular frequency of the current times the self-inductance.

instantaneous current—The value of an alternating current at any given point in a cycle or at any given point in time.

instantaneous voltage—The value of an AC voltage at any given point in a cycle or at any given point in time.

IR drop—Also voltage drop. A drop in line voltage in the conductors of an electric circuit, caused by its resistance to the flow of current, and in AC circuits by the impedance.

kelvin—Pertaining to an absolute scale of temperature with intervals equal to those of the Celsius scale. Zero degrees kelvin equals $-273.16°$ Celsius, or $-459.69°$ Fahrenheit.

kilovolt-ampere—An electrical unit equal to 1000 volt-amperes.

kilowatt—A unit of electrical power equal to 1000 watts.

kilowatt-hour—A unit of electrical energy equal to the energy transferred or expended in one hour by one kilowatt of power.

line of light—A long, straight light source such as fluorescent light fixtures.

line side—The contacts or terminals of an electrical device, designed to be connected to a conductor carrying line voltage from the source; the input side.

line voltage—The voltage supplied by an electrical power line or source, measured at the point of supply or sometimes at the point of ultilization.

load—The electrical demand placed upon a circuit or a system by the utilization equipment connected to it. Also, any piece of electrical utilization equipment of any given rating so connected.

load side—The contacts or terminals of an electrical device, designed to be connected to conductors carrying voltage away from the device to the utilization equipment, or load, or another deivce; the output side.

local light—Also called task or work light. Concentrated light falling upon a small area to illuminate specific tasks to a high level.

lumen—The unit of luminous flux; equal to the luminous flux emitted in a unit solid angle by a point source of one-candle intensity.

luminous flux—The rate of transmission of luminous energy, expressed in lumens.

magnetic field—A region in space in which a magnetic space exists, having intensity and direction at any given point, forming a magnetic circuit roughly analogous to an electric circuit.

main disconnect—The means provided in an electrical system to disconnect all conductors in a building from the service-entrance conductors. The means may consist of one or more but not to exceed six closely grouped switches or circuit breakers.

main entrance panel—A single enclosure housing the main disconnect, main overcurrent protection devices, branch circuit overcurrent devices, and branch circuit and feeder distribution means.

microfarad—A unit of capacitance equal to one-millionth of a farad.

mil—A unit of length equal to 0.001 inch, used in measuring the diameter of wires.

mutual inductance—The ratio of the electromagnetic force in one of two circuits to the rate of change of current in the other circuit.

mutual induction—The production of an electromagnetic force in one circuit by a change in current in another circuit.

National Electric Code—A compendium of articles pertaining to the general electrical wiring field, containing provisions considered essential for the practical safeguarding of persons and property from hazards arising from the use of electricity.

ohm—The mks unit of resistance equal to the resistance in a conductor in which one volt of potential difference produces a current of one ampere. Symbolized by the Greek letter Ω (omega).

open circuit—A discontinuous electrical path or circuit through which no current can flow.

outlet—A point on the wiring system at which current is taken to supply utilization equipment.

overcurrent protection device—A device designed to open an electric circuit automatically at a predetermined overcurrent level.

raceway—Any channel for holding wires, cables, or busbars that is designed expressly for, and used solely for, this purpose.

reactance—The opposition of inductance and capacitance to alternating current; equal to the product of the sine of the angular phase difference between current and voltage and the ratio of the effective voltage to the effective current, symbolized by the letter X.

reactive power—The sine counterpart of the cosine expression of true power, or $P = EI \times \sin(\theta)$.

receptacle—A contact device installed at the outlet for the connection of a single attachemnt plug.

rectifier—An electrical aparatus used for changing an alternating current into a direct current.

resistance—A property of a conductor by virtue of which the passage of current is opposed, causing the conversion of electric energy into heat. It is equal to the voltage across the source divided by the current flow and is measured in ohms.

rms current—The effective current value of an alternating current; equal to the root mean square of the variable values taken throughout one cycle. For a sine wave, it is equal to 0.7071 times I_{PEAK}.

rms voltage—The effective voltage value of an alternating current; equal to the root mean square of the variable values taken throughout one cycle. For a sine wave, it is equal to 0.7071 times E_{PEAK}.

root mean square—The square root of the arithmetic mean of the squares of the numbers in a given set of numbers.

R value—A value of thermal resistance assigned to any given material of a given thickness or density, which is indicative of the ability of that material to resist or retard the flow of heat. R values bear a reciprocal relationship to U values.

secondary—The winding in a transformer or similar electrical apparatus in which a current is induced by a primary winding; output side.

series circuit—An electrical circuit in which all load elements are connected in series with each other and with the poles of the source voltage; "end-to-end" or "along the line."

series-parallel circuit—An electrical circuit in which some of the load elements are connected in series with the poles of the source voltage, and others are connected in parallel with them, in any combination.

service—The conductors and equipment for delivering energy from the electricity supply system to the wiring system of the premises served.

service conductors—The supply conductors that extend from the street main or from the transformers to the service equipment of the premises served.

service drop—The overhead service conductors from the last pole or other aerial support to and including the splices or taps, if any, connecting to the service-entrance conductors at the premises served.

service-entrance conductors—The service conductors between the terminals of the service equipment and a point usually outside the building served, clear of the building walls, where joined by tap or splice to the service drop. Or, the service conductors between the terminals of the service equipment and the point of connection to the service lateral, or the point of entry of the service lateral into the building served.

service equipment—The necessary equipment and apparatus, usually consisting of a circuit breaker, or switch and fuses, and their accessories, located near the point of entrance of the service-entrance conductors to the building or other structure served, or an otherwise defined area, and intended to constitute the main control and means of disconnect of the supply.

service lateral—The underground service conductors between the street main or transformers, including any risers at a pole or other structure, and the first point of connection to the service-entrance conductors in a terminal box or meter or other enclosure, inside or outside the building served. Where there is no terminal box, meter, or other enclosure with taps or splices made therein, the point of connection shall be considered to be the point of entrance of the service conductors into the building served.

service raceway—The raceway that encloses the service-entrance conductors, including protective risers.

short circuit—An abnormal and usually unintentional condition of relatively low resistance between two points of different potential in a circuit, resulting in an excess flow of current.

soffit lighting—A lighting technique that uses diffused light projected downward from a source recessed in a soffit. Often localized, but may be line-of-light. Usually illuminates a specific work area or interest center.

spectrophotometry—A process of making comparisons between parts of light spectra by means of a photometric device in combination with a spectrometer.

stationary appliance—An appliance that is not easily moved from one place to another in normal use.

takeoff—A compiled list of equipment and materials extracted from a set of plans or blueprints, consisting of all of the items needed to complete an installation or job of construction as delineated by the specifications.

tap—A connection of the terminal end of one conductor to another conductor at some point along its run; the process of making such a connection. Also, a connection or connection point brought out of a winding, as in a transformer, at some point between the ends of the winding, for controlling output ratios.

tap conductor—A conductor, usually short but longer than a pigtail, that serves to connect utilization equipment to the serving circuit.

time rating—See *duty cycle*.

tinning—A process of coating a conductor surface with solder.

total connected load—The total current draw in amperes, or power consumption in watts, of all of the utilization equipment connected to all or any given segment of a wiring system.

troffer—An inverted trough serving as a support and reflector for a fluorescent lighting unit.

troffer lighting—A lighting technique that uses diffused light projected downward from troffers recessed in ceiling areas. Usually used for general lighting at either high or low levels.

true power—Also called active power. At the two points of entry of a single-phase two-wire circuit, it is equal to the time average of the values of the instantaneous power taken over one cycle of alternating current; equal to $P = EI \times \cos(\theta)$, where θ is the phase angle between voltage and current.

U value—A coefficient of heat transfer. The amount of heat flow per square foot per hour per degree Fahrenheit temperature difference between inside and outside temperatures of a building section, expressed in Btu. U is equal to 3.412 times the W factor.

valance—A narrow fabric, wood, metal, or plastic frame or facing used as a decorative heading to conceal the tops of drapes and the rods and attaching hardware.

valance lighting—A lighting technique that uses semidirect light projected onto drapes below and ceiling above from a line-of-light source concealed behind an opaque valance.

var—A unit of reactive power equal to an effective value of one ampere in quadrature with an effective value of one volt in an alternating current. (See *reactive power*.)

volt—The mks unit of electromotive force or potential difference; equal to the emf that causes a current of one ampere to flow through a conductor with a resistance of one ohm.

voltage—Electromotive force or potential difference expressed in volts.

volt-ampere—A unit of electric measurement; equal to the product of voltage and amperes. In direct current, one volt-ampere is the equivalent of one watt, and in alternating current it is a unit of apparent power.

volt-ampere reactive—Also reactive volt-ampere. See *var*.

voltage drop—See *IR drop*.

W factor—Similar to U value. A coefficient of heat transfer; equal to the amount of heat flow per square foot per degree Fahrenheit temperature difference between inside and outside temperatures of a building section, and expressed in watts. W is equal to 0.293 times the U value.

wall bracket lighting—A lighting technique similar to valance lighting, using generally diffused light projected on walls and ceilings from a wall-mounted line-of-light source, but concealed behind a faceboard rather than a valance.

watt—The mks unit of power; equal to the rate of work represented by a current of one ampere flowing through a conductor under an electromtive force of one volt.

wire—Metal in the form of a thread or slender rod.

Symbols
and Abbreviations

A, a	ampere
AC; A.C.; ac; a-c; a.c.	alternating current
amp	ampere
ANSI	American National Standard Institute
Awg; AWG	American wire gauge
Btu	British thermal unit
Btuh	British thermal unit per hour
c	cycle
C	capacitance, capacity
C	experience factor (heating)
°C	degrees centigrade or Celsius
CF	construction feature heat loss
cm	circular mils
cps	cycles per second
cy	cycle
DC; D.C.; dc; d-c; d.c.	direct current
DD	degree days
E; e	electromotive force (volt); voltage
emf	electromotive force (volt)
EMT	electrical metallic tubing
E_s	voltage source
F	air heating coefficient
f	frequency (cycles per second)
°F	degrees Fahrenheit
fc	footcandle

GFCI	ground fault current interrupter
GFI	ground fault interrupter
H	henry
HI	infiltration heat loss
HR	heat required
HT	total heat loss of a building
Hz	hertz
I; i	current (ampere)
°K	degrees kelvin
kv; kV	kilovolt
kva; kVA	kilovolt-ampere
kw; kW	kilowatt
kwh; kWh	kilowatt-hour
kwh_T ; kWh_T	total kilowatt-hours
L	inductance
LVC	low-voltage control
ma; mA	milliampere
Mcm; MCM	thousand circular mils
mf; μf; mF; μF; mfd	microfarad
mks; m.k.s.; MKS	meter-kilogram-second
N	number of air changes (heating)
NC	National Coarse (thread)
NEC	National Electrical Code
NF	National Fine (thread)
NFPA	National Fire Protection Association
θ (theta)	angular phase displacement
Φ (phi)	phase; phase angle
P	power (watt)
PF; pf	power factor
QR	quick-recovery
R	resistance (ohm)
R	thermal resistance factor
Rms; RMS; rms	root mean square
TCL	total connected load
TD	design temperature (heating)
T_I	inside temperature (heating)
T_O	outside temperature (heating)
U	coefficient of heat transfer (Btu)
U	reactive power
UL	Underwriters' Laboratories, Inc.
V; v	volt
V	volume
VOM	volt-ohm-milliameter
W	coefficient of heat transfer (watt)
W, w	watt

wg	with ground
X, x	reactance (ohm)
X_C	capacitive reactance
X_L	inductive reactance
Z; z	impedance (ohm)
Ω (omega)	ohm

Index

Index